Diffusion Phenomena

CASES AND STUDIES

Reprinted by permission from Chas Addams.

Diffusion Phenomena

CASES AND STUDIES

SECOND EDITION

Richard Ghez

Dover Publications, Inc.
Mineola, New York

Bibliographical Note

This Dover edition, first published in 2018, is a corrected republication of the work originally published by the Kluwer Academic/Plenum Publishers, New York, in 2001.

International Standard Book Number

ISBN-13: 978-0-486-82832-9
ISBN-10: 0-486-82832-8

Manufactured in the United States by LSC Communications
82832801 2018
www.doverpublications.com

A Margalit qui,
tout en ne sachant pas la Mathématique,
ne cesse de croquer la Pomme.

Preface

This book is a second edition of the one that was published by John Wiley & Sons in 1988. It carries a new title because the former one, *A Primer of Diffusion Problems*, gave the impression of consisting merely of a set of problems relating to diffusion. Nonetheless, my intention was clearly spelled out and it remains the same, namely, to teach basic aspects and methods of solution for diffusion phenomena through physical examples. Again, I emphasize that the coverage is not encyclopedic. There exist already several outstanding works of that nature, for example, J. Philibert's *Atom Movements, Diffusion and Mass Transport in Solids*. My emphasis is on modeling and methodology. This book should thus constitute a consistent introduction to diffusion phenomena, whatever their origin or further application.

This edition has been largely revised. It contains a completely new chapter and three new appendices. I have added several new exercises stemming from my experience in teaching this material over the last 15 years. I hope that they will be instructive to the reader for they were not chosen perfunctorily. Although they are the bane of authors and of readers, I have retained footnotes if they might help the reader's comprehension. Additional, but nonessential material is collected at the end of chapters, and is indicated in the text by superscripts.

A few words are in order to explain why I have added a new chapter, the sixth, on surface rate limitations. By its very nature, diffusion is multidisciplinary; it is also multidimensional. All real physical systems have boundaries, and these are always dominant when the system is "small." I had long felt that an introduction to the influence of boundaries—beyond abstract mathematical statements for the resolution of specific problems—was lacking in most texts on diffusion theory. I can but hope that this additional chapter will help the reader in this regard.

It was my pleasure to have worked with Evelyn Grossberg, editor for Kluwer Academic/Plenum Publishers, whose editorial remarks I gladly accepted.

Jerusalem
April 2001

Preface to the First Edition

Most kinetic processes in ponderable media require the redistribution of mass, energy, momentum, or charge over macroscopic distances. This redistribution occurs through a random process called diffusion, and its macroscopic description rests on a particular differential equation called the diffusion equation. Because this equation is to ponderable media what the wave equation is to vacuum, it is no wonder that its solutions should be relevant to widely different problems: from the heat treatment of alloys to the motion of proteins in growing axon membranes; from the dynamics of galactic structures to the multitude of device-processing steps. Consequently, the literature on diffusion contains at least four different points of view: as a mathematical topic in partial differential equations, as a physicochemical topic in nonequilibrium thermodynamics, as an engineering topic in materials science, and as a topic in stochastic processes. It would be presumptuous to claim familiarity with all these areas and to suggest, here, an equal coverage. And yet, they all derive from the same basic principle, namely, that something or other is *conserved* or can be *balanced*. Conservation is not so much an expression of longevity—as in "this person is well preserved"—but rather it expresses our ability to distinguish and to reckon. Nowhere, to my knowledge, is this point of view better illustrated than in the frontispiece, and I am deeply indebted to Mr. Chas Addams for his permission to reproduce an example of his always penetrating wit.

In spite of this formal unity of diffusion problems, it is my experience that many scientists and engineers remain baffled by the diffusion equation and by its solution in practical cases. Alternatively, they are perhaps trapped between purely mathematical texts and others that deal exclusively with diffusion mechanisms. This book seeks to bridge the gap between physicochemical statements of certain kinetic processes and their reduction to diffusion problems. It also attempts to introduce the reader to the many lines of attack, both analytic and numerical, on the diffusion equation. I have chosen to teach through physically significant examples taken mainly from my experience in the areas of metallurgy and semiconductor technology. The theory is interwoven, I hope, with sufficient rigor so that the tools that evolve do not immediately crumble with use. Nevertheless, I have chosen to write an introductory text, whose only prerequisites are a serious year of calculus, through ordinary differential equations, one semester of thermodynamics, and, if the reader be uneasy with algebraic manipulations, that the fear of calculating be controllable.

As its title implies, this is largely a book about *solving* the diffusion equation. I also wanted a short and lively text. Consequently, unless they provide particular insight, intermediate steps of specific calculations are often only verbally (though completely) expressed. In other words, the beginner would be well advised to arm himself with pad and pencil. Exercises, not so numerous as to paralyze the reader, are sprinkled throughout the text. They are designed either to emphasize a point that has just been made or to suggest some further clarification or extension of that point. These exercises are an integral part of the text, and (a word of encouragement) none is so difficult as to constitute a doctoral thesis.

The text consists of seven chapters and three appendices. These appendices contain more advanced material, whose inclusion within chapters would have detracted from the general flow. References, cited in the text are listed at the end of each chapter, and, in general, equations are numbered decimally within each chapter. I have sometimes labeled equations with an additional Latin letter when they are closely related. It can happen, therefore, that "well ordering" is not everywhere preserved.

I might be faulted on at least one count. My choice of topics is definitely biased, and each specialist will surely be disappointed to find that his particular area of expertise was not covered. This is unavoidable in so short a text, and I will simply observe that presentations of methodology and techniques are useful even though they fail as encyclopedias.

If any text is a slice of the author's life, it is also an image of his relations with others from whom he has benefited. I am grateful to my employer, the IBM Corporation, for allowing me the time and the use of its facilities to complete this book. Moreover, my experience with diffusion would be insignificant were it not for discussions and collaboration with many colleagues in this and other laboratories. Among these, I wish to single out my friends E. A. Giess, M. B. Small, F. M. d'Heurle, T.-S. Kuan, all at IBM, and A. S. Jordan and G. H. Gilmer at AT&T, Bell Laboratories. J. S. Lew and L. Kristianson, also at IBM, are gratefully thanked for their kindness and patience with my many queries. They are holders of increasingly rare skills (mathematical analysis and technical illustration) in a world that too often confuses quality with touch typing. I am also deeply indebted to Bea Shube and to her staff at John Wiley & Sons for their encouragement and help, particularly during those dark periods that every author knows. Finally, I wish to recall the memory of M. M. Faktor (Queen Mary College, London), a true and courageous friend and scientist.

Yorktown Heights, N.Y.
February 1988

Contents

1

The Diffusion Equation

We begin with a model for diffusion: the isotropic one-dimensional random walk.[1-4] It is so simple that the basic physical processes cannot elude us. It also has a continuum limit, the diffusion equation, whose solutions are our main concern here and some of whose properties we then examine. Conversely, this model forms the basis for numerical methods of solution. We then discuss the diffusion equation's form in higher dimensions and other physical instances where that equation offers a realistic description. This chapter ends with a brief account of the origin of conservation principles and constitutive relations that pervade transport phenomena.

1.1. The Isotropic One-Dimensional Random Walk

Consider points on a line, as shown in Fig. 1.1, choose an arbitrary origin, and label these points through integers $i = 0, \pm 1, \pm 2, \ldots$. Attribute particles to each point, henceforth called sites, namely, N_i is the number of particles at site i. Let us assume that each of these particles can jump to adjacent sites with a frequency Γ that does not depend on the site i. Jumps to the right are, on average, as likely as those to the left, so that $\frac{1}{2}\Gamma N_i$ particles jump, per unit time, from site i to each of its nearest-neighbors.[†]

[†]Contrary to certain claims, it is not $N_i/2$ particles that jump on average in each direction, but rather N_i particles that have a transition frequency (or transition probability per unit time) $\Gamma/2$ of carrying out a specific jump. Thus, Γ must be understood as the *total* transition probability. It is also a conditional probability per unit time in the sense that a transition occurs, *given* that an initial site is occupied and that a final site is available. [See remarks (b) and (d), below.]

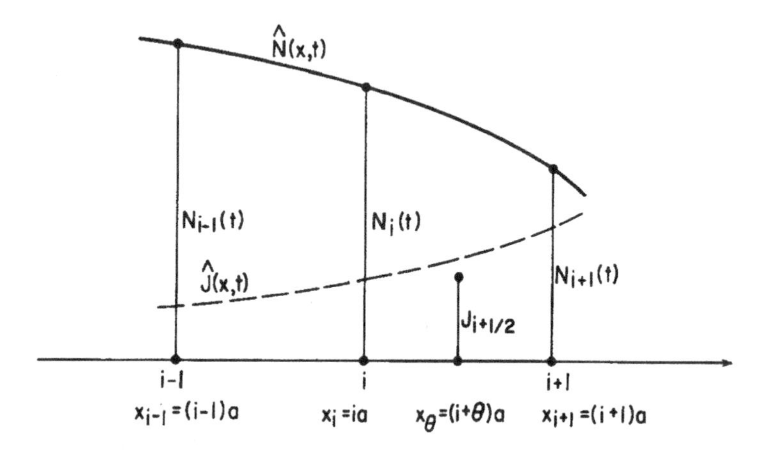

Fig. 1.1. Particle numbers N_i at sites i, the continuous interpolation $\hat{N}$, and the flux $\hat{J}$ derived therefrom. Note the "gap" between $\hat{J}$ and $J_{i+1/2}$, which is a minimum at the midpoint.

Particles that jump from site to site cause the numbers N_i to change in the course of time. To obtain a rate equation, assume that, sitting on the ith site, an observer evaluates all possible transitions to and away from that site

$$\frac{dN_i}{dt} = +\tfrac{1}{2}\Gamma N_{i+1} + \tfrac{1}{2}\Gamma N_{i-1} - \tfrac{1}{2}\Gamma N_i - \tfrac{1}{2}\Gamma N_i.$$

Collecting terms, we get a rate equation for the particle distribution $N_i(t)$

$$\dot{N}_i = \tfrac{1}{2}\Gamma(N_{i+1} + N_{i-1} - 2N_i). \tag{1.1}$$

This equation involves three sites: the observer's station i and his two nearest neighbors. Just as a second-order differential equation can be decomposed into two first-order equations, Eq. (1.1) is decomposed in the following way:

$$\dot{N}_i = -(J_{i+1/2} - J_{i-1/2}), \tag{1.2}$$

where

$$J_{i+1/2} = \tfrac{1}{2}\Gamma(N_i - N_{i+1}) \tag{1.3a}$$

and

$$J_{i-1/2} = \tfrac{1}{2}\Gamma(N_{i-1} - N_i). \tag{1.3b}$$

The quantities $J_{i \pm 1/2}$ represent the rate of exchange, the *net flux*, between adjacent sites. The subscript notation $i \pm \frac{1}{2}$, admittedly somewhat obscure (the N's are defined only at integer sites), will later become clear. For the moment it simply indicates that the particle fluxes (1.3) depend on *two* adjacent sites, in addition to time. These fluxes vanish when $N_i = N_{i+1}$ for all i; in other words, the flow of matter vanishes when mass is evenly distributed among sites. This is nothing but a necessary condition for equilibrium.

Equation (1.1) gives the rate of change in terms of the nearest-neighbor distribution around site i, whereas Eq. (1.2) expresses that change as a difference of fluxes. In a sense, this last expression is the more fundamental, for it is a statement of a *conservation law*. Imagine a cell centered at site i, whose boundaries are somewhere in the intervals $[i-1, i]$ and $[i, i+1]$. Then, Eq. (1.2) merely expresses the change in the cell's content caused by the difference of the "stuff in" $J_{i-1/2}$ and the "stuff out" $J_{i+1/2}$. In other words, this equation is nothing but a statement of (local) mass conservation, and it would still be true even if the fluxes had forms different from (1.3), for example polynomial functions of the differences $N_i - N_{i \pm 1}$, or functions involving next nearest-neighbor jumps. Let us pause here for a few remarks as follows:

Remarks

a. Equation (1.1) is linear in the unknown functions $N_i(t)$. Structurally, it is a linear difference-differential equation, second-order in the discrete variable i and first-order in the continuous variable t. Given an initial configuration [the "initial condition" $N_i(0)$] and boundary conditions (on i), this initial-value problem can be solved analytically. We will not develop this here. However, Eq. (1.1) does provide a numerical scheme for its continuum analog, as Sect. 1.4 will show.

b. The particular simplicity of our random walk rests on the constancy of the transition frequency Γ. Recall that Γ is assumed independent of the site index i and of time. In addition, Γ does not depend on the actual particle distribution in the neighborhood of an ith site. Succinctly, this random walk model is *homogeneous*, an assumption that must often be relaxed, for example in vacancy diffusion: a jump can occur only if adjacent sites are vacant. As pointed out in the first footnote, the model is also *isotropic* because jumps to the right and to the left are equally likely: $\frac{1}{2}\Gamma$ represents either frequency if Γ is the total jump freqency. Later, in Chap. 3, we shall see that external forces introduce a measure of anisotropy.

c. Equations (1.1)–(1.3) are topological in the sense that neither the *distance* between points nor the *shape* of the lattice plays any role whatsoever. Only the *order* of points matters: One must always be able to tell which of two adjacent points is "the first." For example, these equations are valid for points distributed along a space curve—a helix, say—but points of self-intersection, as in a "figure eight," require special attention. In brief, these equations are valid for sites embedded in spaces that are topologically equivalent to the real line.

d. The reader will note the similarity between Eqs. (1.1)–(1.3), applied here to mass transport, and rate equations that describe nuclear decay, chemical kinetics, and stochastic processes. In fact, similar equations describe the time evolution of any property over an ordered list of *states i*. If $P_i(t)$ is that property, and $\Gamma_{i,k}$ is the transition frequency from state i to k, then

$$\dot{P}_i = \sum_k (\Gamma_{k,i} P_k - \Gamma_{i,k} P_i) \tag{1.4}$$

is a generalization of Eq. (1.1), often called a *master equation.* Its probabilistic interpretation is interesting: If P_i is the *a priori* occupation probability of state i and $\Gamma_{i,k}$ is the conditional probability of the $i \to k$ transition (given that i is occupied), then sums such as $\Sigma_i \Gamma_{i,k} P_i$ represent the *total* probability of transition to a given final state k, regardless of the initial state i. Both sums of Eq. (1.4) can be interpreted accordingly.

EXERCISE 1.1. *Steady-state solutions of the rate equations:* As will be emphasized in Chap. 2, the steady state occurs when $\dot{N}_i = 0$ for all sites. Explore the consequences of this condition for Eqs. (1.1)–(1.3). Carefully distinguish equilibrium from steady states. □

1.2. Continuum Equations

In many instances we may be unconcerned with the discrete nature of matter. It is then possible to find a continuum analog of Eqs. (1.1)–(1.3). The essential step requires that the sites i be points of the real line, thereby introducing a distance, or "metric," between points. Assume, then, that the sites are equidistant, a being the lattice (or jump) distance. Therefore, the ith site has the coordinate $x_i = ia$.

Now introduce *any* continuous and sufficiently differentiable function $\hat{N}(x, t)$ that interpolates the previous functions $N_i(t)$, namely, that satisfies the condition

$$\hat{N}(x_i, t) = N_i(t) \tag{1.5}$$

at sites $x = x_i$, but that is quite arbitrary elsewhere. This is shown schematically in Fig. 1.1. Omitting the time variable for the moment, we expand the interpolating function in a Taylor series around the central site x_i. For adjacent sites we get†

$$\hat{N}(x_{i\pm 1}) = \hat{N}(x_i) \pm a \left.\frac{\partial \hat{N}}{\partial x}\right|_{x_i} + \tfrac{1}{2}a^2 \left.\frac{\partial^2 \hat{N}}{\partial x^2}\right|_{x_i} + \mathrm{O}(a^3), \tag{1.6a}$$

which, with the condition (1.5), becomes

$$N_{i\pm 1} = N_i \pm a \left.\frac{\partial \hat{N}}{\partial x}\right|_{x_i} + \tfrac{1}{2}a^2 \left.\frac{\partial^2 \hat{N}}{\partial x^2}\right|_{x_i} + \mathrm{O}(a^3). \tag{1.6b}$$

Introducing this equation into Eq. (1.1), we get

$$\left.\frac{\partial \hat{N}}{\partial t}\right|_{x_i} = \tfrac{1}{2}\Gamma a^2 \left.\frac{\partial^2 \hat{N}}{\partial x^2}\right|_{x_i} + \mathrm{O}(a^4). \tag{1.7}$$

If we agree to neglect terms of fourth order in the jump distance a, and if we assume the truncated equation's validity at *all* points x (and not merely at sites), then Eq. (1.7) reduces to a *single* partial differential equation for the function $\hat{N}$. It replaces the *denumerable* (possibly infinite) coupled system of difference-differential equations (1.1).[N1]

It is somewhat more delicate to analyze the fluxes, for it is *a priori* not at all evident how a continuous analog $J(x, t)$ should be defined.

†The "O" symbol, $\mathrm{O}(a^3)$ in this case, is a shorthand for "plus terms of the order a^3." At least we then know *what* is being neglected, instead of merely writing three little dots.... More precise definitions (but not any more illuminating) are given in Ref. 4.

Again introducing Eqs. (1.6) into Eqs. (1.3), we find that

$$J_{i\pm 1/2} = -\tfrac{1}{2}\Gamma\left(a\frac{\partial \hat{N}}{\partial x} \pm \tfrac{1}{2}a^2\frac{\partial^2 \hat{N}}{\partial x^2}\right)\Bigg|_{x_i} + \mathrm{O}(a^3). \tag{1.8}$$

In other words, both discrete fluxes converge to the *same* value $-\frac{1}{2}\Gamma a\partial\hat{N}/\partial x|_{x_i}$, to first order in a, and the second order terms are equal and opposite. It follows that Eq. (1.8), if blandly inserted into Eq. (1.2), would produce $\partial\hat{N}/\partial t|_{x_i} = 0 + \mathrm{O}(a^2)$: *We would have no continuous analog of the discrete conservation law (1.2)*. Therefore, let us agree to *define* a continuous flux function

$$\hat{J}(x, t) = -\tfrac{1}{2}\Gamma a\frac{\partial \hat{N}}{\partial x} \tag{1.9}$$

in terms of the interpolating function $\hat{N}$. In other words, a flux function can be calculated *for all* x if $\hat{N}(x, t)$ is known, but it is quite unlikely that $\hat{J}$ will ever pass exactly through the discrete fluxes (1.3), no matter where these are placed. The "gap" shown in Fig. 1.1 can be minimized, however, if we measure the difference between $J_{i+1/2}$, say, and expression (1.9) evaluated at arbitrary points in the interval $[i, i+1]$. Specifically, we define an arbitrary intermediate point $x_\theta = (i+\theta)a$, where $0 \leqslant \theta \leqslant 1$, and expand N_i and N_{i+1} around x_θ to form the difference $J_{i+1/2} - \hat{J}(x_\theta, t)$. The result is given in the exercise that follows.

Exercise 1.2. *To justify the notation* $J_{i\pm 1/2}$: Show that the above difference in fluxes is "optimal" when $\theta = \frac{1}{2}$. What else can you say about the series in θ? Then develop the same arguments for the flux $J_{i-1/2}$ in the interval $[i-1, i]$, and show that it has the *same* formal expansion around the appropriate intermediate point as does $J_{i+1/2}$. Finally, form the difference $J_{i+1/2} - J_{i-1/2}$, expand it around the current lattice site x_i, and show that it is equal to $a\partial\hat{J}/\partial x|_{x_i}$ plus fourth-order terms in a. □

It follows from this exercise that the discrete conservation law (1.2) has a continuum analog

$$\frac{\partial \hat{N}}{\partial t} = -a\frac{\partial \hat{J}}{\partial x} + \mathrm{O}(a^4). \tag{1.10}$$

We have now reached our goal, for, if $C = \hat{N}/a$ is the average concentration per cell, and if we introduce the definition

$$D = \tfrac{1}{2}\Gamma a^2 \tag{1.11}$$

for the *diffusion coefficient* (or "diffusivity" for short), then Eqs. (1.9), (1.10), and (1.7) become

$$J(x, t) = -D\frac{\partial C}{\partial x}, \tag{1.12}$$

$$\frac{\partial C}{\partial t} = -\frac{\partial J}{\partial x}, \tag{1.13}$$

$$\frac{\partial C}{\partial t} = D\frac{\partial^2 C}{\partial x^2}. \tag{1.14}$$

To summarize briefly, we obtain the continuum equations (1.12)–(1.14) from the discrete equations (1.1)–(1.3) by: (i) embedding the sites i in the real line; (ii) interpolating the discrete distribution $N_i(t)$ by a continuous function $\hat{N}(x, t)$, which opens up the arsenal of calculus; (iii) agreeing to define $\hat{J}$; (iv) truncating the resulting equations and requiring their validity at all points; (v) defining the concentration C in terms of the interpolating function $\hat{N}$; (vi) defining a diffusivity $D = \frac{1}{2}\Gamma a^2$ which remains constant as $\Gamma \to \infty$ and $a \to 0$.

EXERCISE 1.3. *Next-nearest-neighbor jumps:* Let Γ_1 and Γ_2 be the total jump frequencies to nearest and next-nearest neighbors, respectively. Show that Eq. (1.14) is still valid if the diffusivity is defined appropriately. Can you generalize? This example is of some interest because it shows that distinct difference equations can yield the same continuum equation. Although next-nearest jumps are unlikely on one-dimensional lattices, they are common mechanisms in higher dimensions. □

Remarks:

a. Equations (1.12) and (1.14) are sometimes called Fick's first and second laws.[3] This is but nomenclature. It *is* important to remember, however, that Eq. (1.13), the so-called *continuity equation,* is a precise statement of mass conservation, as we shall soon see. On the other hand, the linear relation (1.12) between the flux J and the "thermodynamic force" $\partial C/\partial x$, sometimes called a *constitutive relation*, does not have the same generality. Its form depends on the details of the diffusion mechanism. Note, nonetheless, that its minus sign means that material must flow from regions of high concentration to regions of lower concentration, a property that we ordinarily associate with diffusion. We will refer, henceforth, to Eq. (1.14) as the *diffusion equation*; it is a consequence of the two previous equations if the diffusivity is a constant. Indeed, any one of Eqs. (1.12)–(1.14) is a consequence of the two other equations.

b. The diffusion equation is sometimes also called the "heat equation,[5,6] because the evolution of energy also obeys a conservation law. From a mathematical point of view, it is the simplest so-called "parabolic" equation (more nomenclature). From the theory of partial differential equations[7] we learn that it must be solved as an initial-value problem: Given initial data on C and boundary conditions on the spatial variable x, we can then find the solution by "marching forward" in time. This will soon become clear in Sects. 1.4 and 1.5 on numerical methods.

c. From Eq. (1.11) we learn that the constancy of D is related to the constancy of the jump distance and frequency. If, for any reason, these are not constants (see the next exercise), then Eq. (1.14) must be replaced by the more general

$$\frac{\partial C}{\partial t} = \frac{\partial}{\partial x}\left(D\frac{\partial C}{\partial x}\right). \tag{1.15}$$

d. Physical dimensions are important, especially if we wish to cast a given problem in as few variables as possible.[N2] We will always denote dimensions by square brackets. Thus, from Eq. (1.11), we learn that $[D] = \text{cm}^2/\text{s}$.† The dimensions of C and J depend on that choice for N_i and $\hat{N}$. For example, if the particles are evenly distributed along planes perpendicular to the direction of diffusion, the usual case of "volume" diffusion, then we have $[N_i] = \text{cm}^{-2}$, $[C] = \text{cm}^{-3}$, and $[J] = \text{cm}^{-2}\,\text{s}^{-1}$. Can you find these dimensions if the particles are

†Purists distinguish between "dimensions" and "units." Dimensions will always be displayed in representative units (often cgs).

evenly distributed along lines ("surface" diffusion) or restricted only to sites ("edge" diffusion)? Note also that mass or molar units, rather than numbers of particles, can be used to measure N_i, with concomitant changes for C and J.

e. Order-of-magnitude estimates of D are instructive. If, for Γ, we take the Debye frequency in a solid, about $10^{13}\,\mathrm{s}^{-1}$, and if we take roughly 2 Å for the jump distance a, then Eq. (1.11) tells us that $D \simeq 2 \times 10^{-3}\,\mathrm{cm}^2/\mathrm{s}$. This is many orders of magnitude larger than what is actually measured in solids.[N3] The moral is simply that not all atomic oscillations lead to successful jumps: Only big ones succeed, as is true in other random walks of life.

f. The careful reader will have noticed that the "hat" has come off the symbol J for the flux somewhere between Eqs. (1.9) and (1.12). This is not an oversight. There are two points to bear in mind regarding fluxes: First, they are directed quantities; the function J must then be the component of a flux *vector* along the jump direction. Second, by a similar construction in more than one dimension it becomes clear that the fluxes, as defined here, are quantities integrated over lines or surfaces transverse to the jump direction. The true flux derives from $\hat{J}$ by dividing by appropriate lengths or areas. This remark is related to remark (d), above; a fuller account is provided in Sect. 1.6 and in Appendix A.

EXERCISE 1.4. *A model for variable diffusivity:* Write first the master equation (1.4) restricted to nearest-neighbor jumps. Then assume that the transition frequencies depend on the occupation of the final sites: $\Gamma_{i,i\pm1} = \frac{1}{2}\Gamma f_{i\pm1}$ for all i, where the dimensionless and arbitrary function $f_{i\pm1}$ depends only on the occupation numbers $N_{i\pm1}$ of adjacent sites. Show that Eq. (1.15) follows from Eq. (1.4), with a concentration-dependent diffusivity $D(C) = \frac{1}{2}\Gamma a^2 f(1 - d\ln f/d\ln C)$. Can you interpret this? (See Ref. 8.). □

1.3. Elementary Properties of the Diffusion Equation

The diffusion equation enjoys several interesting properties[4–7] that are useful, either to characterize the behavior of the solutions or to derive new solutions from known ones. These properties can be subdivided into three categories: conservation, smoothing, and invariance.

A.1 Conservation in Continuous Regions

The conservative nature of Eqs. (1.13) and (1.14)–(1.15), or of their discrete analogs (1.1)–(1.2), has already been emphasized. We are now ready for a disarmingly simple proof. We consider a diffusion process in a stationary "slab," and denote by a and b the coordinates of its left- and right-hand boundaries, respectively. The total mass[†] contained in this region is then

$$M(t) = \int_a^b C(x, t)\, dx. \tag{1.16}$$

Taking its time derivative and using Eq. (1.13), we get

$$\dot{M} = \int_a^b \frac{\partial C}{\partial t}\, dx = -\int_a^b \frac{\partial J}{\partial x}\, dx = J(a, t) - J(b, t). \tag{1.17}$$

This string of equalities can be viewed in two ways: Either the container (the slab's boundary) is impermeable, in which case the two boundary fluxes in the above expression are zero, and the total mass M is constant, or the constancy of M implies, at the very least, that whatever enters the container must also leave. This characterizes the "steady state," to which we shall return in the next chapter. The very same operations can be performed on the discrete form (1.2). Note that they are independent of the constitutive relations (1.3) and (1.12). At the end of this chapter we shall see a converse statement, namely, global mass conservation in continuous regions implies the continuity equation (1.13).

EXERCISE 1.5. *Fluxes are useful:* Consider a semi-infinite region $\{x > 0\}$, initially at the uniform concentration C_∞, into which material (e.g., a dopant) can diffuse from its free surface. Call $M(t)$ the *additional* mass that has accrued at time t. Show that this mass can be calculated as the integral

$$M(t) = \int_0^t J_0(t')\, dt',$$

[†]The word "mass" is used loosely, for the integral (1.16) over concentration depends on the dimensions of C.

where J_0 is the surface flux. Note that this result is independent of the boundary conditions at $x = 0$ and that it also holds for mass depletion by outdiffusion. □

A.2 Conservation at Fixed Surfaces of Discontinuity

The previous calculations assumed that there were no points of discontinuity within the slab. What if that slab consists of two distinct materials or of two distinct phases of the same substance, each characterized by its own diffusivity? If $x = \xi$ is such a point separating the two materials or phases, then it is only necessary that we break up the integral (1.16) into two parts:

$$M = \int_a^\xi C^{(1)}\,dx + \int_\xi^b C^{(2)}\,dx. \tag{1.18}$$

For simplicity, we assume that the fluxes at the external boundaries, $J^{(1)}(a, t)$ and $J^{(2)}(b, t)$, either vanish or are equal and that the total mass is constant.† Then, manipulations similar to those leading to Eq. (1.17) yield‡

$$\dot{M} = 0 = J^{(2)}(\xi^+, t) - J^{(1)}(\xi^-, t). \tag{1.19a}$$

In other words, the flux is always continuous at stationary internal boundaries as long as these boundaries are not the seat of other processes that change the number of particles. The concentration distribution is not smooth, however. With Eq. (1.12) we have

$$D^{(1)} \frac{\partial C^{(1)}}{\partial x}\bigg|_{\xi^-} = D^{(2)} \frac{\partial C^{(2)}}{\partial x}\bigg|_{\xi^+}, \tag{1.19b}$$

which shows that the concentration profile suffers a change in slope if the diffusivities are unequal.

†These assumptions are not necessary as the development in Sect. 1.8 and Chap. 6 will show.

‡The symbols $\xi^\pm$ have their usual mathematical meaning as right and left limits at the point of discontinuity $x = \xi$.

A.3 Conservation at a Moving Boundary

Not all boundaries are fixed. In fact, some of the more interesting ones execute motions, as we will see in the examples of Chaps. 2, 4, and 5. It is then required, in the last calculations, to relax one assumption: The internal boundary point's position is now an arbitrary function of time, $x = \xi(t)$. The integrals (1.18) then have variable limits, and, using Leibniz's rule[N4] for such cases, we easily get a generalization of Eqs. (1.19):

$$\dot{\xi}(C^{(1)}|_{\xi-} - C^{(2)}|_{\xi+}) = J^{(1)}|_{\xi-} - J^{(2)}|_{\xi+}$$
$$= -D^{(1)}\frac{\partial C^{(1)}}{\partial x}\bigg|_{\xi-} + D^{(2)}\frac{\partial C^{(2)}}{\partial x}\bigg|_{\xi+}. \qquad (1.20)$$

Thus, the quantity $\dot{\xi}C - J$ is now continuous across a moving boundary, but the jump in concentration and the flux at that boundary are related through its velocity $\dot{\xi}$. Concentration fields often *do* suffer discontinuities at phase boundaries, although, in general, the temperature field for analogous melting and freezing problems does not.[N5] In that case, one can show[5,7] that the left-hand side of Eq. (1.20) is replaced by $\dot{\xi}L$, where L is the latent heat per unit volume. Relations such as Eq. (1.20) are called *Stefan conditions*; they are, in fact, equations of motion for moving boundaries. Any failure to satisfy Stefan conditions is equivalent to a violation of mass or energy conservation.[N6]

EXERCISE 1.6. *Geometric interpretation of flux continuity:* Show that the flux continuity condition (1.19b) for a stationary boundary leads to a "law of refraction." Discuss the possible signs of the gradients at the interface. How is this modified if the boundary moves according to Eq. (1.20)? □

B. Smoothing Property

The diffusion equation (1.14) has an interesting geometric interpretation if we think of the concentration distribution as a surface $C = C(x, t)$ over the plane of the independent variables (x, t). The time derivative $\partial C/\partial t$ is proportional to the local curvature $\partial^2 C/\partial x^2$ along isochronal lines (we usually call these "profiles") of the surface.

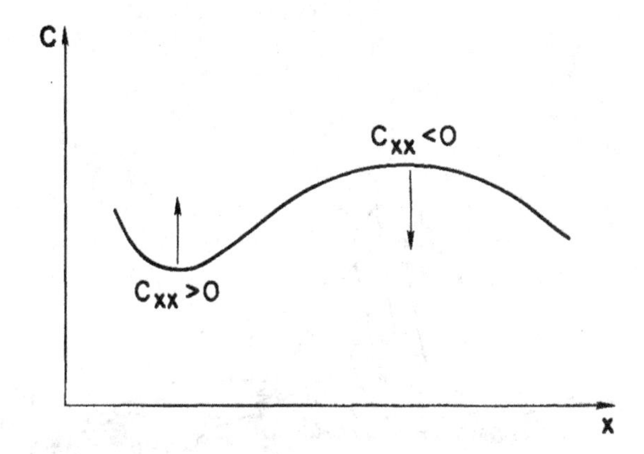

Fig. 1.2. Profile of a "bumpy" concentration surface. It tends to smooth out with time.

Therefore, the distribution's behavior can be inferred from $C(x, t + \Delta t) \approx C(x, t) + \Delta t \partial C/\partial t$. Figure 1.2 shows a "snapshot" of that surface. Now, because the diffusivity is positive, bumps with negative curvature will tend to move down in time, and positive curvature bumps will move up. In other words, a bumpy profile becomes smoother as time progresses.

Figure 1.3 shows a particular concentration surface that we will discuss at length in Chap. 4. The vectors $(0, 1, \partial C/\partial t)$, tangent to lines of constant x and directed toward increasing values of t, indeed point downward where the curvature is negative. Can you make similar statements for Eq. (1.15)?

C.1 Linearity of the Diffusion Equation

Any constant is a "trivial" solution of the diffusion equation (1.14), as is any first-degree polynomial in x. More generally, the diffusion equation satisfies a "principle of superposition": If C_1 and C_2 are two solutions of this equation and if α_1 and α_2 are two arbitrary constants, then $\alpha_1 C_1 + \alpha_2 C_2$ is also a solution, provided that the initial and boundary conditions of the full problem can be decomposed into those of the partial problems for C_1 and C_2. More generally, if $\mathscr{L}$ is a linear operator acting on a solution C, such as differentiation or integration with respect to the independent variables

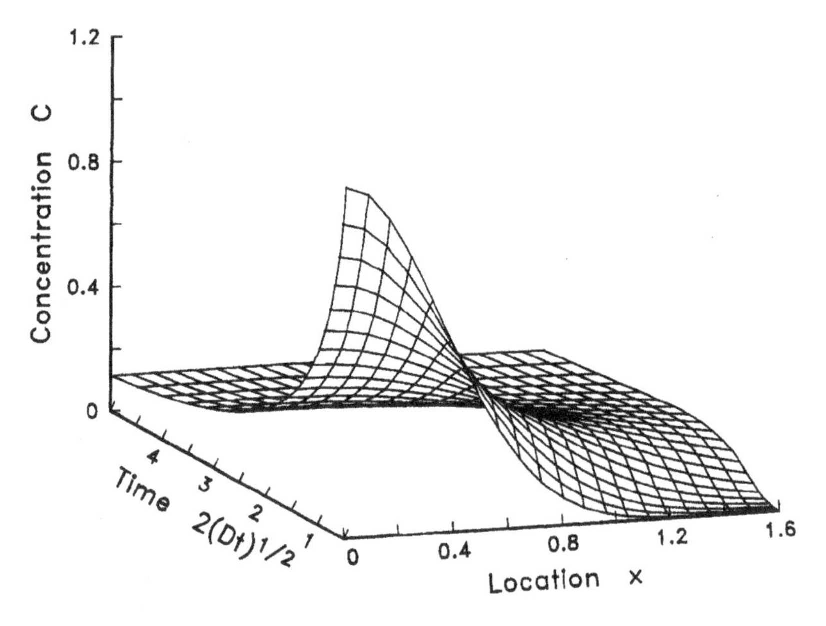

Fig. 1.3. Concentration distribution represented as a surface over the (x, t)-plane. This is, in fact, the Gaussian (4.3) of Chapter 4. Note the essential singularity at $t = 0$. [See Note 2 of Chap. 4.]

(x, t), or with respect to a parameter, then $\mathscr{L}\{C\}$ is also a solution. For example, the flux J, derived from any solution C, is also a solution of the diffusion equation. These facts are easy to verify directly from Eq. (1.14) because the operations in question are linear and they commute. Can you make similar statements for Eq. (1.15)?

EXERCISE 1.7. *Integrating solutions:* Contrary to what was stated above, the integral of a solution is *not* always a solution. Let $u(x, t)$ be any solution of Eq. (1.14), and construct the new function

$$U(x, t) = \int_a^x u(x', t)\, dx',$$

where a is some fixed coordinate. Calculate its derivatives, and find what additional condition one must impose so that U will also be a solution of the diffusion equation (1.14). Similarly, is the time-integral of a solution a possible solution? □

C.2 Symmetries: Invariance Under Reflections and Translations

The diffusion equation does not change its form under reflection operations in space. More precisely, if $C(x, t)$ is a solution and if $x \to \bar{x} = -x$, then $C(\bar{x}, t)$ is also a solution, provided that the boundary conditions are unchanged under that transformation. This is easily verified (using the chain rule) because Eq. (1.14) is second order in x. Similarly, this equation is invariant under translations of the independent variables, i.e., the addition of an arbitrary constant to x and t does not affect its form. A reversal of the time arrow, however, does not preserve the form of the diffusion equation because it is only first order in t. This is the mathematical content of "irreversibility," and diffusion is a prime example of an irreversible process. These results are amplified in Appendix B and in the exercise that follows.

EXERCISE 1.8. *Affine transformations of the diffusion equation:* Consider transformations of the independent variables (x, t) of the form

$$\begin{pmatrix} \bar{x} \\ \bar{t} \end{pmatrix} = \begin{pmatrix} \alpha & \beta \\ \gamma & \delta \end{pmatrix} \begin{pmatrix} x \\ t \end{pmatrix} + \begin{pmatrix} x_0 \\ t_0 \end{pmatrix},$$

where all the coefficients α, β, γ, δ, x_0, and t_0 are constants. What conditions must these coefficients satisfy to preserve the form of the diffusion equation? □

The diffusion equation has many other interesting properties among which are the *maximum principle* and the proof of the *existence* and *uniqueness* of solutions. This last, in particular, is not merely of esthetic appeal. In practice, if we have found a solution to a particular diffusion problem—by whatever means—then we are assured of having found *the* solution of the problem in question. The interested reader will find a full account in Refs. 6 and 7, at least for linear problems, but the properties listed here are sufficient for our purposes. We shall become familiar with them in the following chapters.

1.4. Numerical Methods

In Sect. 1.2 we derived the diffusion equation (1.14) from a model (1.1) for diffusion. Now we do the reverse: Given the equation† $C_t = DC_{xx}$, we derive a so-called "finite-difference" form that can be solved numerically, i.e., on a computer or even with a hand calculator. It is only necessary that we reverse our steps. Whereas interpolation had provided the main tool, we now wish to *sample* the continuous distribution $C(x, t)$.[N7]

Toward that end, we introduce a (constant) rectangular grid on the domain of the independent variables (x, t) by choosing points

$$x_i = i\Delta x \qquad \text{and} \qquad t_n = n\Delta t \tag{1.21}$$

that are labeled by integers i and n. The *a priori* arbitrary mesh sizes Δx and Δt determine the closeness of the finite-difference approximation to the exact solution, but, in general, Δx bears no relation to the jump distance a. Then, as shown in Fig. 1.4, we define a discrete function C_i^n that samples the continuous function $C(x, t)$ at the grid points (1.21), namely, $C_i^n = C(x_i, t_n)$ in a manner similar to Eq. (1.5).‡ We now digress briefly on the approximation of derivatives.

Derivatives are defined in calculus through a limiting process on differences. Since the diffusion equation involves two kinds of partial derivatives, we first consider the concentration profile $C(x)$ at a fixed instant of time, as shown schematically in Fig. 1.5. How does one then estimate the x-derivatives? Similarly to Eqs. (1.6), a Taylor expansion around the current point x_i yields

$$C_{i\pm 1} = C_i \pm \Delta x \left.\frac{\partial C}{\partial x}\right|_{x_i} + \tfrac{1}{2}\Delta x^2 \left.\frac{\partial^2 C}{\partial x^2}\right|_{x_i} + O(\Delta x^3). \tag{1.22a}$$

Hence, taking differences, we get the three possible representations (there are others; see Exercise 1.10, below) for the first derivative:

†The subscript notation for partial derivatives is handy; is saves both time and space. Thus here, e.g., $C_t \equiv \partial C/\partial t$ and $C_{xx} \equiv \partial^2 C/\partial x^2$.

‡To avoid confusion with the space index, note the superscript position of the time index n.

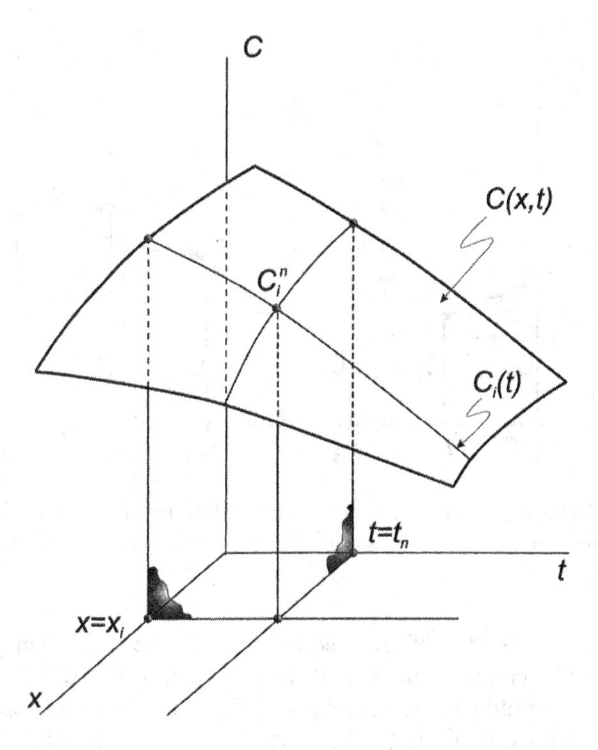

Fig. 1.4. Sampling a concentration surface at grid points. The curve $C_i(t)$ is the intersection of this surface with the plane $x = x_i$.

$$\left.\frac{\partial C}{\partial x}\right|_{x_i} = \begin{cases} (C_{i+1} - C_i)/\Delta x + O(\Delta x), \\ (C_i - C_{i-1})/\Delta x + O(\Delta x), \\ (C_{i+1} - C_{i-1})/2\Delta x + O(\Delta x^2). \end{cases} \tag{1.22b}$$

These representations measure the slope of the chords drawn in Fig. 1.5. The first alternative is called a *forward first difference* because the true derivative is estimated from forward data on the x-axis. The second is called a *backward difference* for similar reasons. Both of these are first-order accurate in Δx. The third alternative, the so-called *central difference*, yields a far better second-order accurate estimate, and it is clear from the figure that the corresponding chord is more nearly parallel to the tangent at the current point x_i.

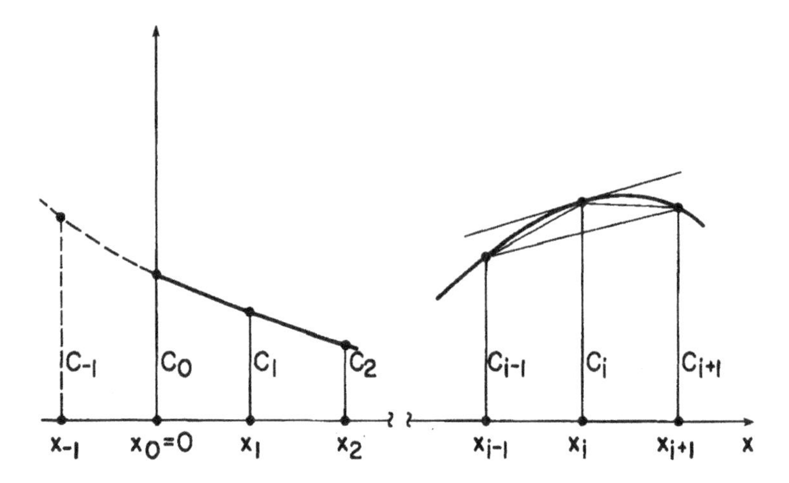

Fig. 1.5. Right-hand side: estimates of derivatives by chords. Left-hand side: a fictitious point x_{-1} for second-order accurate derivative estimate at a boundary.

Returning to the diffusion equation, we note with dismay that it is a *second* derivative in x that is required. But Eq. (1.22a) also provides a formula for estimating C_{xx} that should be reminiscent of the right-hand side of Eq. (1.1), the so-called *central second difference*. Thus, having sliced the concentration surface in Fig. 1.4 by vertical planes $x = x_{i-1}$, $x = x_i$, and $x = x_{i+1}$, we have

$$\dot{C}_i(t) = (D/\Delta x^2)[C_{i+1}(t) + C_{i-1}(t) - 2C_i(t)] + O(\Delta x^2). \quad (1.22c)$$

But how do we estimate the time derivative? As with the space derivatives (1.22b), there are several answers to that question; each provides a distinct discretization scheme for the diffusion equation. Here we choose the simplest, the so-called "explicit (or Euler forward) scheme," in which the left-hand side of Eq. (1.22c) is estimated from data forward in time, i.e., at t_{n+1}. A Taylor expansion in t, similar to Eq. (1.22a) and arrested at first-order terms in Δt, then yields the desired approximation $\dot{C}_i \approx (C_i^{n+1} - C_i^n)/\Delta t$. We thus obtain the finite-difference equation

$$C_i^{n+1} = (1 - 2\lambda)C_i^n + \lambda(C_{i+1}^n + C_{i-1}^n), \quad (1.23)$$

in which

$$\lambda = D\Delta t/\Delta x^2, \tag{1.24}$$

often called the *modulus*, is a dimensionless ratio of time to space mesh sizes. Equation (1.23) represents the diffusion equation if we agree to neglect terms of order $O(\Delta x^2, \Delta t)$, the so-called "local truncation error."

EXERCISE 1.9. *The second difference deserves its name:* For functions f_i over a discrete set of points, one defines[9] the first (forward) difference $\Delta f_i = f_{i+1} - f_i$. Show that the second difference operator $\Delta^2 f_{i-1}$, obtained by iterating the first difference (in complete analogy with second *derivatives*), has exactly the form in square brackets of Eq. (1.22c). □

The finite-difference equation (1.23) recursively generates a solution from initial data. Referring to Fig. 1.6, assume that the concentration is known at the initial time $t_0 = 0$ along the x-axis. This is

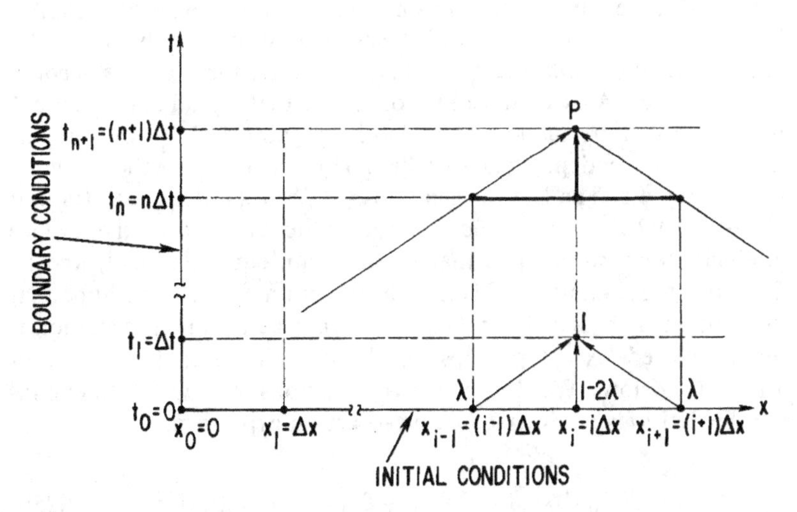

Fig. 1.6. Grid points necessary for finite-difference solutions. With the explicit scheme, evaluation at a current point P depends only on its three immediate predecessors.

known as the *initial condition*. Equation (1.23), written for the time step $n = 0$, then *explicitly* gives the concentration distribution C_i^1 at the first time step $n = 1$ from three previous values of C. The computational pattern (or "computational molecule") has the shape of an inverted "T" with weights attached to its nodes, as shown on Fig. 1.6. The process can then be repeated for $n = 1, 2, \ldots,$ and the solution evidently develops forward in time, from initial data, in lockstep with the above computational procedure. Such methods are therefore sometimes called "marching" procedures.

If the physical system that we are studying is finite in extent, then we must also be careful to respect the *boundary conditions* at each time step. For example, if the concentration is a known function $f(t)$ at the left boundary $x = 0$, say, corresponding to the index $i = 0$, then Eq. (1.23) takes the form

$$C_1^{n+1} = (1 - 2\lambda)C_1^n + \lambda[C_2^n + f(t_n)] \tag{1.25a}$$

at the first interior point $i = 1$.

Another important boundary condition arises when the gradient is known, for instance, at the left-hand boundary, $\partial C/\partial x|_{x=0} = g(t)$. We could represent the gradient through the expression $(C_1^n - C_0^n)/\Delta x$, but the first equation in (1.22b) shows that it is only first-order accurate in Δx, while the finite-difference equation (1.23) is second-order accurate. A useful device for obtaining better gradient conditions rests on the introduction of a fictitious point $i = -1$ outside the domain of the independent variables. This is shown on the left-hand side of Fig. 1.5. The last of equations (1.22b) then demonstrates that $(C_1^n - C_{-1}^n)/2\Delta x$ is a second-order accurate estimate of the surface gradient. Next, we assume that the diffusion equation (1.14) and its finite-difference analog (1.23) hold *at* the boundary, i.e., we hope that its solution there is sufficiently smooth. In practice this expectation is not misplaced, except perhaps for the first time step (see Exercises 1.11–1.13, below). We thus have two equations from which to eliminate the fictitious quantity C_{-1}^n and to get an expression

$$C_0^{n+1} = (1 - 2\lambda)C_0^n + 2\lambda[C_1^n - \Delta x g(t_n)] \tag{1.25b}$$

for the boundary values.

EXERCISE 1.10. *One-sided derivatives at boundaries:* The first derivative at a boundary can also be represented *without* the introduction of fictitious points. Consider, for example, the data at the three points $x_0 = 0$, x_1, and x_2 in Fig. 1.5. Interpolate the concentration by a second-degree polynomial in x and find its coefficients in terms of the data. Then estimate the derivative at the boundary $x = 0$. Is this a first- or a second-order estimate? □

The previous discretization scheme paid no attention to possible restrictions on the values of Δx and Δt except for an intuitive sense that they should be "small." In other words, we expect the numerical solution of Eq. (1.23) to converge toward the analytic solution of the diffusion equation if the mesh sizes are small. This can indeed be proved and is known as the *consistency* requirement of the finite-difference scheme.[9-12] In addition, we require that this convergence occur without spurious oscillations or overshoot. Thus we are also interested in the *stability* of the above finite-difference scheme. There are many ways to analyze this question: perhaps the most compelling is due to von Neumann. We will not consider this here because of space limitations and because the method is well documented elsewhere.[9,10,12] We observe, however, that Eq. (1.23) cannot have oscillatory solutions if all its coefficients are of the same sign. Now the modulus λ, like D, is positive, and therefore the coefficient of C_i^n must also be positive. Hence, the modulus must obey the inequalities $0 < \lambda < \frac{1}{2}$, and from the definition (1.24) we get the important restriction

$$\Delta t < \frac{\Delta x^2}{2D}. \tag{1.26}$$

This is the crux of the matter, for we need a fine space mesh to resolve spatial gradients. Equation (1.26) then imposes an upper bound, *quadratic in* Δx, on the size of the time mesh. Small Δt's, in turn, may mean prohibitively large running times and possibly the accumulation of roundoff errors.†

EXERCISE 1.11. *A finite-difference calculation—Constant surface concentration:* Consider diffusion in a region bounded by two parallel

†The end of Appendix D suggests analytic procedures that alleviate these problems.

planes $x = 0$ and $x = L$ (a "slab" region), under the initial condition $C(x, 0) = C_L$ (a constant) for all x inside that slab, and under the two boundary conditions: (a) $C(0, t) = C_0$ (also a constant); (b) $C(L, t) = C_L$. Both boundary conditions hold for all $t > 0$. It is advised to sketch these conditions in the (x, t)-plane and to note the discontinuity at the origin. The problem depends on four dimensional parameters: D, L, C_0, and C_L. Introduce the new dimensionless variables $\bar{x} = x/L$, $\bar{t} = Dt/L^2$, and $u = (C - C_L)/(C_0 - C_L)$. Why is this advantageous? Choose a spatial mesh consisting of five equal parts, and choose the modulus $\lambda = 0.2$. What are the corresponding values of $\Delta\bar{x}$ and $\Delta\bar{t}$? Write the (explicit) finite difference equation (1.23), and carefully specify the values of i and n for which it applies. Prepare an (i, n)-matrix and fill in the initial and boundary values of u. What value would you attribute to u_0^0? Solve first by hand for the first few time steps. You will then develop a "feeling" for the solution, in lockstep with the algorithm. Only then, write a general purpose computer program. Experiment with various values of λ and $\Delta\bar{x}$. Does your solution tend to a steady state? Is mass conserved? □

EXERCISE 1.12. *A finite-difference calculation—Variable surface concentration:* The same as the previous exercise, except that the "near" boundary concentration is a given, exponentially decreasing function of time: $C(0, t) = C_L + (C_0 - C_L)e^{-t/\tau}$, where τ is constant. The physical meaning of such a condition is explained in Sect. 4.5 of Chap. 4. The initial condition and the "far" boundary condition remain unchanged. There are now two timescales and thus two distinct sets of dimensionless variables. Discuss the suitability of each set. □

EXERCISE 1.13. *A finite-difference calculation—Constant surface flux:* Same problem as Exercise 1.11, except that the "near" boundary flux $-D\partial C/\partial x = J_0$ is a given constant. The other conditions remain unchanged. □

The explicit scheme (1.23) is still very popular, in spite of its limitations. As mentioned earlier, there are several other ways to discretize Eq. (1.22c), even as there are several distinct finite-difference schemes that yield the same continuum equation (see Exercise 1.3). Some of these do not suffer the limitation (1.26). An obvious candidate is the "fully implicit (or backward) scheme," in which the time derivatives of Eq. (1.22c) are estimated through $(C_i^n - C_i^{n-1})/\Delta t$, namely, the time analog of the second Eq. (1.22b). We then get the following set of linear equations

$$(1 + 2\lambda)C_i^n - \lambda(C_{i+1}^n + C_{i-1}^n) = C_i^{n-1} \tag{1.27}$$

for the unknown quantities at time t_n. It can be shown that this scheme is unconditionally stable: λ is unrestricted and the inequality (1.26) can be violated. The penalty for such freedom is the need to solve a (possibly large) linear system, and computer methods are definitely advisable. Fortunately, the coefficient matrix of Eqs. (1.27) has a very simple structure, and is a prime example of a "sparse" matrix: Only the main, upper, and lower diagonals are filled; the rest is nothing but a bunch of zeros. Very powerful methods exist for the solution of such "tridiagonal" systems. These questions, as well as the discussion of other popular discretization schemes, such as the one due to Crank and Nicholson, can be found in standard books on numerical analysis.[9,11,13†]

EXERCISE 1.14. *Matrix form of the implicit method:* Cast Eqs. (1.27) in matrix form and observe that all the unknowns at time t_n must be obtained *simultaneously* from earlier data. Draw the computational molecule. Verify that the matrix is tridiagonal. How would you handle a boundary condition such as $C(0, t) = f(t)$? □

A few words regarding the information flow are in order. Looking back at Fig. 1.6 and at the computational pattern for the explicit scheme (1.23), it is clear that the concentration's value C_i^{n+1} at the current point P depends on all previous values *inside* the sector (or pyramid, in more than one space dimension) with an apex at P. Values outside this sector are irrelevant to the value at P. This "corporate" structure is, in fact, nonphysical, because it implies that "signals" with velocity greater than $\Delta x/\Delta t$ (the inverse of the sector's inclination) cannot affect the current point's concentration.[N8] On the other hand, the implicit scheme (1.27) solves simultaneously for *all* concentration values along a time row n, and it uses all the information available below that row. Thus, it is both rather more democratic and more physical than the explicit scheme.

†By now it should be obvious that the correspondence between continuum equations and their finite-difference analogs is not one-to-one.

Numerical methods are very powerful because they are *not* limited to linear problems, typical of constant D. For example, the right-hand side of the nonlinear equation (1.15), if discretized, becomes $\Delta x^{-2}\{D_{i+1/2}[C_{i+1}(t) - C_i(t)] - D_{i-1/2}[C_i(t) - C_{i-1}(t)]\}$, a form that should remind us of Eqs. (1.2) and (1.3). The diffusivities must be evaluated at midpoints $i \pm \frac{1}{2}$ to ensure an optimal truncation error. The reader is encouraged to write out such finite-difference schemes in both explicit and implicit form, and to think of ways to evaluate $D_{i\pm 1/2}$.

1.5. A Numerical Example

In electrostatics, a point charge at the origin creates the simplest potential distribution. Similarly for diffusion problems, one may ask how a unit mass influences an infinite medium if it is initially placed at the origin. This is called a *point source*. Mathematically, the initial condition is $C(x, 0) = \delta(x)$, or in finite-difference terms[†]

$$C_i^0 = \delta_{i,0} \qquad i = \ldots, -2, -1, 0, 1, 2, \ldots \tag{1.28a}$$

This problem has no boundary conditions at any finite distance from the origin.[‡] If we now choose a value for the modulus λ, then we know from the last section that Eq. (1.23) allows us to compute the concentration values at the first time step. We easily see from Fig. 1.7 and Eq. (1.23) that only C_{-1}^1, C_0^1, and C_1^1 are nonzero. The process can be repeated for the next time step, and we note that the point source yields nonzero concentration values within the sector of unit slope and apex at the origin in the (i, n)-plane.

The number of arithmetic operations for this simple problem (the so-called "operational count") can be roughly halved if we use its symmetry intelligently. This will be a first application of Property C.2 of Sect. 1.3, for both the diffusion equation and the boundary and initial conditions are symmetric under a reversal of the x-axis. This

[†] The Dirac delta-function $\delta(t)$ should be familiar from quantum mechanics. We will see more of it later on in Chap. 7. Its discrete analog is the Kronecker delta $\delta_{i,k}$, which has the value 1, if $i = k$, and 0 otherwise.

[‡] There are boundary conditions at infinity: the solution should decay sufficiently rapidly to ensure the convergence of sums over the concentration distribution, such as the total mass (see Appendix F).

means that

$$C_i^n = C_{-i}^n, \tag{1.28b}$$

and we only need be concerned with the concentration values for positive index i, say. But what about the concentration along the line of symmetry $i = 0$? The answer is evident from Eq. (1.23); evaluating it at that position and taking Eq. (1.28b) into account,

$$C_0^{n+1} = (1 - 2\lambda)C_0^n + 2\lambda C_1^n. \tag{1.28c}$$

The complete solution is displayed in Table 1.1, which provides first a simple *APL* program, called *EXPLICIT*, and then some sample ouput. The arguments *NF* and *L* are the final value of the time index n (so that you won't run all night) and of the modulus λ, respectively. A few points are worth mentioning. First, we note that the output for $\lambda = 1$ is nonsensical because this case violates the stability criterion (1.26). The output for $\lambda = 0.5$ is somewhat better and is sometimes called the case of marginal stability. At least the concentration does not go negative. The cases $\lambda = 0.2$ and 0.1 are definitely more reasonable, and they also show the improvement that accrues with smaller moduli: Since $\Delta t = \lambda \Delta x^2/D$ by definition, it follows that t_n is proportional to λ for constant space mesh and diffusivity. Thus, for example, we compare directly the output rows $n = 1$ and 2 of the case $\lambda = 0.2$ with the rows $n = 2$ and 4 of the case $\lambda = 0.1$. Yet more refinement would indicate, heuristically at least, that the solution converges to something reminiscent of a Gaussian. It is recommended to plot these data.

The last case, $\lambda = -0.1$, is also of some interest, for it demonstrates the basic irreversibility of the diffusion equation. In fact, negative moduli would represent either a negative flow of time (because $\Delta t < 0$)—and we saw that the diffusion equation is *not* invariant under time reversal—or that the diffusivity D is negative, which was excluded in Property B of Sect. 1.3, for then the diffusion process cannot be stable. This is a general feature of all irreversible processes: The choice of what one means by the *positive* flow of time is related to the positivity of the transport coefficient D.

This is as good a point as any to introduce the notion of *Green's function*. We have just described the numerical solution to the finite-difference equation (1.23) when the initial disturbance is a point

Table 1.1: Listing and Execution of an *APL* Program for a Point Source

```
        ∇ NF EXPLICIT L;N;C;C1
[1]     N←0
[2]     C←1,(1+NF)ρ0
[3]     'MODULUS IS ',⍕L
[4]     'I =   ',(9 0)⍕0,⍳NF
[5]    PRINT:'N = ',(⍕N),(9 3)⍕¯1↓C
[6]     C1←C[1]+2×L×C[2]-C[1]
[7]     C←C+L×(1⌽C)+(¯1⌽C)-2×C
[8]     C[1,NF+2]←C1,0
[9]     →(NF≥N←N+1)/PRINT
        ∇

4 EXPLICIT 1   MODULUS IS 1
I =            0          1          2          3          4
N = 0      1.000       .000       .000       .000       .000
N = 1     ¯1.000      1.000       .000       .000       .000
N = 2      3.000     ¯2.000      1.000       .000       .000
N = 3     ¯7.000      6.000     ¯3.000      1.000       .000
N = 4     19.000    ¯16.000     10.000     ¯4.000      1.000

4 EXPLICIT .5  MODULUS IS 0.5
I =            0          1          2          3          4
N = 0      1.000       .000       .000       .000       .000
N = 1       .000       .500       .000       .000       .000
N = 2       .500       .000       .250       .000       .000
N = 3       .000       .375       .000       .125       .000
N = 4       .375       .000       .250       .000       .063

4 EXPLICIT .2  MODULUS IS 0.2
I =            0          1          2          3          4
N = 0      1.000       .000       .000       .000       .000
N = 1       .600       .200       .000       .000       .000
N = 2       .440       .240       .040       .000       .000
N = 3       .360       .240       .072       .008       .000
N = 4       .312       .230       .093       .019       .002

4 EXPLICIT .1  MODULUS IS 0.1
I =            0          1          2          3          4
N = 0      1.000       .000       .000       .000       .000
N = 1       .800       .100       .000       .000       .000
N = 2       .660       .160       .010       .000       .000
N = 3       .560       .195       .024       .001       .000
N = 4       .487       .214       .039       .003       .000

4 EXPLICIT -.1  MODULUS IS -0.1
I =            0          1          2          3          4
N = 0      1.000       .000       .000       .000       .000
N = 1      1.200      ¯.100       .000       .000       .000
N = 2      1.460      ¯.240       .010       .000       .000
N = 3      1.800      ¯.435       .036      ¯.001       .000
N = 4      2.247      ¯.706       .087      ¯.005       .000
```

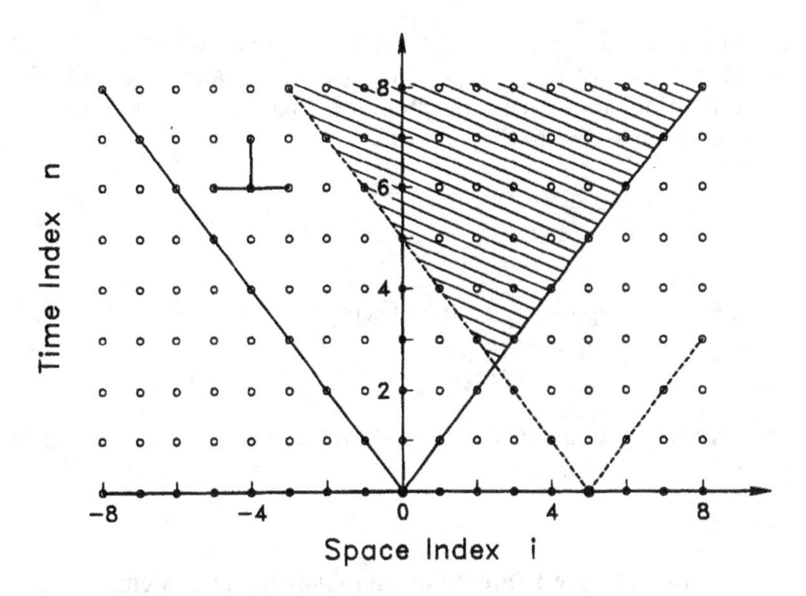

Fig. 1.7. Sectors of influence for initial point sources. Concentration values are additive in the hatched region. The computational molecule for the explicit scheme is shown.

source. Simply call *that* solution the Green's function (or "elementary solution") G_i^n.† Now, if that point source had been placed at any other point $x_k = k\Delta x$, then it is easily seen from the computational procedure that the solution would develop exactly as before, *with exactly the same concentration values,* inside the sector with apex at $i = k$, also shown (dashed, for $k = 5$) in Fig. 1.7. Thus $G_i^n = G_{i-k}^n$ has the translational symmetry of this problem.[N9] Further, the linearity of Eq. (1.23) means that the effect of a unit source at $i = k$ simply adds to the effect of our initial source at the origin $i = 0$ (see again Property C.1 of Sect. 1.3). These effects are additive in the hatched region of Fig. 1.7, namely, in the intersection of the two sectors.

We can now use these properties to compute the effect of an *arbitrary* initial condition $C(x, 0) = \varphi(x)$, or its finite-difference analog:

$$C_i^0 = \sum_{k=-\infty}^{\infty} \varphi_k \delta_{i,k}. \tag{1.29a}$$

†Strictly speaking, we have a *family* of functions G that depends on the modulus λ.

Again, because of the linearity property of the diffusion equation, each partial disturbance φ_k propagates independently, and each gives rise to a multiple of Green's function for that source point k. The concentration at any time t_n is thus

$$C_i^n = \sum_{k=-\infty}^{\infty} \varphi_k G_{i-k}^n, \tag{1.29b}$$

namely, a superposition of the effects of point sources that were distributed along the x-axis.

EXERCISE 1.15. *An initial dipole:* Assume that your initial condition is $C_i^0 = \delta_{i,k} - \delta_{i,-k}$. Discuss. □

1.6. Higher Dimensions and Coordinate Systems

The world is never simple and rarely one-dimensional. Sometimes, however, it can display a measure of symmetry. In this section we give, without much proof, the forms of Eqs. (1.12)–(1.15) for such cases. A thorough treatment can be found in References 4, 5, and 7.

We first observe that the concentration C is a scalar quantity. Thus, we expect the left-hand sides of Eqs. (1.13)–(1.15) to remain unchanged under coordinate transformations. The flux is another matter. In fact, the development in Sects. 1.1 and 1.2 indicates that it is a directed quantity, as is the partial derivative C_x. The following calculation is designed to demonstrate heuristically the vectorial nature of these quantities. It should be remembered that our random-walk model was independent of a coordinate system, i.e., the linear lattice can have an arbitrary orientation with respect to a fixed frame. This is shown schematically in Fig. 1.8. In that case, the lattice sites i are the end-points of position vectors $\mathbf{x}_i$. Taylor's expansion applied to the interpolating function $\hat{N}(\mathbf{x}_i, t)$ again leads to

$$N_{i\pm 1} = N_i \pm \mathbf{a}\cdot\nabla\hat{N}|_{\mathbf{x}_i} + O(a^2), \tag{1.30}$$

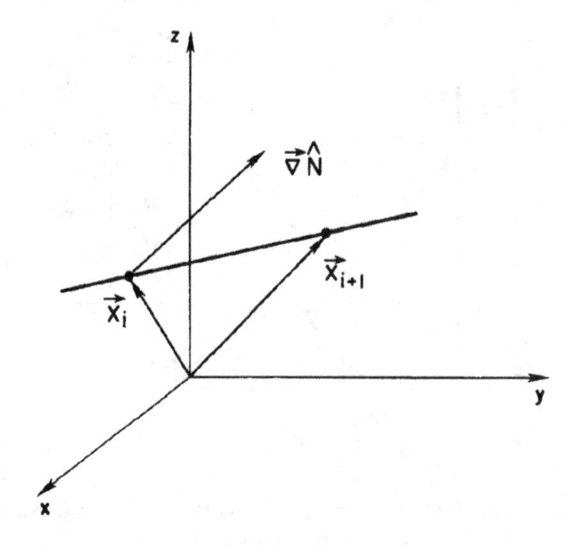

Fig. 1.8. Lattice points in space referred by their location vectors $\mathbf{x}_i$. The concentration gradient $\nabla\hat{N}$ need not be collinear with the lattice.

where now the jump *vector* $\mathbf{a} = \mathbf{x}_{i+1} - \mathbf{x}_i$ is dotted into the gradient $\nabla\hat{N}$.† Thus, expression (1.3a) of the flux $J_{i+1/2}$, which represents mass flow along the lattice in the direction of increasing i, is proportional to this scalar product. It is therefore related to the projection of a flux *vector* **J** onto the lattice. A more rigorous treatment (see Appendix A) would show that Eq. (1.12) can be generalized to give

$$\mathbf{J}(\mathbf{x}, t) = -D\nabla C \tag{1.31}$$

if it is further assumed that the diffusivity is isotropic.[N10]

Next, given that the flux is a vector quantity, we will show in Sect. 1.8 that the continuity equation (1.13) reads more generally:

$$\frac{\partial C}{\partial t} = -\nabla \cdot \mathbf{J}. \tag{1.32}$$

†Recall that the "del" or "nabla" operator ∇ has the form $\mathbf{i}\partial/\partial x + \mathbf{j}\partial/\partial y + \mathbf{k}\partial/\partial z$ in Cartesian coordinates (x, y, z), and that it serves to form the gradient ∇f and the Laplacian $\nabla^2 f$ of a scalar function f. The divergence and curl of a vector field $\mathbf{v}$ are then $\nabla \cdot \mathbf{v}$ and $\nabla \times \mathbf{v}$, respectively.

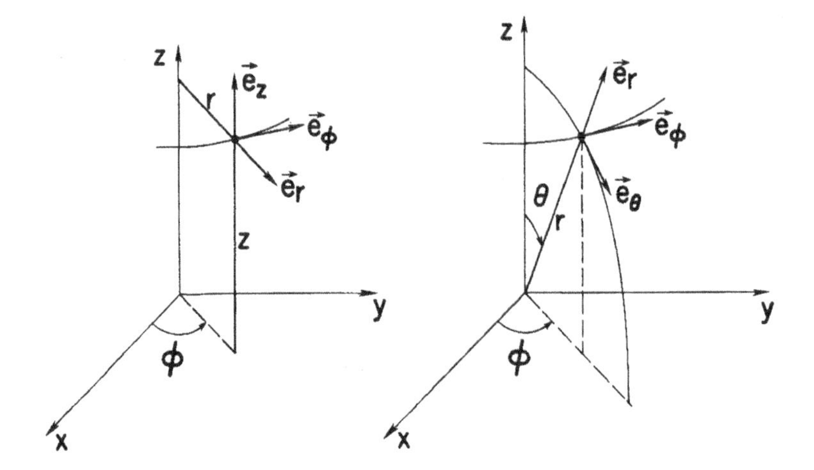

Fig. 1.9. Cylindrical (left) and spherical (right) coordinate frames and their unit vectors.

Finally, the last two equations, together, yield the diffusion equation

$$\frac{\partial C}{\partial t} = D\nabla^2 C \tag{1.33}$$

if diffusion occurs in a region of more than one space dimension, and an obvious generalization of Eq. (1.15),

$$\frac{\partial C}{\partial t} = \nabla \cdot (D\nabla C), \tag{1.34}$$

if the diffusivity is variable.

Now, a good many physical problems have, or are assumed to have, certain spatial symmetries. Just as it is never wise to fit a square peg into a round hole, we should take advantage of a body's symmetry to simplify the preceding Eqs. (1.31)–(1.34). Coordinate systems that mimic this symmetry are most helpful, mainly because the boundary conditions then assume simple forms. This will become evident in the following chapters. Of these, the cylindrical and spherical systems are the most commonly used. They are illustrated in Fig. 1.9, and we record below the expressions for the flux, divergence, and Laplacian in these coordinates:

CYLINDRICAL COORDINATES (r, ϕ, z)	SPHERICAL COORDINATES (r, θ, ϕ)

Coordinate transformations:

$$x = r\cos\phi \qquad x = r\sin\theta\cos\phi$$
$$y = r\sin\phi \qquad y = r\sin\theta\sin\phi \tag{1.35a}$$
$$z = z \qquad z = r\cos\theta$$

Expressions for the flux $\mathbf{J} = -D\nabla C$:

$$J_r = -D\frac{\partial C}{\partial r} \qquad J_r = -D\frac{\partial C}{\partial r}$$
$$J_\phi = -\frac{D}{r}\frac{\partial C}{\partial \phi} \qquad J_\phi = -\frac{D}{r\sin\theta}\frac{\partial C}{\partial \phi} \tag{1.35b}$$
$$J_z = -D\frac{\partial C}{\partial z} \qquad J_\theta = -\frac{D}{r}\frac{\partial C}{\partial \theta}$$

Expressions for the divergence $\nabla \cdot \mathbf{J}$:

$$\frac{1}{r}\frac{\partial}{\partial r}(rJ_r) + \frac{1}{r}\frac{\partial}{\partial \phi}J_\phi + \frac{\partial}{\partial z}J_z \qquad \frac{1}{r^2}\frac{\partial}{\partial r}(r^2 J_r) + \frac{1}{r\sin\theta}\frac{\partial}{\partial \phi}J_\phi + \frac{1}{r\sin\theta}\frac{\partial}{\partial \theta}(J_\theta \sin\theta) \tag{1.35c}$$

Expressions for the Laplacian $\nabla^2 C$:

$$\frac{1}{r}\frac{\partial}{\partial r}\left(r\frac{\partial C}{\partial r}\right) + \frac{1}{r^2}\frac{\partial^2 C}{\partial \phi^2} + \frac{\partial^2 C}{\partial z^2} \qquad \frac{1}{r^2}\frac{\partial}{\partial r}\left(r^2\frac{\partial C}{\partial r}\right) + \frac{1}{r^2\sin^2\theta}\frac{\partial^2 C}{\partial \phi^2} + \frac{1}{r^2\sin\theta}\frac{\partial}{\partial \theta}\left(\frac{\partial C}{\partial \theta}\sin\theta\right) \tag{1.35d}$$

It must be emphasized that the components of the flux vector $\mathbf{J}$ are evaluated in the *orthonormal* frames $(\mathbf{e}_r, \mathbf{e}_\phi, \mathbf{e}_z)$ and $(\mathbf{e}_r, \mathbf{e}_\theta, \mathbf{e}_\phi)$ of Fig. 1.9.[†] Each unit $\mathbf{e}$-vector is tangent to a given coordinate line, and it points in the direction of increasing values of that coordinate.

Problems in several space dimensions can be forbidding, as is illustrated by the full equations (1.35). Fortunately, a notable simplification accrues when problems are "rotationally symmetric," namely, when the concentration depends only on the distance from a fixed axis or fixed point. Then, the angular and axial parts of the Laplacian, Eqs. (1.35d), vanish identically, and the radial part appears to take a regular and simple form. This is indeed a general result.

EXERCISE 1.16. *Laplacian in rotationally symmetric systems:* If $C = C(r)$, where $r = (x^2 + y^2 + \cdots)^{1/2}$, and d is the number of dimensions of the rotationally symmetric subspace, show that

$$\nabla^2 C = \frac{1}{r^{d-1}} \frac{\partial}{\partial r}\left(r^{d-1} \frac{\partial C}{\partial r}\right) = \frac{\partial^2 C}{\partial r^2} + \frac{d-1}{r} \frac{\partial C}{\partial r}.$$

[Hint: Evaluate the Laplacian as the divergence of a gradient in rectangular coordinates.] □

It follows that Eqs. (1.32)–(1.34) take the form

$$\frac{\partial C}{\partial t} = -\frac{1}{r^{d-1}} \frac{\partial}{\partial r}(r^{d-1} J_r) = \frac{1}{r^{d-1}} \frac{\partial}{\partial r}\left(r^{d-1} D \frac{\partial C}{\partial r}\right) \tag{1.36}$$

for such cases where the concentration has neither angular nor axial dependence. Finally, in the spherically symmetric case $d = 3$, it is a curious fact that the function $u = rC$ satisfies the 1-d equation (1.14), with r substituted for x. This result is generalized in Exercise 1.17.

EXERCISE 1.17. *Simplifications of Eq. (1.36):* Assume that the concentration in a rotationally symmetric system has the form $C(r,t) = f(r)u(r, t)$. What can you say about the form of the functions f and u that simplify Eq. (1.36)? [Hint: Substitute and discuss the differential equations in f.] □

[†]These components are called the "physical" components, to be distinguished from covariant and contravariant components that must be considered in nonorthonormal frames.

1.7. Other Instances of the Diffusion Equation

A mere description of hopping on lattices would hardly justify a study of the diffusion equation. We hinted, however, that "conservation" or "balance" of certain physical quantities was the main issue. Although a complete treatment is outside our scope—it requires the arsenal of continuum mechanics—we now briefly consider other situations where the diffusion equation, in one guise or another, is a useful description of physical processes.[14,15] We write these in 1-d; the generalization to several dimensions is either trivial, as in heat conduction, or subtle, as in fluid dynamics.

A. Heat Conduction in a Stationary Solid

This case is closest to mass diffusion. It differs, however, in that we must introduce *two* related transport coefficients. First, there is a continuity equation,

$$\rho c \frac{\partial T}{\partial t} = -\frac{\partial J}{\partial x}, \tag{1.37a}$$

that expresses energy conservation. Here, T is the body's temperature, ρ and c are its mass density and heat capacity per unit mass, respectively, and J is the heat flux. The relation between temperature and flux is then expressed by Fourier's law of heat conduction,

$$J = -K \frac{\partial T}{\partial x}, \tag{1.37b}$$

in strict analogy with Fick's first law (1.12) and with no greater degree of validity. It defines the *thermal conductivity* K. Together, these equations produce the form (1.14) if we introduce the *thermal diffusivity* $\kappa = K/\rho c$, which indeed has the dimensions (cm^2/s) of D. It is now evident that Eq. (1.37a) represents energy conservation because $\rho c T$ is nothing but the internal energy density of a solid body. On the other hand, temperature is not a conserved quantity. Consequently, as mentioned above, two coefficients, K and κ, are required to describe temperature flow, rather than one, D, in the case of mass flow. The dimensions of J (W/cm^2) serve to find those of K ($W/cm\ K$).

B. Hydrodynamic Flow Past an Infinite Flat Plate

Consider the following, sometimes known as "Stokes's first problem." A semi-infinite region, filled with some incompressible fluid at rest, is bounded by a horizontal plate normal to an x-axis that points into the fluid. If the plate suddenly begins to move in its plane, then its translational motion imparts a shear motion to the fluid, and momentum is thus transferred. The fluid's velocity field has a single nonzero component v parallel to the plate, and one then shows that momentum conservation requires that it obey

$$\rho \frac{\partial v}{\partial t} = -\frac{\partial \tau}{\partial x}, \tag{1.38a}$$

where the shear stress τ serves to define the *viscosity* η if Newton's law

$$\tau = -\eta \frac{\partial v}{\partial x} \tag{1.38b}$$

holds. Note that the shear stress has the dimensions of a momentum flux, $[\tau] = (\mathrm{g\,cm/s})/\mathrm{cm}^2\,\mathrm{s}$. Equations (1.38), together, again yield an expression similar to Eq. (1.14) if one defines the *kinematic viscosity* $\nu = \eta/\rho$, which also has the dimensions of D. Again, the momentum density ρv is the conserved quantity, rather than the velocity.

C. Quantum Mechanics

Schrödinger's equation is also a diffusion equation, as is evident from

$$i\hbar \frac{\partial \psi}{\partial t} = \mathscr{H}\psi, \tag{1.39a}$$

with

$$\mathscr{H} = \frac{p^2}{2m} + V \quad \text{and} \quad p = -i\hbar \frac{\partial}{\partial x}. \tag{1.39b}$$

We now have an expression for the probability density ψ that is formally similar to a one-dimensional diffusion equation with a source[N11] related to the potential energy V, and where the momentum p plays the role of flux. The analogy ends there, however, because the related diffusivity $i\hbar/2m$ is purely imaginary. Nonetheless, some methods of scattering theory are formally similar to the solution by Green's functions of the diffusion equation.

EXERCISE 1.18. *Conservative form:* Rewrite Eqs. (1.37)–(1.39) in the form of Eqs. (1.12)–(1.13) when all the coefficients are constants. Interpret the "concentration" and the "flux" in each case. What are the "diffusivities"? □

D. Semiconductor Equations

The macroscopic transport of electrons and holes in semiconductors is also described by similar equations. If n and p denote their concentrations (i.e., their *number* densities), then one shows that they are conserved in the sense of Eq. (1.13):

$$\frac{\partial n}{\partial t} = -\frac{\partial J_n}{\partial x} \tag{1.40a}$$

and

$$\frac{\partial p}{\partial t} = -\frac{\partial J_p}{\partial x}. \tag{1.40b}$$

The fluxes, now, have a form more complex than (1.12) because both electrons and holes can be affected by electric fields E. This is a prime example of "drift" or "convective" terms in the flux. Because charged particles are constantly being accelerated between collisions, they achieve an average velocity, and hence give rise to a flow that is proportional to the field. An elementary account will be given in Chap. 3. We shall find that the total fluxes are

$$J_n = -D_n \frac{\partial n}{\partial x} - \mu_n E n, \tag{1.40c}$$

$$J_p = -D_n \frac{\partial p}{\partial x} + \mu_p E p, \tag{1.40d}$$

where the *mobilities* μ_n and μ_p characterize the particles' response to the field. Note that the drift terms come with opposite signs to ensure that both mobilities will be positive. This example also illustrates the possible complexities that arise in charged systems, for charges are also sources of the electric field. In fact, the three unknown functions n, p, and E are related through Gauss's law of electrostatics

$$\epsilon \frac{\partial E}{\partial x} = e(p - n), \tag{1.40e}$$

where e is the *magnitude* of the electronic charge and ϵ is the semiconductor's dielectric constant. The solution of any such problem is evidently highly nonlinear because of the products En and Ep in Eqs. (1.40c) and (1.40d). Ionic transport in solutions or in dielectrics is also governed by similar equations.

E. Current Flow Through a Conductor

Free electrons in a conductor are neither created nor destroyed. It can be shown that their density n obeys Eq. (1.40a), above, but that the diffusive component of their flux J_n is negligible compared to the drift component. If $\rho = -en$ represents the electron *charge* density and if it is recalled that the electric field derives from an electrostatic potential ϕ such that $E = -\phi_x$, then we have a continuity equation

$$\frac{\partial \rho}{\partial t} = -\frac{\partial J}{\partial x} \tag{1.41a}$$

that requires a current density

$$J = -eJ_n = -\mu_n en \frac{\partial \phi}{\partial x}. \tag{1.41b}$$

This last is nothing but Ohm's law, with *conductivity* $\sigma = \mu_n en = -\mu_n \rho$. The potential, however, satisfies Laplace's equation $\phi_{xx} = 0$ (in 1-d) because the free electrons in a conductor are compensated by a background sea of positive charges. The right-hand side of Gauss's law (1.40e) thus vanishes. These problems are linear, in contrast to semiconductor problems, because finding the electrostatic potential distribution is decoupled from the calculation of current density.

1.8. General Balance Laws and Constitutive Relations

There are many other manifestations of the diffusion equation: in biology, in queueing theory, in the theory of option pricing, and in the description of telephone traffic and other bottlenecks. At this point, the conclusion of the first chapter, we might ask why this is so. The answer rests on the observation that there are quantities in nature that are conserved or, at least, that can be balanced in some grand accounting scheme.[14–16] Such a quantity—call it Q—is additive in the sense that its value for a collection of subsystems is the algebraic sum of its values for each subsystem. Examples include the mass, charge, momentum, angular momentum, various forms of energy, and entropy of any macroscopic system. One then assumes that matter is continuous and that there exists a density function, or *field* q (per unit mass)[N12] such that

$$Q(t) = \int_{\mathscr{D}} \rho(\mathbf{x}, t) q(\mathbf{x}, t)\, dV. \tag{1.42}$$

Thus, an additive quantity is expressed as an integral over a domain $\mathscr{D}$ (a specified open and connected region of space) that contains a given physical system. Here, ρ is again the mass density, itself a field. In former days, additive quantities were known as "extensive," probably to distinguish them from fields, which were known as "intensive" quantities. Some books make much of this distinction, but in fact Eq. (1.42) is all there is to say.

To begin our balancing act requires merely that we distinguish what causes Q to change with time. There are exactly two reasons for change: either because there are internal sources or sinks σ distributed inside the system,[N11] or because there are exchanges, represented by a vector field $\mathbf{J}_{\text{tot}}$, with the world at large. The first are internal to $\mathscr{D}$, while the second act through its closed boundary $\partial\mathscr{D}$, as shown in Fig. 1.10. Thus we have the ledger

$$\dot{Q} = \int_{\mathscr{D}} \sigma\, dV - \oint_{\partial\mathscr{D}} \mathbf{J}_{\text{tot}} \cdot \mathbf{n}\, dA,$$

where $\mathbf{n}$ is the outward pointing normal to the system's boundary and dA is its area element.

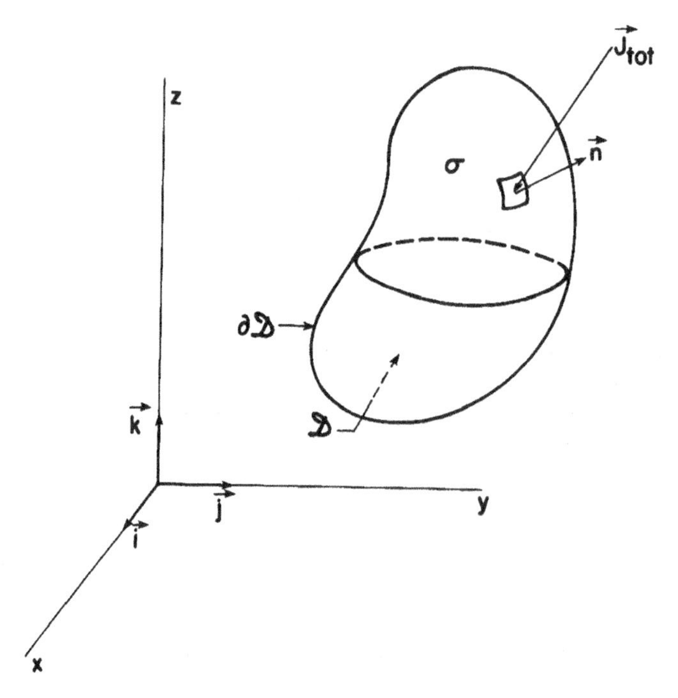

Fig. 1.10. The domain $\mathscr{D}$ of any physical system, its boundary $\partial\mathscr{D}$, and the unit exterior normal $\mathbf{n}$. The flux and (internal) source fields are shown.

It is now a simple matter to use the definition (1.42), to carry out the differentiation, and to use the divergence theorem[N13]

$$\oint_{\partial\mathscr{D}} \mathbf{J}_{\text{tot}} \cdot \mathbf{n}\, dA = \int_{\mathscr{D}} \nabla \cdot \mathbf{J}_{\text{tot}}\, dV. \tag{1.43}$$

We then find[N14]

$$\frac{\partial \rho q}{\partial t} = -\nabla \cdot \mathbf{J}_{\text{tot}} + \sigma, \tag{1.44}$$

an equation already reminiscent of the continuity equation (1.32). It remains only to recognize that a given field q can also be carried (or "convected") by an overall fluid motion $\mathbf{v}$, so that the decomposition of the total flux

$$\mathbf{J}_{\text{tot}} = \rho q \mathbf{v} + \mathbf{J} \tag{1.45}$$

defines the true flux $\mathbf{J}$ owing to causes other than convection. In particular, the total mass is an additive quantity with density ρ. Moreover, mass can be neither created nor destroyed, and its total flux is merely that due to convection. Thus, putting $q = 1$, $\sigma = 0$, and $\mathbf{J}_{\text{tot}} = \rho \mathbf{v}$ in the above Eq. (1.44), we get the (mass) continuity equation

$$\frac{\partial \rho}{\partial t} = -\nabla \cdot (\rho \mathbf{v}). \tag{1.46}$$

Inserting Eq. (1.45) into Eq. (1.44) and using Eq. (1.46), we obtain the general balance equation for any field†

$$\rho \frac{\partial q}{\partial t} + \rho \mathbf{v} \cdot \nabla q = -\nabla \cdot \mathbf{J} + \sigma. \tag{1.47}$$

This is the point of departure for a full development and justification of the special cases, and others as well, that were mentioned in the previous section. Note that Eq. (1.47) is first order in time and that it contains the continuity equation (1.32) as a special case when the velocity field and source term vanish.

Here, we are content to note that the balance equation (1.47) is an exact statement in the sense that all one does is distinguish causes and effects. Nevertheless, for a given source σ and impressed velocity field $\mathbf{v}$, that equation relates the *two* unknowns: field q and flux $\mathbf{J}$. To solve such an equation still requires a relation between these unknowns. Such a relation is called "constitutive" or "phenomenological" because it depends on the system's basic building blocks and because it must ultimately find its justification in a microscopic model of the transport process. For example, our random walk for mass diffusion was precisely such a finer-scale model. Indeed, it yielded a *linear* relation between concentration gradient and mass flux, and that relation, Fick's first law (1.12) or (1.31), was characterized by a single *transport coefficient*, the diffusivity, which was also estimated [see Eq.

†This calculation requires an important result of vector analysis: $\nabla \cdot (\rho q \mathbf{v}) = q \nabla \cdot (\rho \mathbf{v}) + (\rho \mathbf{v}) \cdot \nabla q$.

(1.11)] in terms of jump frequencies and jump distances. Experimental observation shows that such linear relations are not limited to mass transport. The laws of Fourier (1.37b), Newton (1.38b), and Ohm (1.41b) are often obeyed when systems are "close to equilibrium." The following heuristic argument may serve.

At equilibrium the various fields q that characterize the state of a given physical system are generally (but not always) independent of the space coordinates (x, y, z). The fluxes $\mathbf{J}$ of these fields also vanish at equilibrium. Therefore, when a system is disturbed, it is plausible that the flux $\mathbf{J}$ of any such field should be functionally related to the *gradients* of q. A Taylor expansion around the equilibrium state then yields[N15]

$$\mathbf{J} = -L\nabla q, \tag{1.48}$$

which defines the transport coefficient L. This relation has an immediate interpretation, for the equation $q(\mathbf{x}) = \text{const.}$ defines level surfaces (isoconcentration surfaces, isotherms, equipotentials, etc.) The corresponding fluxes are then orthogonal to these surfaces if the medium is isotropic (L is then a scalar). The gradients ∇q are sometimes called "thermodynamic forces" because they appear to cause the corresponding fluxes or currents $\mathbf{J}$, and, in the idiom of electrical circuits, the transport coefficient L has the character of a conductance.[N16] Finally, the minus sign and the positivity of L in Eq. (1.48) are related to the stability of transport processes, much as in our elementary discussion in Sects. 1.3 and 1.5. Contemporary folklore couches these considerations in terms of "minimum entropy production" or "dissipative structures." For our purposes, however, these would be mere slogans, and we shall merely say that physical systems evolve, most often, toward states of lower energy. In a nutshell, *physical quantities coast downhill all the way.*[17]

EXERCISE 1.19. *Invariance of balance laws under affine transformations:* Consider, for simplicity, the 1-d form of Eqs. (1.47) and (1.48). Eliminate the flux in favor of the field to get a "diffusion equation," and recall the affine transformation of Exercise 1.8. What conditions must the transformation coefficients satisfy to preserve the form of these equations? It is assumed that the field q is a scalar under that transformation. How must the source σ and the velocity v transform? □

Notes

1. How good an approximation is Eq. (1.7) to the discrete rate equation (1.1)? That question is easily answered if we assume that $\hat{N}$ is an analytic function of x. We can then carry out the Taylor expansions (1.6) to all orders. Substituting into the right-hand side of Eq. (1.1), all terms of odd order in a vanish identically, and we get

$$\frac{\partial \hat{N}}{\partial t} = \Gamma \sum_{n=1}^{\infty} \frac{a^{2n}}{(2n)!} \frac{\partial^{2n} \hat{N}}{\partial x^{2n}}. \tag{1}$$

We attempt a plane-wave solution $\hat{N}(x, t) = A_k(t) \exp(ikx)$, where $i = \sqrt{(-1)}$. From the theory of Fourier transforms, any solution satisfying given initial and boundary conditions is a linear superposition of such plane waves with wave vectors k, so that there is no loss of generality. Introducing the plane-wave solution into Eq. (1) we get a differential equation

$$\dot{A}_k / A_k = \Gamma \sum_{n=1}^{\infty} (-1)^n \frac{(ak)^{2n}}{(2n)!} = -\Gamma(1 - \cos ak) \tag{2}$$

for the amplitudes, which is easily solved

$$A_k(t) = A_k(0) e^{-2\Gamma t \sin^2(ak/2)}. \tag{3}$$

The exponent is always negative, which means that we have decaying (i.e., stable) solutions. Inserting the plane-wave solution into the truncated expression (1.7) and comparing to Eq. (3), we see that the continuum approximation is valid if $ak \ll 1$. For any of the usual boundary conditions (that determine the "spectrum" of k), the jump distance must be much smaller than the system's characteristic dimensions.

2. The random walk (1.1)–(1.3) depends on a single physical quantity, the jump frequency, and Γ^{-1} can serve as a time-scale. There are no distance scales for such problems, even in bounded regions, $0 < i < i_{max}$, say, because the site indices are pure numbers (indeed, integers). On the other hand, the continuum equations (1.12)–(1.15) do admit a distance scale when the system is bounded $0 < x < L$, and the diffusivity then serves to define a time scale L^2/D.

3. Representative values of D in various phases, in units of cm^2/s:

Gases	Liquids	Solids
10–0.1	10^{-4}–10^{-6}	10^{-8}–10^{-20}

It is striking that diffusivities of gases and liquids are bounded in tight ranges and indeed are weak functions of temperature. On the other hand, diffusivities in solids vary over a wide range, both from material to material and as a function of temperature for a given material. An explanation is offered in Chap. 3.

4. This rule of calculus reads

$$\frac{d}{dt}\int_a^{\xi(t)} f(x, t)\, dx = \int_a^{\xi(t)} \frac{\partial}{\partial t} f(x, t)\, dx + \dot{\xi} f[\xi(t), t]. \tag{4}$$

Thus, the derivative with respect to a parameter (t in this instance) of an integral, whose upper limit is also a function of time, is the sum of two terms: (i) an integral evaluated as if the boundary does not move; (ii) a term proportional to the boundary's velocity, as if the boundary "carries" a "density" f. Leibniz's rule is intuitively obvious, as a quick sketch will show.

5. This can be understood if, for example, processes at the moving boundary occur fast enough for *local* equilibrium to obtain. Equilibrium then implies the continuity of the chemical potential and of the temperature field across that boundary. But equal chemical potentials μ generally imply a discontinuity in concentration, because the relation $\mu(C)$ varies with every material or phase. These questions will be addressed in Sect. 2.4, Appendix C, and Chap. 6.
6. This statement is strictly true only for planar boundaries on which the number of particles remains constant. For example, adsorption phenomena or surface reactions require that one consider concentration changes of the "surface phase," and then Eqs. (1.19)–(1.20) must be modified accordingly. Chapter 6 is devoted to some of these questions.
7. The continuum standpoint takes the view that the diffusion equation is a faithful representation of reality, although we know full well that a discrete random walk gives a more detailed description of diffusion. (see Note 3 in Chap. 6).
8. In contrast to the wave equation, it can be shown that the diffusion equation, being only first order in time, implies that signals must propagate with infinite speed. This will be obvious from the gaussian solution in Chap. 4 (see Note 2 in that chapter). For effects based on finite propagation speed, see "Finite Speed of Propagation in Heat Conduction, Diffusion, and Viscous Flow," by H. D. Weymann, *Amer. J. Phys.* **35**, 488 (1967), and "Heat Waves," by D. D. Joseph and L. Preziosi, *Rev. Mod. Phys.* **61**, 41 (1989); *addendum, ibid.* **62**, 375 (1990).
9. Call $G_{i,0}^n$ the function that we have just calculated numerically, where the second subscript, 0, is to remind us that the disturbance was placed at the origin. Formally, G is invariant under the translation $i \to \bar{\imath} = i - k$, which implies $G_{i,0}^n = G_{\bar{\imath},\bar{0}}^n = G_{i-k,k}^n$. In an infinite region, Green's function depends

spatially only on the difference between the "observation point" i and the "source point" k. The second subscript k is thus superfluous.

10. Isotropy, here, means that the concentration gradient is collinear with the flux. In crystals, for example, this may not be the case. Then Eq. (1.31) must be generalized to give

$$J^\alpha = -\sum_\beta D^{\alpha\beta} \frac{\partial C}{\partial x^\beta}, \tag{5}$$

where Greek superscripts label cartesian components (see Appendix A).

11. The notion of a distributed source (or sink) is best illustrated through an example: If a binary gas undergoes a homogeneous chemical reaction $A \to B$, then the mole density of each species changes as a function of time. The reaction *rate* is then a source-like term, because it describes the removal of moles of 'A' from the left-hand side of the above reaction and the creation of moles of 'B' on its right-hand side.
12. A "field" is generally understood to be *any* scalar-, vector-, or even tensor-valued function of the space and time variables $(\mathbf{x}, t)$. Not all fields, however, are density functions, i.e., functions such that their integral over all space has a physical meaning. For example, the temperature is a field, but its integral over space means nothing at all.
13. This theorem is attributed to Green or to Ostrogradskii (or to both); it was surely known to Gauss. It also illustrates a minor semantic difficulty associated with the word "flux": In potential theory, one often defines the flux of a vector field as an expression similar to the left-hand side of Eq. (1.43); it corresponds to our *integrated* flux. Some authors thus prefer the word "current" for the vector quantity $\mathbf{J}$ representing the flow of a physical quantity q. Besides his theorems and functions, very little is known about George Green except, perhaps, that his education was mercifully unconventional and that he sired at least seven illegitimate children. See the remarkable article by J. Gray, "Green and Green's Functions," in *The Mathematical Intelligencer* **16**, No. 1, 45 (1994).
14. The derivation rests critically on the assumption that the domain $\mathscr{D}$ be continuous. If not, as in a system of several phases, that domain must be broken up into disjoint pieces and each integral considered separately, much as was done for Properties A.2–A.3 of Sect. 1.3.
15. The total flux (1.45) cannot be thus functionally related because the resulting constitutive relation would not be galilean invariant, a requirement of all laws of nonrelativistic physics.
16. By now it should be clear that linear constitutive relations such as Eq. (1.48) do not have the same degree of generality as the balance laws (1.47). There exist physical systems and phenomena where the former do not hold, e.g., viscoelastic and porous media, hysteresis, and spinodal decomposition. Furthermore, multicomponent systems can exhibit "uphill"

diffusion, so that the diffusivity appears to be negative in certain regions. Although the system's energy can increase locally in these regions, the overall energy must decrease with time.

References

1. Lord Rayleigh, "On James Bernoulli's Theorem in Probabilities," *Phil. Mag.* **47**, 246 (1899). [Reprinted in *Scientific Papers* **4**, p. 370 (Dover, New York, 1964).]
2. S. Chandrasekhar, "Stochastic Problems in Physics and Astronomy," *Revs. Mod. Phys.* **15**, 1 (1943). [Reprinted in *Selected Papers on Noise and Stochastic Processes*, N. Wax, Ed. (Dover, New York, 1954).]
3. P. G. Shewmon, *Diffusion in Solids* (McGraw-Hill, New York, 1963).
4. C. C. Lin and L. A. Segel, *Mathematics Applied to Deterministic Problems in the Natural Sciences* (MacMillan, New York, 1974).
5. H. S. Carslaw and J. C. Jaeger, *Conduction of Heat in Solids*, 2nd ed. (Oxford University Press, London, 1959).
6. D. V. Widder, *The Heat Equation* (Academic Press, New York, 1975).
7. A. N. Tikhonov and A. A. Samarskii, *Equations of Mathematical Physics* (Pergamon Press, Oxford, 1963). [Reprinted by Dover, New York.]
8. R. Ghez and W. E. Langlois, "More on the Concentration Dependence of Fick's Laws," *Amer. J. Phys.* **54**, 646 (1986).
9. E. Isaacson and H. B. Keller, *Analysis of Numerical Methods* (John Wiley, New York, 1966).
10. G. D. Smith, *Numerical Solutions of Partial Differential Equations: Finite Difference Methods*, 3rd ed. (Oxford University Press, Oxford, 1985).
11. R. Courant, K. Friedrichs, and H. Lewy, "Über die Partiellen Differenzengleichungen der Mathematischen Physik," *Math. Ann.* **100**, 32 (1928). [Translated and reprinted in *IBM J. Res. & Dev.* **11**, 215 (1967).]
12. R. D. Richtmeyer and K. W. Morton, *Difference Methods for Initial Value Problems*, 2nd ed. (Interscience, New York, 1967).
13. G. G. O'Brian, M. A. Hyman, and S. Kaplan, "A Study of the Numerical Solution of Partial Differential Equations," *J. Math. & Phys.* **29**, 223 (1950).
14. R. B. Bird, W. E. Stewart, and E. N. Lightfoot, *Transport Phenomena* (John Wiley, New York, 1960).
15. J. C. Slattery, *Momentum, Energy, and Mass Transfer in Continua*, 2nd ed. (Robert E. Krieger, New York, 1981).
16. S. R. de Groot and P. Mazur, *Non-Equilibrium Thermodynamics* (North-Holland, Amsterdam, 1962). [Reprinted by Dover, New York.]
17. L. Woolf, *Downhill All the Way: Autobiography of the Years 1919–39* (Hogarth Press, London, 1967).

2

Steady-State Examples

This chapter introduces the notion of steady state through several physically significant examples. Indeed, there exist many time-dependent problems, among which are problems of phase growth, whose solution can be "almost" at steady state. Such is the case for oxidation kinetics, precipitate growth, and crystal growth under conditions of constant supersaturation. Along the way, the equilibrium properties of binary systems and of "small" phases will be reviewed.

2.1. The Steady State Is *Not* the Equilibrium State

The steady state must be carefully distinguished from the equilibrium state, even though the latter is a subcase of the former. We first should recall that equilibrium obtains when the total energy of a body is at a minimum under certain constraints (e.g., constant total entropy, volume, and total masses of various species) that characterize a given physical system macroscopically. With this variational principle, one then *proves* that the fields "conjugate" to the constraints—temperature, pressure, and chemical potentials—are independent of the space coordinates within each phase.[N1] These fields are said to be *constant*. On the other hand, we have the

DEFINITION. *Any field* $q(\mathbf{x}, t)$ *is at steady state if and only if* $\partial q/\partial t = 0$.

Therefore, a steady-state field q can depend only on position $\mathbf{x}$ and on any other coordinates that characterize the physical system's internal state: Measurements, today, of a steady-state quantity yield the same value as any later measurement, as long as all external conditions remain constant in time.

The simple 1-d equations (1.12)–(1.14), where the diffusivity D is a constant, will serve to illustrate this difference. If $\partial C/\partial t = 0$, then the flux (1.12) is also independent of time. Furthermore, the continuity equation (1.13) shows that the flux is in fact independent of position x, as well: $J(x) = J_0$, say, is a "first integral" of this set of equations. Finally, Eq. (1.12) is now a simple differential equation whose solution is evidently $C(x) = \alpha - (J_0/D)x$, where α is another constant. The flux is thus constant with respect to the space coordinate x, and the concentration is a linear function of that coordinate. This is characteristic of the steady state in one dimension (for constant D's). The same holds true for the random walk equations (1.1)–(1.3), as we saw in Exercise 1.1. In contrast, equilibrium is that subcase where the flux is not only constant but strictly vanishes, which implies that the particle distribution is then constant.[†]

EXERCISE 2.1. *Steady-state thermal conduction through a composite slab:* Assume constant conductivity K_i in each distinct region (x_{i-1}, x_i), for $(i = 1, 2, \ldots, M)$, and temperature continuity at the phase boundaries. The temperatures at the external boundaries x_0 and x_M are given. Find the temperature distribution and interpret the result in terms of "thermal resistances," "current," and "potential drop." □

It must be emphasized that a linear concentration distribution, derived above in steady state, is the exception rather than the rule. Nonlinear steady-state distributions always occur in more than one space dimension, and, even in 1-d, nonconstant diffusivities cause nonlinearities, as the following exercises and examples will show.

EXERCISE 2.2. *Steady-state mass diffusion when $D = D(C)$:* Repeat the argument above when the diffusivity depends exclusively on concentration. In other words, Eqs. (1.12)–(1.13) are assumed valid but Eq. (1.15) rather than (1.14) holds. □

[†] In general, a nonconstant concentration distribution is consistent with equilibrium because only the constancy of the electrochemical potential is required. This will be illustrated in Chap. 3 with a short introduction to double layers.

It follows from the definition of the steady state and from the general balance law (1.47) that the steady-state values of any density function q must satisfy

$$\rho\mathbf{v}\cdot\nabla q = -\nabla\cdot\mathbf{J} + \sigma. \tag{2.1}$$

If, in addition, the velocity field and the source term are negligible, then it follows that the flux $\mathbf{J}$ must be divergence free:

$$\nabla\cdot\mathbf{J} = 0. \tag{2.2a}$$

If, further, field and flux are related by the constitutive relation (1.48),

$$\mathbf{J} = -L\nabla q, \tag{2.2b}$$

repeated here for convenience, then the field q must satisfy

$$\nabla\cdot(L\nabla q) = 0. \tag{2.2c}$$

Finally, if the transport coefficient L is independent of location, then the field q obeys Laplace's equation

$$\nabla^2 q = 0. \tag{2.2d}$$

These are the equations that we intend to illustrate and solve in this chapter.

EXERCISE 2.3. *Power-law steady-state solutions in a slab:* Find all physical solutions $C(x)$ for steady-state diffusion in a slab $0 < x < L$, without sources or convective motion, if $D(C) = KC^{\nu}$, where K is a positive constant and the exponent ν is arbitrary. Assume that the boundary condition at $x = 0$ is either (i) $C(0) = C_0$ or (ii) $J(0) = J_0$, where C_0 and J_0 are arbitrary constants. Assume that $C(L) = 0$ in both cases. Calculate the slopes $C'(x)$. Are there any restrictions on the possible values of C_0, J_0, and ν? Is the "existence theorem" violated? □

2.2. General Solutions for Rotationally Symmetric Cases

When the concentration depends only on the distance to a fixed point or axis, Eq. (1.36) shows that Eqs. (2.2a) and (2.2b) reduce to

$$\frac{d}{dr}(r^{d-1}J) = 0, \tag{2.3a}$$

$$J = -D\frac{dC}{dr}, \tag{2.3b}$$

where, for brevity, the subscript r of the radial flux component has been dropped.† The first equation (2.3a) yields immediately

$$J = \beta/r^{d-1} \tag{2.4a}$$

for all points other than the origin. With this "first integral," the second equation (2.3b) gives

$$C = \begin{cases} \alpha - [\beta/D(2-d)]r^{2-d} & \text{for } d \neq 2, \\ \alpha - (\beta/D)\ln r & \text{for } d = 2, \end{cases} \tag{2.4b}$$

if, for whatever reason, the diffusivity is spatially constant. Here, α and β are integration constants. Note also that Eqs. (2.4) are singular at the origin $r = 0$, except in the case of one space dimension.

We now apply these equations to a particularly simple, yet useful case: *diffusion through a long hollow cylinder*. This method is of practical interest since it has often been used in the past to measure diffusivities or thermal conductivities,[1,2] even when these depend on concentration or temperature. Consider, therefore, a long circular cylinder of length H and of radius R_2 made of some material whose diffusive properties we wish to investigate. Then bore a hole of radius R_1 along its axis, as shown schematically in Fig. 2.1. This has two functions: To physically remove the singularity of the previous equations, and to allow the passage of a substance whose diffusive properties through the cylinder we wish to investigate. For example, one could pass a reactive substance such as water vapor through the wall

†Recall that d is the dimension of the rotationally symmetric subspace.

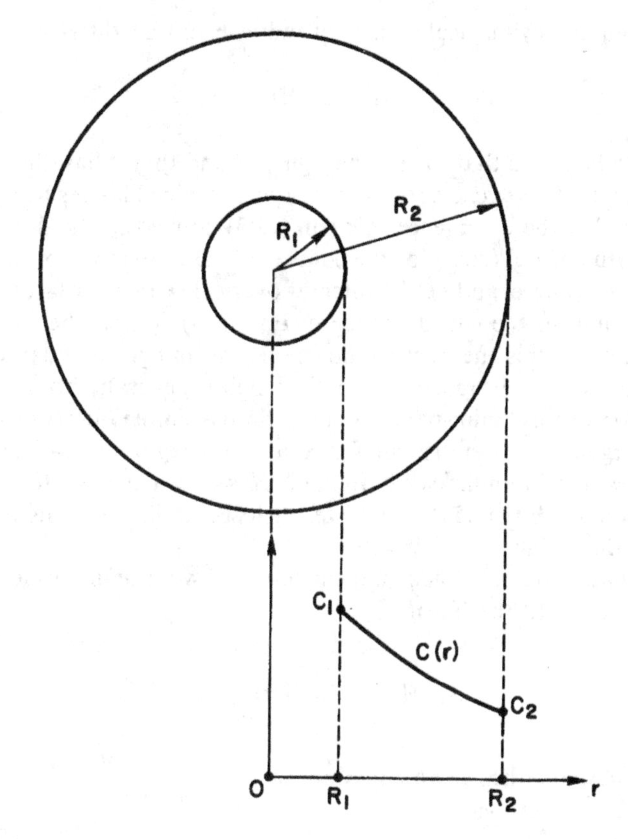

Fig. 2.1. Schematic representation of the hollow-cylinder problem.

of a metal cylinder to study its permeability or oxidation kinetics, or an impurity to modify its electrical or mechanical properties. We assume for simplicity that the concentration of the diffusing species is fixed at the inner and outer cylindrical surfaces, namely, $C(R_1) = C_1$ and $C(R_2) = C_2$. We also assume that the cylinder is so long that the concentration does not depend appreciably on axial position z. Consequently, the concentration distribution $C(r)$ is purely radial, and the previous relation (2.4a) holds in each transverse section of two dimensions. A short discussion of this problem follows.

Applying Eq. (2.4a) for $d = 2$, we get $rJ(r) = \text{const.}$ for all r. Therefore, the total amount of substance that passes, in a time interval

t, through *any* cylindrical surface of radius r and height H is

$$M = 2\pi r H t J. \tag{2.5a}$$

It should be noted that M has the sign of J and that it has dimensions of grams if $[J] = \text{g/cm}^2\text{s}$. Now for a quick aside. This result appears to contradict the divergence theorem (1.43) applied to the flux vector **J**: By virtue of Eq. (2.2a) (or the equivalent 2.3a), that vector ought to be divergence-free, and yet its integral over cylindrical surfaces is here proportional to the left-hand side of Eq. (2.5a). Thus, whereas blithe application of this theorem would predict no integrated flux, we find a quantity that is proportional to M, which is generally nonzero. The resolution of this "mini-paradox" rests on the simple observation that the divergence theorem is true for continuous regions $\mathscr{D}$ only, i.e., for regions whose boundaries $\partial\mathscr{D}$ *do not enclose singularities*. Here, in the derivation of Eq. (2.5a), we have enclosed a line of singularities, namely the cylinder's axis $r = 0$.

Equation (2.5a) leads immediately to two useful results if we remember Eq. (2.3b). Then,

$$M = -2\pi r H t D \frac{dC}{dr} \tag{2.5b}$$

can be viewed either as an equation for D

$$D(r) = -\frac{M}{2\pi r H t C'}, \tag{2.6a}$$

or as a differential equation for C. In the first case, measurements (e.g., by sectioning) of a concentration profile $C(r)$, under conditions of known constant mass flow rate $\dot{M}$, give the position dependence of the diffusivity in the annular region $R_1 < r < R_2$. From the measured profile we can then construct the dependence $D(C)$. In the second case, variables separate in Eq. (2.5b) to give

$$M \ln(R_2/R_1) = 2\pi H t \int_{C_2}^{C_1} D(C)\, dC. \tag{2.6b}$$

Since the above integral is nothing other than the average diffusivity

$\langle D \rangle$ times the concentration drop $(C_1 - C_2)$ over the annulus, Eq. (2.6b) provides a rapid estimate of the diffusivity if we can measure all other quantities accurately. The computation of the concentration distribution is the subject of the exercise that follows.

EXERCISE 2.4. *Concentration distribution in a long hollow cylinder:* Assume that the diffusivity is constant and that the concentrations are constants on the inner and outer cylindrical surfaces. Find the function $C(r)$ either by determining the integration constants in Eq. (2.4b) or by integrating the equivalent Eq. (2.5b). Discuss the case of a "thin" annulus.[N2] [Hint: Use dimensionless variables such that $(r - R_1)/R_1 \ll 1$.] □

2.3. The Thermal Oxidation of Silicon

In the semiconductor industry, there is perhaps no single processing step more important than the formation of SiO_2 on silicon. As is well known,[3] these overlayers form protective "masks" against the diffusion of impurities, and they also provide electrical insulation and dielectric properties necessary for the fabrication and operation of semiconductor devices. These facts alone warrant the study of SiO_2 growth kinetics. From our point of view, however, this physical system is ideally suited to show how one can develop a physical model: what assumptions must be retained and which ones should be discarded, how to set up the problem mathematically, and how to test the validity of the model.

First, some experimental evidence. Silicon is a highly reactive element. In particular, a stable amorphous oxide forms on its surface if it is exposed to an oxygen-bearing ambient. This can be, typically, a mixture of inert gases (e.g., H_2 or N_2) and oxidants O_2 or H_2O. A thin ($\simeq 10$ Å) layer of oxide then forms, even at room temperature.† But the oxidation reaction is an *activated process*, which means that it is greatly accelerated at elevated temperatures ($> 650°C$) to promote the formation of much thicker oxide layers that are then useful. These layers are also *conformal* and *adherent*, in the sense that they follow exactly the topography of the underlying Si substrate. Other shreds of

†This so-called "native" oxide causes processing engineers much grief because it covers what should be an otherwise clean Si surface. The considerations of this section do *not* apply to the formation of native oxides because chemisorption steps are involved.

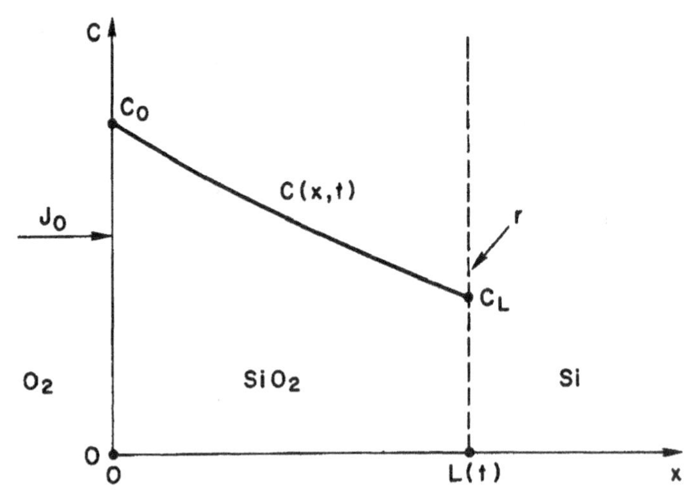

Fig. 2.2. Three-phase oxidation system and excess O_2 distribution in the oxide.

experimental evidence include the following: The oxidation *rate* is not a constant; it decreases roughly as $t^{-1/2}$ for long enough oxidation times, and it is approximately proportional to the oxygen partial pressure p. This suggests that the oxidation process is mainly due to the migration of *oxygen molecules* to the Si/SiO_2 interface where they react, at the silicon substrate's expense, to form fresh SiO_2. Silicon does *not* appear to migrate significantly. These observations form the basis of a model for oxidation kinetics that was popularized by Deal and Grove,[4] and which we now develop in 1-d.

Figure 2.2 shows schematically a planar layer of SiO_2 during its formation. It is sandwiched between an O_2 ambient, with partial pressure p, and the Si substrate. This is clearly a three-phase system and a first example of a *phase change*. We seek an expression for the oxide's thickness L as a function of time and of other experimentally controllable parameters. We assume that:

a. Oxygen dissolves molecularly at a rate J_0 into the existing SiO_2 at the outer interface (taken as the origin of coordinates), and its concentration there is given by Henry's law,

$$C_0 = Kp, \tag{2.7a}$$

where the *solubility* K depends only on temperature T.

b. Oxygen diffuses molecularly through the existing SiO_2, and it does not react *homogeneously* with the glass network.[N3] This assumption has been verified experimentally.[5] If $C(x, t)$ is the O_2 concentration in *excess* of what is contained in the network, then we assume that its evolution can be described by the diffusion equation (1.14).

c. Oxygen molecules react *heterogeneously* at the inner interface according to the overall reaction

$$Si + O_2 \rightarrow SiO_2. \tag{2.7b}$$

The chemical *reaction rate* r must thus be first order in the oxygen concentration C_L that exists at the Si/SiO_2 interface. Expressed mathematically,

$$r = kC_L, \tag{2.7c}$$

i.e., the reaction rate is linear in the O_2 concentration and its rate constant k depends only on temperature. Note that r has the dimensions of flux and that the position of the Si/SiO_2 interface $L(t)$, in our coordinate system, is also the instantaneous oxide thickness.

d. Diffusion occurs at steady state, so that the diffusion equation reduces simply to $C_{xx} = 0$. This statement clearly requires elaboration because oxidation is evidently a time-dependent process the (oxide thickness L is a nonlinear function of time). In fact, this example provides us with our first nontrivial problem: How can we model a physical system whose boundary moves? A complete answer, in this instance, requires the numerical solution of the full time-dependent problem. It is useful, however, to assume that the concentration field C is approximately at steady state, namely that

$$C \approx \alpha + \beta x, \tag{2.7d}$$

in accordance with the results of Sect. 2.1, but where the "constants" α and β can depend weakly on time. This is sometimes called the assumption of *quasi*-steady state, to which we will return later on.

To solve this problem demands that we take mass conservation somewhat more seriously, because the inner interface $x = L(t)$ is the seat of physical processes other than the mere exchange of mass by diffusion. We must thus refine the reasoning implicit in Properties A of Sect. 1.3. For definiteness, let us say that all concentrations are measured in units of number densities, i.e., $[C] = \mathrm{cm}^{-3}$. Then the total number of excess O_2 molecules in the system,[N4] per unit of substrate area, is

$$M_{O_2}(t) = \int_0^{L(t)} C(x, t)\, dx. \tag{2.8a}$$

Its time derivative must be equal to any known reason for change, namely, an accretion by the inward flux J_0 from the gas phase minus a loss due to the reaction rate r of Eq. (2.7b). Since we have no reaction at the outer interface that changes the number of moles (such as in the case of dissociative adsorption; see Exercise 2.7, below), it follows that the flux must be continuous there, exactly as in Eqs. (1.19). Carrying out the indicated differentiation, using Leibniz's rule (recall Note 4 of Chap. 1), and balancing the fluxes at $x = 0$, we find a modified Stefan condition

$$D \left.\frac{\partial C}{\partial x}\right|_L + r + C_L \dot{L} = 0 \tag{2.8b}$$

that must hold at the inner interface. It remains to relate the reaction rate to the movement of that interface. This is easily done by considering the total number of SiO_2 molecules in our system. The concentration C_{SiO_2} is evidently a constant and is the reciprocal of the molecular volume Ω of SiO_2. Thus

$$M_{SiO_2}(t) = \Omega^{-1} L(t), \tag{2.8c}$$

whose time derivative can only be the gain of SiO_2 though the right-hand side of the reaction (2.7b). Thus, we get the desired relation

$$r = \Omega^{-1} \dot{L}. \tag{2.8d}$$

We are now almost done with model building. It simply remains to solve the above equations. With Eq. (2.7c), eliminate the reaction rate r from Eqs. (2.8b) and (2.8d):

$$DC_x|_L + kC_L(1 + \Omega C_L) = 0, \tag{2.9a}$$

$$\dot{L} = \Omega k C_L. \tag{2.9b}$$

These are boundary conditions that must hold at the Si/SiO_2 interface. The first determines the yet unknown concentration C_L at that interface, and the second, a Stefan condition, gives that interface's rate of motion. On the other hand, the assumed steady-state distribution (2.7d) can also be written

$$C(x) = C_0 - (C_0 - C_L)x/L \tag{2.10}$$

in terms of the concentrations at both interfaces. Inserting this expression into Eq. (2.9a) gives a quadratic equation for C_L, and the problem would be vastly simpler if we could neglect the second term in the parenthesis of Eq. (2.9a). This is indeed the case, as the following numerical estimate shows.

The atomic volume of SiO_2 is equal to its molecular mass (60.1 g/mole) divided by its density (2.27 g/cm^3) and by Avogadro's number (6.02×10^{23} mole^{-1}). Thus, $\Omega = 4.4 \times 10^{-23}$ cm^3. On the other hand, the concentration C_L is certainly less than C_0 that is imposed at the gas/SiO_2 interface. Now, one finds experimentally that the solubility K in Henry's law (2.7a) is of the order 7.5×10^{16} atm^{-1} cm^{-3}, so that the dimensionless term we are considering, ΩC_L, is less than $\Omega C_0 = 3.4 \times 10^{-6}$ at a pressure of 1 atm. This is definitely negligible in comparison with unity in Eq. (2.9a). Using that approximate boundary condition, we find that

$$C_L = \frac{C_0}{1 + kL/D}, \tag{2.11a}$$

and, with Eq. (2.9b), for the inner interface's rate of motion we have

$$\frac{dL}{dt} = \frac{\Omega k C_0}{1 + kL/D}. \tag{2.11b}$$

This last is a differential equation for $L(t)$ that is easily solved (the variables separate) under the initial condition $L(0) = 0$. We thus get the required thickness–time relation

$$L/k_L + L^2/k_P = t, \tag{2.12}$$

where the so-called *linear* and *parabolic* rate constants k_L and k_P are

$$k_L = \Omega k C_0 = \Omega k K p, \tag{2.13a}$$

$$k_P = 2\Omega D C_0 = 2\Omega D K p. \tag{2.13b}$$

Discussion and Critique

a. The oxidation law (2.12) is called *linear-parabolic* for obvious reasons. This kinetic behavior was recognized by various workers[6] well before Deal and Grove. The linear term on the left-hand side dominates for short enough oxidation times, while the quadratic term weighs more heavily for longer times. Most experimental observations confirm this behavior. In particular, the oxidation rate $\dot{L}(t)$ goes from a constant, for short times, to a $t^{-1/2}$ behavior, for long times. However, the experimental determination of the rate constants k_L and k_P is no easy matter, because the transition from "pure" linear to quadratic behavior occurs over several orders of magnitude of time. The reader is encouraged to peruse Ref. 7, an experimental study using an automated ellipsometer, to understand just how difficult an unambiguous determination of these rate constants can be. These considerations are amplified in the exercise that follows.

EXERCISE 2.5. *Characteristics of the linear-parabolic law:* Equation (2.12) describes a parabola in the (t, L) plane. Solve for the function $L(t)$ and characterize this parabola (axis, branches, physical and nonphysical parts). Then expand that function in an appropriately chosen dimensionless time variable and define times t_L and t_P such that linear kinetics hold for $t < t_L$, while parabolic kinetics dominate for $t > t_P$. The interval of time between t_L and t_P is called the region of *mixed kinetics*; it is also cause for much confusion in the literature. □

b. Both of Eqs. (2.11) involve the denominator $1 + kL/D$. In the language of equivalent circuits (see Exercise 2.1), this can be understood as the sum of two resistances in series, k^{-1} and L/D, that contribute to the overall rate limitation of the oxidation process. For example, Eq. (2.11a) expresses the total "potential drop" C_0 as the sum of a potential C_L across the surface resistance k^{-1} and a potential $C_0 - C_L$ that develops across the diffusional resistance L/D. The dimensionless ratio kL/D measures the relative dominance of these two resistances. Such a decomposition into surface and bulk rate limitations is of fundamental importance in any rate process, and Chap. 6 is entirely devoted to this phenomenon.

c. The measured rate constants k_L and k_P allow us to determine the primary constants of the model, k and D, *if* the solubility K is known precisely from independent measurements. If not, Eqs. (2.13)

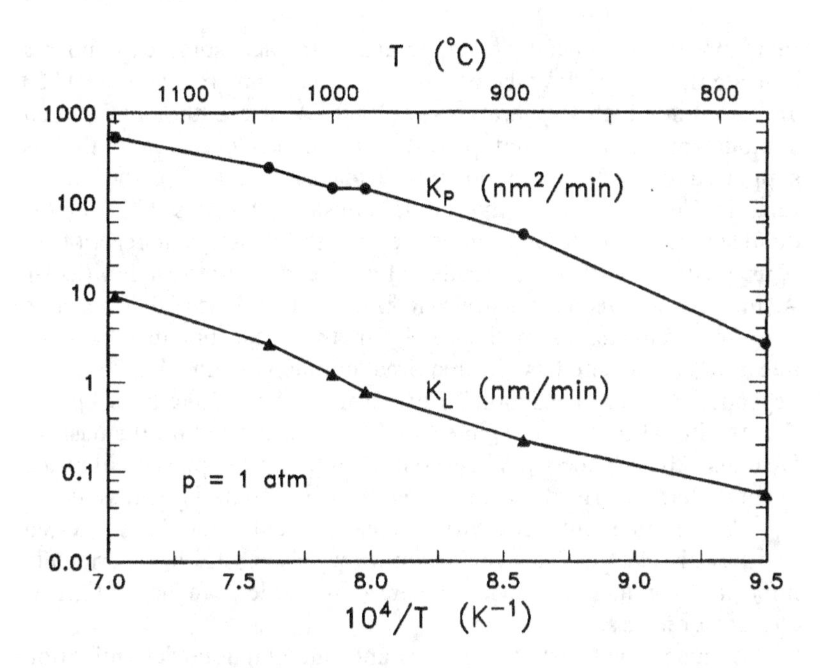

Fig. 2.3. Arrhenius plot of the linear and parabolic rate constants for silicon oxidation at 1 atm of oxidant. Data from Fig. 2 in Irene and Dong;[8] reprinted by permission of the publisher, *The Electrochemical Society, Inc.*

show that data-fitting oxidation experiments (at quasi-steady state) can only yield the *permeabilities* Kk and KD.

d. An example of the linear and parabolic rate constants' temperature dependence is given in Fig. 2.3, adapted from Ref. 8. We note that they do *not* have a strict Arrhenius behavior, proportional to $\exp(-E/k_BT)$, where E is an activation energy, k_B is Boltzmann's constant (8.62×10^{-5} eV/K), and T is the absolute temperature. Therefore, we realize immediately that the simple model of Deal and Grove[4] cannot tell the whole story of silicon oxidation: The postulated kinetic steps cannot be *elementary steps* in the sense of chemical kinetics.[9]

e. From Eqs. (2.13) we learn that both rate constants should be proportional to the oxygen partial pressure. While this has been properly assayed for the parabolic constant k_P, and thus gives credence to the model, the linear constant has a power dependence on p

that appears to depend on temperature. In fact, some experiments indicate that $k_L \propto p^{\alpha(T)}$, where $\alpha \to 0.5$ for low temperatures ($<750°C$) and $\alpha \to 1$ for high temperatures ($>1000°C$). The author has offered a model of oxidation that provides this dependence[10] and that is supported by further experimental evidence[11] regarding the curvatures in Fig. 2.3. Briefly, this model considers the possibility of O_2 dissociation at the Si/SiO_2 interface and the subsequent reaction of *atomic* oxygen with Si, in parallel, of course, with the reaction (2.7b). Again, one obtains linear-parabolic kinetics, with essentially the same long-time behavior as in the Deal–Grove model, but in which the linear rate constant has the required nonintegral and temperature-dependent "order of reaction." Furthermore, it has long been known that the initial oxidation regime cannot be understood on the basis of Deal and Grove's theory. The initial oxidation rate appears enhanced beyond that described by the linear rate constant k_L. Massoud *et al.*[12] has carried out very careful measurements, and he suggested that built-in electric fields can account for this effect. This is critically important for modern devices where gate oxides can be as thin as 25 Å, or even less.

As mentioned earlier, a precise and unambiguous determination of oxidation rate constants is very difficult, and possible extensions of the Deal–Grove model must thus be regarded as tentative. Nevertheless, we are in possession of sufficient qualitative evidence to rule out certain processes, shown, for example, in the two exercises that follow.

EXERCISE 2.6. *The impossibility of homogeneous reactions:* Work out the kinetics of SiO_2 growth if O_2 can react homogeneously with the "glass" network, and show that they are incompatible with the experimental evidence. [Hint: Include a sink term $-\kappa C$ in the diffusion equation; the other equations remain unchanged.] □

EXERCISE 2.7. *The impossibility of surface dissociation*: Assume that O_2 dissociates at the gas/SiO_2 interface and that *atomic* oxygen diffuses and combines with Si to form fresh SiO_2. Define the appropriate reactions and find the boundary conditions from mass balance. Show that this scheme is incompatible with the experimental evidence. □

These exercises indicate the nature of model building: It behooves anyone who proposes a given physicochemical process to work out and discuss the consequences of any other imaginable scheme.

f. In its original formulation,[4] the Deal–Grove model neglected, from the outset and without explanation, the term ΩC_L in Eq. (2.9a). Although this term is numerically small, its neglect in the boundary conditions is equivalent to a violation of mass conservation. The Deal–Grove model, as well as its formulation here, suffers from another logical inadequacy. It will be recalled that the origin of coordinates was chosen at the gas/SiO_2 interface (it could have been chosen at the other interface as well). But oxidation experiments are conducted in a laboratory frame such that the silcon substrate is at rest. In this frame *both* interfaces appear to move, and the diffusion equation for O_2 in SiO_2 must then contain convective terms that reflect these motions. These correction terms are likely to be negligible for oxidation processes, but they are certainly vital for the accurate description of impurity redistribution in oxides.

g. The model we have discussed is one-dimensional. The fabrication of device structures, however, presents problems[13,14] that cannot be understood without 2-d or 3-d modeling. Solutions are necessarily numerical,[15] rather than analytic. In spite of its long history, the kinetics of SiO_2 formation is not a closed field, and is still being actively investigated.[16]

h. Let us return and examine the assumption of a quasi-steady state more closely. From Eqs. (2.10), (2.11a), and (2.12) it is evident that the concentration distribution depends on time *through* the schedule $L(t)$, i.e., $C = C[x, L(t)]$. This situation should be familiar from the "adiabatic approximation" in mechanics. For example, the period of a pendulum *does* change with time as the pendulum slowly changes its altitude above sea level, say, in a balloon. This period, however, changes only through the change in the local gravitational acceleration. Similarly, many problems of phase change occur slowly enough that the fields (here, the concentration) adjust to their steady-state values corresponding to the instantaneous value of a forcing term (here, the oxide thickness). The rate of that forcing term (here, $\dot{L}$) is then determined by the overall mass balance, as in Eq. (2.9b). The "goodness" of this approximation cannot be assessed in 1-d, as for this example, without detailed numerical work. The following sections, however, will provide an example (growth of spherical precipitates) for which there also exists an exact time-dependent solution. The desired comparison will be found in Chap. 5.

2.4. A Few Considerations Concerning Capillarity

The next example deals with nucleation and growth. Problems of this type require a clear understanding of a curved surface's effect on local thermodynamic variables. These variables are all fields in the sense of Sect. 1.8 and Note 10 of Chap. 1. In this section we review some *equilibrium* laws for small systems.[17–21] These laws follow from basic thermodynamics, which is summarized in Appendix C. The *kinetic* problem of nucleation and growth will be taken up in the next section.

Consider, for simplicity, a system of two phases, one of which might (but does not necessarily) contain the other, and each of which contains c components.[N5] This system is represented schematically in Fig. 2.4. Assume that each phase can be completely described by its temperature T, pressure p, and chemical potentials μ_i $(i = 1, 2, \ldots, c)$. This is equivalent to an assumption of fluid-like behavior, even though, later, we shall apply the results of this section to solids. These variables are not independent; their variations, *for whatever reason*, must satisfy the Gibbs–Duhem relation:

$$Cs\,dT - dp + \sum_{i=1}^{c} C_i\,d\mu_i = 0. \tag{2.14a}$$

For definiteness, let us say that the concentrations C_i are measured here in units of number density. The total concentration is simply the sum over components,

$$C = \sum_{i=1}^{c} C_i, \tag{2.14b}$$

and it follows that the entropy density s is defined per particle. Equations (2.14) must hold for each phase α $(\alpha = 1,2)$. Therefore, *all* the above variables come "dressed" with an additional superscript to indicate which of the two phases they characterize. For example, $C_1^{(2)}$ is the concentration of the first component in the second phase.

Gibbs[17] showed long ago that the equilibrium of *each* phase requires the constancy of the T-, p-, and μ-fields over the domains $\mathscr{D}^{(1)}$ and $\mathscr{D}^{(2)}$. For example, $\nabla p^{(2)} = 0$ in the second phase if it is at equilibrium.[N1] Moreover, equilibrium *between* phases demands the

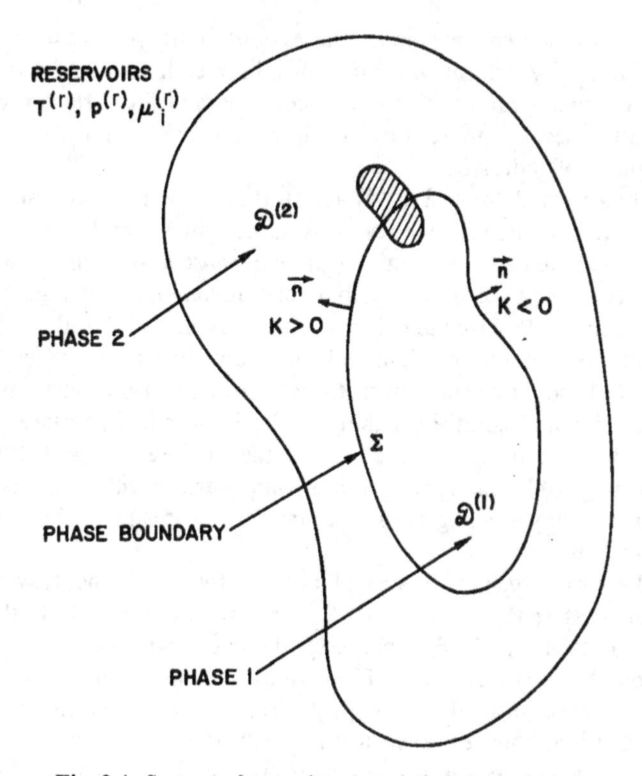

Fig. 2.4. System of two phases and their boundaries.

following equalities at the phase boundary Σ:

$$T^{(1)} = T^{(2)}, \tag{2.15a}$$

$$p^{(1)} - p^{(2)} = \gamma K, \tag{2.15b}$$

$$\mu_i^{(1)} = \mu_i^{(2)} \qquad (i = 1, 2, \ldots, c). \tag{2.15c}$$

We see that the temperature and chemical potentials are continuous across the phase boundary, but that the jump in pressure is proportional to the surface tension γ and to the local curvature K. Equation (2.15b) is called Laplace's law of capillarity, but it was certainly known earlier to Young. The "curvature" in this equation is the *mean curvature* of a surface as defined in differential geometry, i.e., the sum of inverse radii of curvature of curves formed by the intersection of the surface with the two "principal" planes containing the unit normal **n**.[N6] Referring to Fig. 2.4, the unit normal is directed from phase 1 to phase 2, and the mean curvature is reckoned positive if the surface is convex toward the second phase. For example, $K = 2/R$ or $1/R$ for a sphere or cylinder of radius R.

Equations (2.14) and (2.15) are all that is required to describe the "equilibrium of heterogeneous substances," in general. Toward that end, it is convenient to imagine our two-phase system immersed in a large external phase with which it can reversibly exchange heat at temperature $T^{(r)}$, mechanical work at pressure $p^{(r)}$, and masses at chemical potentials $\mu_i^{(r)}$. This idealization requires comment. The external phase represents interactions with the world at large, namely, the experimental conditions that can be imposed. This phase is also assumed large; therefore, changes in the enclosed phases, 1 and 2, induce negligible changes within it, and surface effects at its outer boundaries play a negligible role. Such a phase is often called a system of "reservoirs."

We now consider a special case: a two-component system at constant external temperature $T^{(r)}$ and pressure $p^{(r)}$. Call the two components A and B. We then seek the chemical potentials' dependence on the curvature K.[N7] The situation is now very simple if we mentally travel inward, so to speak, from the external phase toward phase 1. First, constant experimentally imposed $T^{(r)}$ implies constant $T^{(2)} = T^{(r)}$ at the second phase's external boundary, because a relation of the type (2.15a) must hold there. Since the temperature $T^{(2)}$ is a constant field in phase 2, its value $T^{(r)}$ "propagates," unchanged, throughout that phase. Likewise, $T^{(1)} = T^{(r)}$ by the same continuity arguments. Therefore, we have $dT^{(r)} = 0 = dT^{(1)} = dT^{(2)}$. Then, $dp^{(r)} = 0 = dp^{(2)}$ for similar reasons. However, $dp^{(1)} = d(\gamma K) \neq 0$ because of Laplace's law (2.15b). In other words, the phase boundary Σ is a barrier to the propagation of $p^{(r)}$ into phase 1.

The four chemical potentials cannot all be prescribed without violating the phase rule.[N7] Nevertheless, at equilibrium, they are constant fields over their respective domains. By virtue of Eqs. (2.15c), their values, for each component, must be equal throughout the two-phase system. If μ_A and μ_B are these values common to both phases, and if $d\mu_A$ and $d\mu_B$ are their variations, then the above considerations show that Eqs. (2.14) and (2.15) reduce to

$$-d(\gamma K) + C_A^{(1)}\,d\mu_A + C_B^{(1)}\,d\mu_B = 0, \tag{2.16a}$$

$$C_A^{(2)}\, d\mu_A + C_B^{(2)}\, d\mu_B = 0, \tag{2.16b}$$

from which, for example, we can eliminate $d\mu_A$:

$$-d(\gamma K) + (C_B^{(1)} - C_A^{(1)}C_B^{(2)}/C_A^{(2)})\, d\mu_B = 0. \tag{2.16c}$$

If, now, phase 2 is a dilute solution of component B, then the second term in parentheses must be small. If, further, the concentration of that component in phase 1 does not vary much with the curvature of that phase, then Eq. (2.16c) can be integrated to give

$$\mu_B(K) - \mu_B(0) = \gamma K/C_B^{(1)}. \tag{2.17a}$$

It must be understood that the integration is carried out from the state for which the surface is flat ($K = 0$, which means that phase 1 is effectively of infinite extent) to the state for which the curvature is finite. Moreover, that integration requires no assumption whatsoever regarding the surface tension γ's dependence on curvature or composition.

Chemical potentials are fine things, but one tends to measure concentrations instead. We can easily find the equilibrium concentration's dependence on curvature from a known $\mu(C)$ relation in the second phase. In a solution (solid, liquid, or gas), we know that[22]

$$\mu_B^{(2)} = \mu_B^0(p^{(2)}, T^{(2)}) + kT^{(2)} \ln a_B^{(2)}, \tag{2.17b}$$

where μ_B^0 is the chemical potential of component B if the solution is pure, $a_B^{(2)}$ is the activity of that component in solution, and k is Boltzmann's constant. Inserting Eq. (2.17b) into (2.17a), we find an exponential dependence of activity on curvature. In particular, for very dilute solutions of component B, the activity tends toward a multiple of that component's mole fraction, and we get the final result

$$C_B^{(2)}(K) = C_B^{(2)}(0)\, e^{\gamma K/kTC_B^{(1)}}, \tag{2.17c}$$

attributed to Ostwald.[N8] The quantity $\gamma/kTC_B^{(1)}$ is positive, has the dimensions of length, and is called the *capillary length* λ.† Therefore,

†The temperature T in these and the next equations is obviously controlled by the reservoir's $T^{(r)}$.

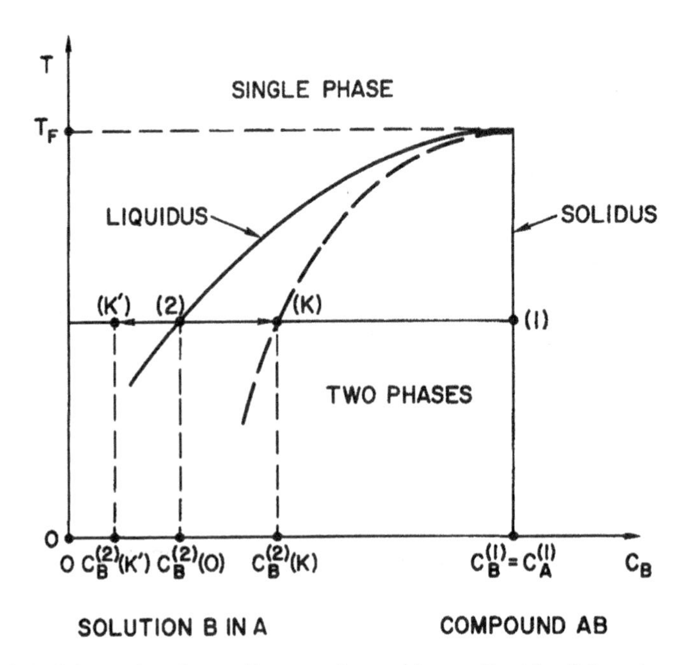

Fig. 2.5. Schematic phase diagram for a binary liquid-solid system. The capillarity-induced liquidus shift is shown dashed.

Eq. (2.17c) shows that the equilibrium concentration of the minor component above a convex surface element ($K > 0$) is larger than what would prevail over a flat element, and *vice versa* for concave surface elements.

The application of Eq. (2.17c) to equilibrium phase diagrams is of the utmost importance. Figure 2.5 shows a simple diagram that corresponds to the equilibrium of a solid compound AB (phase 1) with a liquid solution of B in A (phase 2). A concrete example might be the crystallization of GaAs from a liquid solution of As in Ga. Here, the solidus is essentially a vertical line at the stoichiometric AB composition, while the liquidus is curved and depends exponentially on temperature. The calculation of this liquidus for GaAs assumes that the solution is regular and that both phases are infinite in extent.[22,23] This phase diagram tells us that, at a given temperature (and pressure), a *unique pair* of solid and liquid compositions is in equilibrium. The one-to-one corresponding points are labeled (1) and (2) in Fig. 2.5.[N9]

The region above the liquidus, the so-called *single-phase region*, represents thermodynamic states of the liquid that are not in equilibrium with any finite solid. We are then free to choose the solution's temperature and concentration arbitrarily, as is evident to any good cook. On the other hand, the region of Fig. 2.5 bounded by the solidus and liquidus lines represents metastable states of the liquid; it is called the *two-phase region* for the following reason. We assume that the solid compound is of small extent and, therefore, that its curvature is, on the whole, large and positive. It follows from Eq. (2.17c) that this solid would be in equilibrium with a liquid whose concentration $C_B^{(2)}(K)$ in the minor component B is *larger* than it would be for an infinite solid. This is true at any temperature below the compound's temperature of fusion T_F. Thus, a given positive curvature K of the solid phase 1 causes a *liquidus shift* (dashed) to the right. Reciprocally, given an arbitrary point in the two-phase region [labeled (K) in Fig. 2.5], then there surely exists a solid, and one only, small enough (i.e., of positive and large enough curvature K) that is in equilibrium with the liquid of composition $C_B^{(2)}(K)$.† In the next section we will see that these small solid particles are metastable in the sense that there is a critical dimension for these particles, related to the capillary length, below which the solid dissolves and above which it grows.

EXERCISE 2.8. *Justification of Ostwald's equation:* Carefully justify all the steps leading from Eq. (2.16c) to Eq. (2.17c). [Hint: Recall that the mole fraction of component B, in either phase, is defined as $x_B = C_B/C$, where $C = C_A + C_B$; therefore $x_A + x_B = 1$. For example, take an activity of the form $a = \alpha x + \beta x^2$.] What can you say if the solution is *not* dilute? Consider, first, the case of a stoichiometric AB compound, then a pure solid phase 1 composed of B only. In all cases, also discuss the $C_A(K)$ dependence of the major component. □

EXERCISE 2.9. *Application to GaAs:* Take the case of GaAs, a prototypical binary III–V system. Compute the capillary length at 700°C and 1000°C, knowing that $\gamma \simeq 860$ ergs/cm^2, that the density of solid GaAs is 5.2 g/cm^3, and that its molecular mass is 72.3 g/mole. With these values and Eq. (2.17c), compute the corresponding enhancement of As in solution in equilibrium with 0.01- and 1-μm-diameter spheres. □

†All these considerations are compatible with the phase rule (C.6) for small systems.

Finally, we note that the derivation of Ostwald's equation (2.17c) was predicated on the implicit assumption that phases 1 and 2 are in *global* equilibrium. This is not necessary: Consider an *arbitrary* domain (shown hatched in Fig. 2.4) that "straddles" both bulk phases and their phase boundary. If that system is in thermodynamic equilibrium, then one can show that Eqs. (2.14) and (2.15) hold *locally* because the domain is arbitrary. Consequently, Eq. (2.17c) also holds locally. This point is particularly useful when considering kinetic processes, as we shall see shortly. Although a system of two bulk phases and a phase boundary can be far from equilibrium—and we would then wish to compute its rate of relaxation toward equilibrium—it is often useful to assume that the phase boundary itself is locally at equilibrium with the adjacent bulk phases.

Equation (2.17c) applies regardless of the sign of the mean curvature. A corrugated solid body has surface elements with negative as well as positive curvature. If a surface element has a *negative* mean curvature K', then $C_B^{(2)}(K')$ is the concentration of the liquid phase that is in equilibrium with that element. This causes a local liquidus shift to the *left*, labeled (K′), in Fig. 2.5. These considerations are key to the understanding of morphological features on surfaces.

2.5. The Precipitation of Spherical Particles

Problems of nucleation and growth occur throughout our experience. They are also crucial to our ability to manufacture parts having the required mechanical and electrical properties. Whether rain precipitates from clouds or carbon precipitates from steel depends, in large measure, on the surface properties of small phases. These phases can exchange matter and energy with a large phase—the *matrix* or *parent* phase. The small phases are called *nuclei* because, as they grow at the expense of the matrix, they are the cause of larger, macroscopically observable phases. The origin, shape, and thermodynamic properties of nuclei are still open to conjecture, although molecular-dynamics computer simulations[24] do shed some light on these points. In this section, we solve a problem that is again mathematically very simple, but that will provide us with a clear understanding—and a dynamic one at that—of the critical radius of nucleation.

We consider again a parent phase 2, in any state of aggregation (gas, liquid, or solid), that is a *homogeneous* mixture of two chemically

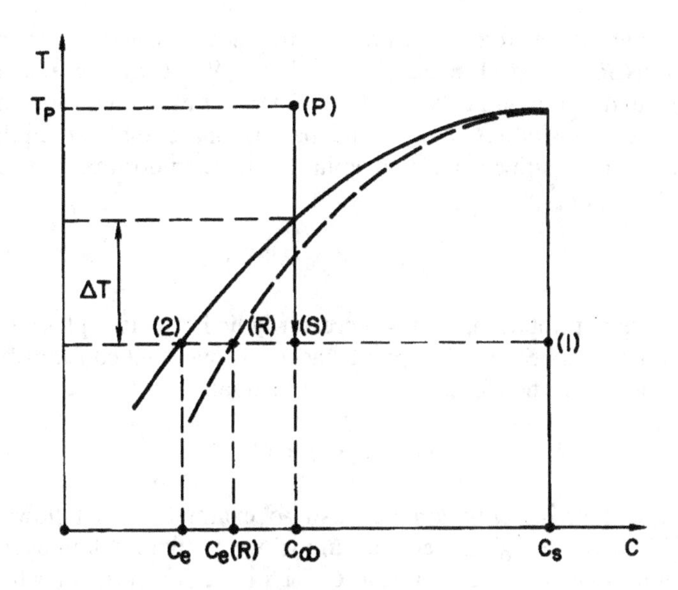

Fig. 2.6. Same as Fig. 2.5 in simplified notation. The undercooling ΔT is equivalent to the supersaturation measures σ or σ_r.

distinct species A and B. We then ask if this mixture at a given temperature, pressure, and composition can promote the formation of another phase 1, the nucleated *precipitate*, containing these same species. In other words, can the initially homogeneous mixture become heterogeneous, and if so, at what rate? Without inquiring into the origin of nuclei, it is clear that, once formed, they will grow (or decay) because of mass or heat transport; and transport implies diffusion over macroscopic distances.

Figure 2.6 is a phase diagram similar to Fig. 2.5. Again, the diagram is drawn for a system of liquid parent phase and solid precipitate, but the considerations that follow are independent of the system's state of aggregation. We also simplify the notation of the last section by putting $C_s = C_B^{(1)}$ and $C_e = C_B^{(2)}(0)$ for the equilibrium concentration of B in a large precipitate and the parent phase, respectively.† At a given temperature and pressure, we know that the compositions C_e and C_s are corresponding equilibrium compositions

†The notation $C_B^{(2)}(0)$ means the phases are *large*: $R \to \infty$, which means $K = 0$.

if both phases are large in extent. Consider, now, a *spherical* precipitate of radius R immersed in the parent phase. We also know that there exists, under the same conditions of temperature and pressure, a certain composition $C_e(R)$ of the parent phase that would be in equilibrium with this (small) precipitate. That composition is given by Eq. (2.17c), or

$$C_e(R) = C_e\, e^{2\gamma/RkTC_s} \tag{2.18}$$

in the present notation. It must necessarily lie in the two-phase region of the diagram because the sphere has a positive (and constant) mean curvature $2/R$. Therefore, we have the inequalities[N10]

$$C_e < C_e(R) < C_s. \tag{2.19}$$

The dynamic situation we wish to examine is as follows: We imagine that at a given temperature T we prepare a homogeneous solution of constant composition C_∞. Its state S can lie anywhere in the "half-strip" $0 < C < C_s$ and $T > 0$ of the phase diagram in Fig. 2.6; here, it is marked in the two-phase region. Then, we assume that the Lord shows his hand through the creation, somewhere in the bulk of this solution, of a solid spherical nucleus of radius R_0. We assume, also, that the surface of this nucleated precipitate is active enough for exchanges of mass to occur rapidly with the solution. The concentration in solution C_R, right next to the precipitate's surface, is then the equilibrium concentration $C_e(R)$ corresponding to its instantaneous radius.[†] Since, in general, $C_e(R) \neq C_\infty$ (remember that the solution's overall composition was arbitrary), it follows that diffusion must occur, the local concentration C of component B in solution must vary, and thus the precipitate must grow or dissolve.

The mathematical problem, in its full generality, is quite complex because we have, here again, a problem of phase change with a moving boundary. However, the quasi-steady-state approximation provides a simple solution, much as in the case of silicon oxidation. Figure 2.7 shows the concentration profile in the parent phase that corresponds to the conditions on Fig. 2.6. Diffusion *in* the solid precipitate is often negligible.

[†] This is equivalent to the assumption of "local equilibrium" that was mentioned at the end of the last section. See Exercise 2.12, however, where that condition is relaxed.

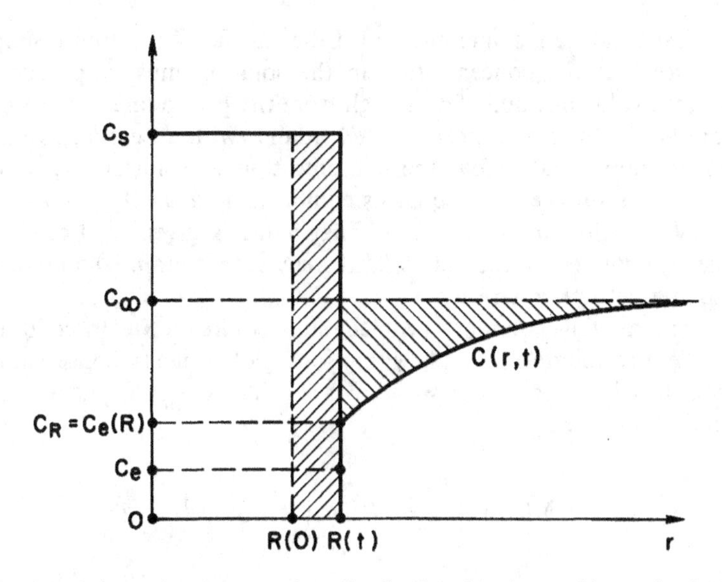

Fig. 2.7. Schematic concentration distribution for the problem of spherical precipitate growth.

In the absence of other information we assume spherical symmetry around the precipitate's center, which is also the origin of our coordinate system. In that system, the precipitate's boundary moves according to a schedule $r = R(t)$ that we must find. Toward that end, we recall the diffusion equation (1.36) for the minor component B in solution,

$$\frac{\partial C}{\partial t} = \frac{D}{r^2}\frac{\partial}{\partial r}\left(r^2 \frac{\partial C}{\partial r}\right), \tag{2.20}$$

written in spherically symmetric coordinates.[N11] We must find its solution $C(r, t)$ subject to the boundary conditions

$$C(\infty, t) = C_\infty \tag{2.21a}$$

and

$$C_R \equiv C(R, t) = C_e(R), \tag{2.21b}$$

and an appropriate initial condition that need not concern us here.

The first condition, sometimes called the "far-field" condition, simply indicates that the concentration in the solution must approach its prepared value at points far enough from the precipitate. The second condition is our real concern. In accordance with an earlier assumption, it simply states that the concentration in solution C_R at the precipitate's surface must equal its equilibrium value, *compatible with the sphere's instantaneous radius.* That value is given by Eq. (2.18), which is nonlinear in the radius R. Moreover, that radius changes with time in a yet unknown way.

Any moving-boundary problem still requires a Stefan condition for the precipitate's boundary motion; global mass conservation, again, provides the "royal way." We consider the total mass of the mobile component B,

$$M(t) = \int_0^{R(t)} C_s \, dV + \int_{R(t)}^{\infty} C(r, t) \, dV,$$

where the volume element is $dV = 4\pi r^2 \, dr$. The calculation is now very simple because there is no reason for M to change. With the condition $\dot{M} = 0$ and the use of Leibniz's rule (recall Note 4 in Chap. 1) and Eq. (2.20), we get†

$$D \frac{\partial C}{\partial r}\bigg|_{r=R} = \dot{R}(C_s - C_R), \tag{2.21c}$$

with an evident geometric interpretation: Any increase of the precipitate's total mass must be strictly accounted for by a depletion of the component B in the parent phase. In other words, the volumes corresponding to the two hatched areas of Fig. 2.7 must be equal.

The solution of this problem is very simple in the quasi-steady-state approximation. We then assume that the left-hand side of Eq. (2.20) is small; its right-hand side thus reduces to an ordinary differential equation whose general solution is

$$C[r, R(t)] = C_\infty - (C_\infty - C_R)R/r, \tag{2.22}$$

in accordance with the general result (2.4b). Here, we anticipate that

†We assume that the flux at infinity decreases faster than r^{-2} (proved in Appendix F).

the concentration's time dependence will occur again through the boundary's schedule, as in our earlier oxidation problem. This schedule is easily determined by inserting Eq. (2.22) into the Stefan condition (2.21c). Thus

$$\frac{dR}{dt} = \frac{D}{R(C_s - C_R)}(C_\infty - C_R). \tag{2.23a}$$

Finally, the yet unknown surface concentration C_R must be compatible with the equilibrium condition (2.18), and we get the differential equation

$$\frac{R}{D}\frac{dR}{dt} = \frac{C_\infty - C_e e^{2\lambda/R}}{C_s - C_e e^{2\lambda/R}}, \tag{2.23b}$$

where $\lambda = \gamma/kTC_s$ is again a convenient definition of the capillary length. The problem is, at least formally, solved: The solution of this *nonlinear* differential equation (2.23b), together with the initial condition $R(0) = R_0$, yields the schedule $R(t)$ and then the surface concentration C_R because of Eq. (2.18). The full concentration distribution $C(r, t)$ follows from Eq. (2.22).

Let us look at the qualitative features of the differential equation (2.23b). It gives the precipitate's rate of growth in terms of purely thermodynamic variables on its right-hand side. Observe first that the denominator of that right-hand side is always positive because of the inequalities (2.19). The numerator can change sign, however, and the precipitate's size for which this occurs is called the *critical radius of nucleation* R^*. Thus, $C_e(R^*) = C_\infty$ implies

$$R^* = \frac{2\lambda}{\ln(C_\infty/C_e)} = \frac{2\gamma}{kTC_s \ln(C_\infty/C_e)}, \tag{2.24}$$

and this radius corresponds to a *turning point* of the differential equation.[N12] The precipitate's rate of growth vanishes when $R = R^*$. In addition, inspection of Eq. (2.23b) shows that if $R < R^*$ then $\dot{R} < 0$, and the precipitate *dissolves*. Likewise, *growth* occurs if $R > R^*$. Thus, although the precipitate is in equilibrium with a solution of given temperature and composition when its radius R is exactly equal to the critical radius R^*, that equilibrium is *metastable*: Any deviation from the equilibrium condition produces further changes that drive the system farther from equilibrium.[†]

Another way of expressing this same result arises if we observe from Eqs. (2.23) that $\dot{R}$ has the sign of $C_\infty - C_e(R)$. Thus, if the parent phase's bulk state (S) lies to the right of state (R) in Fig. 2.6, then the precipitate will grow. In the language of thermodynamics, the liquid is *supersaturated* with respect to a solid of radius R. This is true for *all* radii larger than the critical radius R^*. Therefore, the prepared state (S) of the liquid is supersaturated with respect to all states (R) lying to its left. The converse statements regarding states to the right of (S) then define the concept of *undersaturation* in a natural way. Of course, if state (S) lies to the left of the liquidus itself, then the liquid will always be undersaturated with respect to solids, no matter what their (positive) radii of curvature might be. In that context, one sometimes finds these notions expressed equivalently in terms of temperature differences, and these are useful experimentally. The supersaturated state (S), shown in Fig. 2.6, can be achieved by carefully cooling a solution of prepared composition C_∞ from a temperature T_P above the liquidus, down through the liquidus to the final state S. The maximum amount of available supersaturation $C_\infty - C_e$ is equivalent to the *undercooling* ΔT shown on the diagram. Both are measures of the maximum driving force of the precipitation process.

EXERCISE 2.10. *Undercooling and critical radius:* Discuss a prepared solution's ability to withstand a given undercooling. Assume that $C_e = C_0 e^{-\Delta H/kT}$ and $\partial\gamma/\partial T < 0$. In particular, how does R^* change with temperature? Express R^* in terms of a chemical potential difference. □

We now briefly turn to the integration of Eqs. (2.23). First, we see that capillarity plays a meager role when R is large enough. In fact, a precipitate always grows if a given nucleus has an initial radius $R_0 > R^*$. If, in addition $R_0 \gg 2\lambda$, then $C_e(R) \approx C_e$ for all subsequent radii. Then, it follows that Eq. (2.23b) reduces to

$$\frac{R}{D}\frac{dR}{dt} = \frac{C_\infty - C_e}{C_s - C_e}. \tag{2.25a}$$

[†]In contrast to arguments usually presented for the existence of R^*, the calculation here is dynamical.

The right-hand side is constant and is called the *absolute* supersaturation σ, reserving the words *relative* supersaturation for the ratio $\sigma_r \equiv \ln(C_\infty/C_e)$ that occurs in the definition of the critical radius (2.24). The solution of Eq. (2.25a) is evidently the parabolic law

$$R(t) = (R_0^2 + 2\sigma Dt)^{1/2}, \tag{2.25b}$$

and it represents the asymptotic behavior of a precipitate's increase in size if it had the good fortune to be "launched" with a large enough radius. It is interesting to note that both relations (2.25) hold *exactly* if the surface tension γ vanishes. If the previous conditions are not met, then one must integrate the nonlinear equation (2.23b). This we will not do, because such calculations would carry us far afield into the kinetics of precipitation. To compute these properly would still require an estimate of the *rate* at which nuclei are formed and estimates of their initial size distribution– and these are *not* diffusion problems.

□

EXERCISE 2.11. *Supersaturation measures and length scales:* Discuss the physical meaning of both supersaturation measures, σ and σ_r. In particular, what is σ_r if it is "small"? Note that there are *two* characteristic length scales for the rate equation (2.23b): R^* and 2λ. The first measures deviations from metastability; the second measures finite-size effects. Discuss that equation for all possible initial radii R_0 relative to these lengths. Then discuss the solution (2.25b) for both positive and negative σ. □

EXERCISE 2.12. *Effect of finite surface kinetics:* Assume that exchanges at the precipitate's surface with the adjacent bulk phases occur at a finite rate. Specifically, assume that the solid–liquid phase transformation can be described by a "reaction rate" $k[C_R - C_e(R)]$, where k is a "reaction constant." Discuss its meaning, and find the appropriate boundary and Stefan conditions. [Hint: Consider mass balance in each phase.] Next, find the rate equation that generalizes Eq. (2.23b). What is then the critical radius of nucleation? □

There are many other questions relating to the growth of precipitates. For example, we assumed a spherical shape that was preserved during growth or dissolution. Is this possible? Is the sphere the only "shape-preserving" configuration? These questions are generally difficult to answer in full. We refer to the paper by Ham[25] for a partial answer. In brief, there is a class of "quadric" surfaces that maintain their shape under diffusion if capillarity effects are neglected. The *stability* of these surfaces, on the other hand, rests crucially on such effects. The "morphological stability" of growing crystals was first studied in the early 1960s by Mullins and Sekerka (see Ref. 26 for an excellent review), and these questions have since been actively pursued in the larger context of "pattern formation."[27] Indeed, one may ask how a snowflake gets its orderly shape, even as it is born out of an unordered and essentially homogeneous phase. The scientific and technological importance of these questions cannot be underestimated.

EXERCISE 2.13. *Precipitation of cylindrical particles:* Discuss equilibrium over cylindrical surfaces and find the analog of Eq. (2.18). Compute the kinetics of long cylindrical precipitates and estimate the critical radius of nucleation. □

Another problem that deserves mention is "Ostwald ripening." We assumed, in effect, that there were few nucleation events in the parent phase. If, however, many nuclei are produced, then the diffusion fields around each precipitate can overlap and compete for the available amount of supersaturation. Big precipitates grow at the expense of smaller ones, and the bulk composition C_∞ is no longer constant: it decreases with time as the population of precipitates changes. One says that the solution "ages."[28] The calculation of this time dependence and the precipitates' ultimate size distribution is the aim of a theory proposed by Lifshitz and Slyozov,[29] and later by Wagner.[30] Voorhees[31] provides more recent developments.

Finally, this section has dealt only with problems of *homogeneous* precipitation, i.e., when nucleation occurs in the interior of a parent phase. This must be contrasted with *heterogeneous* precipitation, a subject in its own right. Then, nucleation events and precipitate growth *at* phase boundaries are the main concern.[32,33]

2.6. Crystal Growth According to the Theory of Burton, Cabrera, and Frank

The years 1949–1951 witnessed a truly remarkable development in our understanding of why crystals grow at all.[34] Indeed, a dilemma

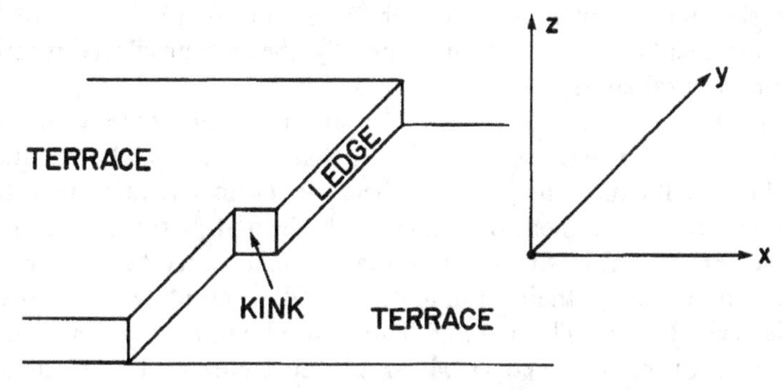

Fig. 2.8. Schematic TLK model of vicinal surfaces.

arises if we consider the structure of crystal surfaces. To understand this question, we observe that there are essentially two types of crystal surfaces, those whose orientiation lies or does not lie near a close-packed direction. The former surfaces, called *vicinal*, are mostly smooth on an atomic scale. They consist of close-packed surface elements called *terraces* that are separated by risers called *steps*. These steps are necessary to produce a surface with a given average orientation. The steps themselves are not simple structures because of thermal fluctuations. Indeed, it had long been known[35] that they roughen to produce re-entrant corners called *kinks*. Figure 2.8 shows how a vicinal surface is composed of terraces and steps, and how steps are composed of ledges and kinks. With simple bond models one can show that kinks are privileged surface sites: Adding or subtracting molecules at these sites cannot change the total cohesive energy of a large crystal. Consequently, the binding energy at a kink site must be related to the equilibrium chemical potential.

From a dynamic point of view, molecules from the parent phase can attach to terraces and diffuse toward ledges. Once attached to ledges, they diffuse toward kinks, where they find their ultimate resting place. In that process, the kink position moves in the negative y-direction of Fig. 2.8, and when that kink reaches the crystal's edge, the whole step will have progressed by one unit in the positive x-direction. In turn, when a step moves toward a crystal edge, then the crystal as a whole will have grown by one unit in the positive z-direction. This model is therefore called the "terrace-ledge-kink" (TLK) model of crystal growth; it was formulated long ago by Kossel,

Stranski, and Volmer (see Volmer[36]). In this sense, kinks are the ultimate agents of growth. More generally, they are privileged reaction sites on crystal surfaces.

However, if kinks arise from fluctuations in steps, then what is the origin of steps? First, it can be shown that steps *cannot* be produced by thermal fluctuations, at least, below a critical temperature that characterizes two-dimensional melting.[34] Second, if these steps were due to intentional misorientation, then their advance to the crystal's edge would cause their annihilation, and, eventually, the growth process would stop. Third — and here is the dilemma — although steps can be produced by the edges of disk-like nuclei on terraces, it can also be shown that the supersaturation required for such heterogeneous nucleation is often much greater than is observed experimentally. Before addressing this question, let us note that crystal surfaces whose average orientation is far from close-packed are necessarily *rough* on an atomic scale. Each surface element contains many kink sites, but steps and terraces are no longer identifiable. Each of these kinks is a nucleation site, and the energy required for nucleation is far lower than on close-packed surface elements. There is then no appreciable barrier to crystal growth. For example, the surface of the spherical precipitate of the last section was implicitly assumed to be rough.

Thus, we rephrase our question: What is the origin of steps on vicinal surfaces? In 1949, F. C. Frank, one of the founders of modern dislocation theory, suggested that a screw dislocation (more precisely, a dislocation with a screw component) that intersects a crystal surface can provide an inexhaustible source of steps. This built-in imperfection bypasses the need to overcome a heterogeneous nucleation barrier. The structure of steps, their motion, and the connection with the growth of crystals was then explored in a now classic paper by Burton, Cabrera, and Frank.[34] This theory is known as the BCF model of crystal growth, and it ranks high among twentieth-century achievements in physics. It must be remembered that these authors postulated a mechanism that was beyond the experimental observations of the time, but one that has been amply confirmed since.

A complete exposition of the BCF theory is beyond our scope. It does provide, however, a very interesting problem in diffusion theory. Consider a single step produced by an emerging screw dislocation. This step reaches out from the dislocation core to an edge of the crystal, and one shows that it is straight, on average, if the crystal is in equilibrium with the parent phase. Assume, now, that the supersaturation in the parent phase is "switched on." Molecules from that phase are then transferred by 3-d diffusion to the terraces; on terraces these adsorbed molecules execute 2-d jumps toward ledges, and, once attached to ledges, the molecules diffuse one-dimensionally toward the 0-d sites, which are kinks. We thus have a nested set of diffusion problems that are distinguished by their dimensionality.† As explained earlier in connection with the TLK picture, the dislocation step will move. But it is also pinned at its core, which means that regions of the step far from the core will move at a velocity controlled essentially by the supersaturation, while the parts near the core will move less. Therefore, the dislocation step must wind itself around the core to produce a spiral. The dynamics of this spiral were explored in depth by BCF. Here, suffice it to say that its shape is essentially determined by the critical radius R^* of heterogeneous nucleation, which determines the minimum radius of curvature that prevails at the dislocation core. In effect, as with homogeneous nucleation, any element of the dislocation step with a radius of curvature *less* than R^* would recede rather than advance.

The spiral, in polar coordinates (r, φ), was shown by BCF to be essentially archimedean

$$r \propto R^* \varphi. \tag{2.26a}$$

Therefore, the distance l between successive arms of the spiral is constant and is proportional to R^*. Since the critical radius for heterogeneous nucleation is also inversely proportional to the relative supersaturation $\sigma_r = \ln(C_\infty/C_e)$ (see Eq. 2.24), we get the important relation

$$l \propto 1/\sigma_r \tag{2.26b}$$

that relates step distance to supersaturation. The constant of proportionality in this equation depends on such things as the edge free energy of a 2-d nucleus, the temperature, and the interaction between competing spirals.

The full diffusion problem outlined above is evidently complex. It is good to remember, however, what we really want, namely an

†It can also be shown that "direct" processes, such as parent-phase → kink exchanges, are unlikely.

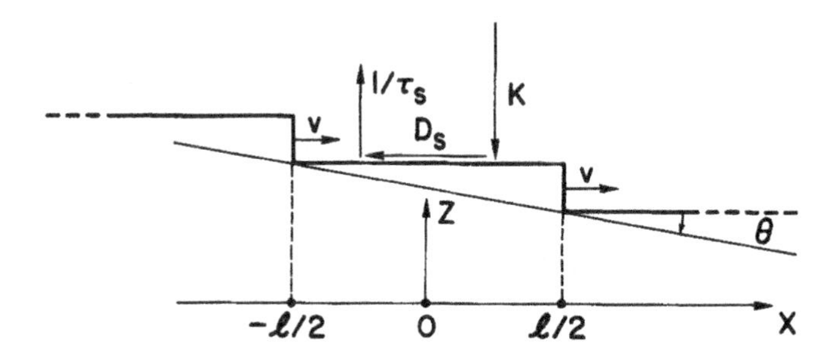

Fig. 2.9. Section through a vicinal surface. Note the adsorption, desorption, and diffusive processes on the surface.

expression for the rate of growth V of a crystal surface. Figure 2.9 shows a normal section of this surface and two of its equidistant steps. The growth rate normal to a given close-packed orientation is equal to the velocity v of a step times the tangent of the misorientation angle θ. If h denotes the step height, then $\tan\theta = h/l$. Further, if J_s denotes the surface flux along terraces, then mass-balance considerations dictate that $vh/\Omega = 2J_s|_{\text{step}}$, where Ω is the volume of a molecule. The factor 2 comes from the two fluxes on either side of a given step. Thus we get an expression for the growth rate,

$$V = (2\Omega/l)J_s|_{\text{step}}, \tag{2.27}$$

and the whole calculation reduces to the computation of molecular fluxes at steps. Following BCF, we now make the simplifying assumptions:

(a) Diffusion in the parent phase and along ledges is fast enough not to be rate-limiting. The problem is thus reduced to the consideration of transport processes on terraces.
(b) The kinks on a given step are so close together that, in view of the assumption of fast diffusion on ledges, steps are effectively line sinks.
(c) Exchanges at kinks are effective enough to maintain the concentration at steps close to its equilibrium value.
(d) The steps are so far from the dislocation core that they can be considered straight and equidistant. This implies a 1-d diffusion field in the x-direction on terraces.

(e) The motion of steps is slow enough, compared to mass redistribution, that the quasi-steady state obtains, and Eq. (2.27) holds.

With these assumptions, the concentration $C_s(x, t)$ of adsorbed molecules on a given terrace must obey

$$0 \approx \frac{\partial C_s}{\partial t} = -\frac{\partial J_s}{\partial x} + F, \tag{2.28a}$$

which is nothing but a balance law (1.47) (on a surface) with a "source" equal to the net flux F of molecules exchanged with the bulk parent phase. The flux F is then expressed as the detailed balance of an adsorption rate kC_∞ and a desorption rate C_s/τ_s, where k is the adsorption rate constant, C_∞ is the (constant) bulk concentration, and τ_s is the lifetime or residence time of an adsorbed molecule on the terrace. Thus, we have

$$F = kC_\infty - C_s/\tau_s. \tag{2.28b}$$

It remains only to express the surface flux according to a phenomenological relation

$$J_s = -D_s \frac{\partial C_s}{\partial x}, \tag{2.28c}$$

which defines the surface diffusivity D_s.†

EXERCISE 2.14. *Balance laws for vicinal surfaces:* Choosing appropriate domains on the surface, prove that mass conservation implies Eq. (2.28a), and find the general Stefan condition for steps that leads to Eq. (2.27). □

The problem is now very simple. Observe, first, that we can relate the equilibrium concentrations in the bulk and surface phases because $F = 0$ at equilibrium. With Eq. (2.28b) we then get a "mass-action"

†Note that the dimensions of the various quantities are $[C_s] = \text{cm}^{-2}$, $[J_s] = \text{cm}^{-1}\text{s}^{-1}$, but that D_s has the dimensions cm^2/s characteristic of all diffusivities. [See remark (d) of Sect. 1.2.]

relation

$$C_{s,e} = \tau_s k C_e, \tag{2.29a}$$

i.e., the equilibrium surface concentration in terms of the equilibrium bulk value. Then, we noted in Exercise 2.11 that C_∞ is given approximately by

$$C_\infty \approx C_e(1 + \sigma_r) \tag{2.29b}$$

if the relative supersaturation is small. Inserting Eqs. (2.28b), (2.28c), and (2.29) into Eq. (2.28a), we get the differential equation

$$D_s \tau_s \frac{d^2 C_s}{dx^2} - C_s + C_{s,e}(1 + \sigma_r) = 0, \tag{2.30a}$$

in which the supersaturation is evidently a forcing term. This equation also shows that the combination of physical quantities

$$x_s = \sqrt{D_s \tau_s} \tag{2.30b}$$

must be a fundamental length scale. It is called the *mean surface diffusion distance*, and it plays the same role here as does the mean free path in other kinetic theories.

Now, what about boundary conditions? By virtue of assumption (c), the concentration at steps must equal its equilibrium value. Since the balance equation (2.30a) holds on terraces between any two adjacent steps, and because this equation is symmetric under reversals of space, it behooves us to choose a coordinate system in Fig. 2.9 whose origin is placed symmetrically, right between adjacent steps. The boundary conditions then read

$$C_s(\pm l/2) = C_{s,e}. \tag{2.30c}$$

Equation (2.30a) is a second-order differential equation with constant coefficients; by standard methods, its solution is

$$\frac{C_s(x) - C_{s,e}}{\sigma_r C_{s,e}} = 1 - \frac{\cosh x/x_s}{\cosh l/2x_s}. \tag{2.31a}$$

Hence, we compute the flux (2.28c) and then the growth rate (2.27)

$$V(l) = (C_{s,e}\sigma_r\Omega/\tau_s)(2x_s/l)\tanh(l/2x_s). \tag{2.31b}$$

The growth rate is thus the product of three terms. The first, $C_{s,e}\sigma_r\Omega/\tau_s$, is due to the net flux from the parent phase, and is the maximum possible growth rate. In fact, when $l \ll 2x_s$ the two other terms go to unity because the whole surface, effectively, becomes a sink.† The second term, $2x_s/l$, shows that only those molecules that impinge in a band of width $2x_s$ (the "catchment" area or capture cross section) about the step are effective for growth. Molecules adsorbed at a distance greater than x_s from a step will have a large probability of desorbing before they reach kinks. The third term, $\tanh(l/2x_s)$, arises because adjacent steps compete for available adsorbed molecules.

EXERCISE 2.15. *Case of an isolated step:* Find the adsorbed concentration distribution around a single *isolated* step, and find the step's velocity. [Hint: Either solve the differential equation *da capo* or take the limit of Eqs. (2.31) and (2.32) for large l.] □

Equation (2.31b), as derived, makes no use of spirals. It is valid for any array of parallel equidistant straight steps, no matter how they are produced. There is yet another way of expressing growth rates if we invoke the spiral mechanism. It involves only the two macroscopic observables: supersaturation and equilibrium bulk concentration. We simply recall that the spiral growth mechanism implies the inverse relation (2.26b). Accordingly, we write

$$l/2x_s = \sigma_1/\sigma_r, \tag{2.32a}$$

where σ_1 is a constant. Then, using the equilibrium relation (2.29a) and the definition (2.30b), we get the normal growth rate in the form

$$V(\sigma_r) = (\Omega\sigma_1 kC_e)(\sigma_r/\sigma_1)^2\tanh(\sigma_1/\sigma_r). \tag{2.32b}$$

This relation, first derived by BCF, is called the "linear-parabolic" law of crystal growth because $V \approx \sigma_r^2$ for small supersaturations (relative to σ_1), whereas that dependence is linear for large supersaturations.

† Recall that $z^{-1}\tanh z = 1 + O(z^2)$.

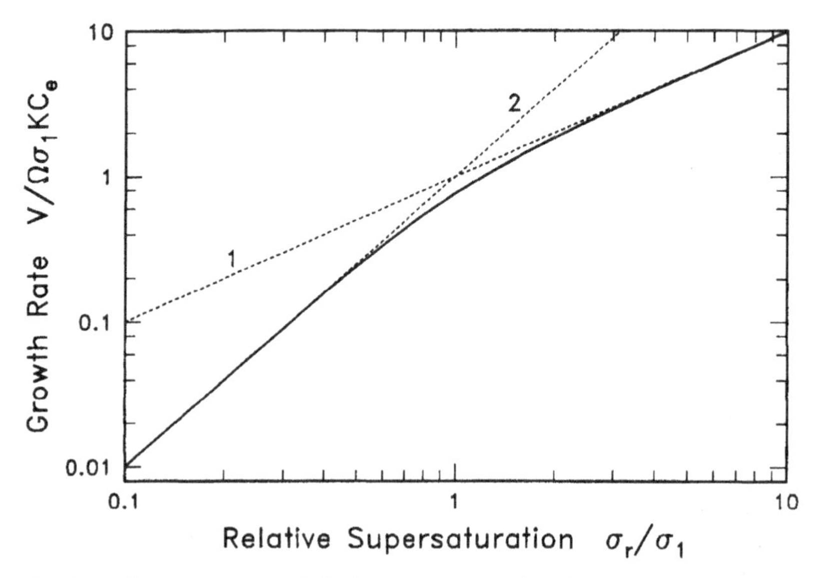

Fig. 2.10. Growth rate *vs.* (relative) supersaturation for the BCF model of crystal growth. Note the limiting linear (1) and quadratic (2) behaviors, shown dashed.

Figure 2.10 shows the behavior of Eq. (2.32b), and it is again clear, as in the case of silicon oxidation, that the transition from pure parabolic to pure linear behavior occurs over one order of magnitude in the supersaturation. Thus, the experimental verification of this BCF result is also not easy.[37]

There are evidently many other interesting diffusion problems related to the BCF model of crystal growth. In one way or another, these simply relax some of the assumptions (a–e) above. Chernov's excellent review article[38] and textbook,[39] and our own more recent paper[40] summarize these extensions. Finally, the exact shape of growth spirals and their dynamics have been investigated by Monte Carlo simulations.[41,42]

One last comment: The BCF theory, in fact, mixes concepts and calculations on a microscopic scale (surface structure) with macroscopic diffusional computations. In that sense it plays the same part in our understanding of crystal growth as the Boltzmann equation plays in the kinetic theory of gases. There, the total time derivative (a macroscopic concept) of the single-particle probability density is equal to the "collision integral," which must be computed from scattering theory.

Notes

1. This statement is strictly true only for isotropic phases at rest and in the absence of force fields, internal or external to the system.
2. This exercise suggests that diffusion through any (not necessarily circular) cylindrical shell behaves approximately like 1-d diffusion when the shell's thickness is small with respect to the local radius of curvature. Generalization is possible for arbitrary shells where both radii of curvature can vary.
3. Recall that homogeneous reactions occur *inside* a given phase. They are described mathematically by source terms in balance laws, and they characterize chemical transformations at interior points. On the other hand, heterogeneous reactions occur at phase *boundaries*. They are described mathematically by boundary conditions. Here, the reaction (2.7b) occurs at the Si/SiO_2 interface, assumed to be a geometric surface and not a 3-d transition region. Some measurements indicate that this may be but a useful idealization. For example, see "X-ray Photoelectron Spectroscopy of SiO_2-Si Interfacial Regions: Ultrathin Oxide Films" by S. I. Raider and R. Flitsch, *IBM J. Res. & Dev.* **22**, 294 (1978), and "Microscopic Structure of the SiO_2/Si Interface" by F. J. Himpsel, F. R. McFeely, A. Taleb-Ibrahimi, J. A. Yarmoff, and G. Hollinger, *Phys. Rev.* **B38**, 6084 (1988). Note, also, that the "back-reaction" and the equilibrium concentration in Eqs. (2.7b,c) are assumed negligible.
4. It is assumed here that no oxygen dissolves into the Si substrate. Recent studies have shown that this assumption is open to question, as is the matter of Si dissolution into the oxide and into the substrate itself (as self-interstitials). See "Interstitial Kinetics near Oxidizing Silicon Interfaces" by S. T. Dunham, *J. Electrochem. Soc.* **136**, 250 (1989), and "A Note on the Linear-Parabolic Law of Phase Formation" by S.-L. Zhang and F. M. d'Heurle, *Phil. Mag.* A **64**, 619 (1991).
5. Some of these components may be absent in one of the phases, in which case one must put their composition variables equal to zero. By *phase* one generally means a region of space throughout which *all fields are continuous.* These fields are, in general, continuously varying within a given phase, but at least one of them must suffer a discontinuity at phase boundaries.
6. The mean curvature obeys an interesting invariance property: K can be computed from *any two* orthogonal sections of a surface. Mathematical works on differential geometry generally call $2H$ the mean curvature, reserving K for the "gaussian" curvature.

7. The thermodynamic variables that can be held constant and those whose variations we seek must be compatible with the "phase rule," see Appendix C. According to Eq. (C.6) of that appendix, here we have three degrees of freedom because $\varphi = 2$ and $c = 2$. From the nine-member list (C.5) we choose fixed values of $T^{(2)}$ and $p^{(2)}$ by maintaining the heat and work reservoirs at the constant values $T^{(r)}$ and $p^{(r)}$. We then inquire, for example, into the functional dependence of the chemical potentials on the mean curvature K. The external values $\mu_A^{(r)}$ and $\mu_B^{(r)}$ must be left free to adjust.
8. Results of this type—shifts in equilibria due to curvature—are generically known as "Gibbs-Thomson effects." For example, the effect of curvature on vapor pressure is attributed to Lord Kelvin, who, after all, was also W. Thomson. For convenience, some more of these results are collected in Appendix C.
9. This is compatible with the phase rule (C.4) of Appendix C for large systems. Because $f = 2$, we can arbitrarily choose experimental conditions of constant T and p. Then, however, all other quantities, such as equilibrium compositions, are fully determined.
10. In passing, we note that the impossibility of finding a nucleus in the single-phase region (the region of good cooks, above the liquidus), and thus $C_e(K) \leqslant C_e$, is related to the impossibility of finding *finite* bodies of constant negative mean curvature.
11. The following calculations are valid only when convection in the liquid is negligible. In particular, note that convection can be caused by density changes in the liquid-solid transition. See "On the Dynamics of Phase Growth" by L. E. Scriven, *Chem. Engng. Sci.* **10**, 1 (1959); *ibid.* **27**, 1753 (1972), for a full account of these effects.
12. The situation should be familiar from mechanics. For example, the position(s) of a mass point at which its velocity vanishes, and therefore at which the motion reverses itself, are the turning points of its equations of motion. These are second order in time, in contrast to the first order equation, above.

References

1. R. P. Smith, "The Diffusivity of Carbon in Iron by the Steady-State Method," *Acta Met.* **1**, 578 (1953).
2. H. S. Carslaw and J. C. Jaeger, *Conduction of Heat in Solids*, 2nd ed., pp. 188–193 (Oxford University Press, London, 1959).
3. W. G. Oldham, "The Fabrication of Microelectronic Circuits," *Scientific American* **237**, pp. 110–128 (September 1977). [That whole issue is devoted to microelectronics. The reader can gauge the accelerating pace

of technological change, since then, by consulting the October 1986, February 1990, and June 1993 issues of *Physics Today*.]
4. B. E. Deal and A. S. Grove, "General Relationship for the Thermal Oxidation of Silicon," *J. Appl. Phys.* **36**, 3770 (1965).
5. E. Rosencher, A. Straboni, S. Rigo, and G. Amsel, "An ^{18}O Study of the Thermal Oxidation of Silicon in Oxygen," *Appl. Phys. Lett.* **34**, 254 (1979).
6. U. R. Evans, *The Corrosion and Oxidation of Metals* (Edward Arnold Ltd., London, 1960).
7. E. A. Irene and Y. J. van der Meulen, "Silicon Oxidation Studies: Analysis of SiO_2 Film Growth Data," *J. Electrochem. Soc.* **123**, 1380 (1976).
8. E. A. Irene and D. W. Dong, "Silicon Oxidation Studies: The Oxidation of Heavily B- and P-Doped Single Crystal Silicon," *J. Electrochem. Soc.* **125**, 1146 (1978).
9. K. J. Laidler, *Chemical Kinetics*, 2nd ed. (McGraw-Hill, New York, 1965).
10. R. Ghez and Y. J. van der Meulen, "Kinetics of Thermal Growth of Ultra-thin Layers of SiO_2 on Silicon. Part II: Theory" *J. Electrochem. Soc.* **119**, 1100 (1972).
11. E. A. Irene, "Evidence for a Parallel Path Oxidation Mechanism at the Si-SiO_2 Interface," *Appl. Phys. Lett.* **40**, 74 (1982).
12. H. Z. Massoud, J. D. Plummer, and E. A. Irene, "Thermal Oxidation of Silicon in Dry Oxygen: Growth-Rate Enhancement in the Thin Regime, Parts I and II," *J. Electrochem. Soc.* **132**, 2685 and 2693 (1985).
13. E. Bassous, H. N. Yu, and V. Maniscalco, "Topology of Silicon Structures with Recessed SiO_2," *J. Electrochem. Soc.* **123**, 1729 (1976).
14. R. B. Marcus and T. T. Sheng, "The Oxidation of Shaped Silicon Surfaces," *J. Electrochem. Soc.* **129**, 1278 (1982).
15. L. O. Wilson, "Numerical Simulation of Gate Oxide Thinning in MOS Devices," *J. Electrochem. Soc.* **129**, 831 (1982).
16. Proceedings of "Workshop on Oxidation Processes," *Phil. Mag.* **55**, February and June issues (1987).
17. J. W. Gibbs, "On the Equilibrium of Heterogeneous Substances," *Trans. Conn. Acad.* (1875–1878). [Reprinted in *Scientific Papers*, Vol. 1 (Dover, New York, 1961).]
18. R. Defay and I. Prigogine, *Tension Superficielle et Adsorption* (Editions Desoer, Liège 1951). [Engl. transl. by D. H. Everett (John Wiley, New York, 1966).]
19. F. P. Buff, "The Theory of Capillarity," in *Handbuch der Physik*, Vol. 10, pp. 281–304 (Springer Verlag, Berlin, 1960).
20. E. A. Guggenheim, *Thermodynamics, An Advanced Treatment for Chemists and Physicists* (North-Holland, Amsterdam, 1986).
21. R. Ghez, "A Theoretical Crystal Grower's View of Phase Equilibria," in *Crystal Growth in Science and Technology*, Proceedings of the *International*

School of Crystallography, Erice, 1987, NATO Advanced Study Institute, Vol. **B-210**, pp. 1–26 (Plenum Publishing, New York 1989).
22. E. A. Guggenheim, *Mixtures* (Oxford University Press, London, 1952).
23. A. S. Jordan and M. E. Weiner, "The Effect of the Heat Capacity of the Liquid Phase on the Heat of Fusion Liquidus Equation of Compound Semiconductors," *J. Phys. Chem. Sol.* **36**, 1335 (1975).
24. J. D. Honeycutt and H. C. Anderson, "Small System Size Artifacts in the Molecular Dynamics Simulation of Homogeneous Crystal Nucleation in Supercooled Atomic Liquids," *J. Phys. Chem.* **90**, 1585 (1986).
25. F. S. Ham, "Shape-Preserving Solutions of the Time-Dependent Diffusion Equation," *Quart. Appl. Math.* **17**, 137 (1959).
26. R. F. Sekerka, "Morphological Stability," in *Crystal Growth: An Introduction*, pp. 403–443, P. Hartman, Ed. (North-Holland, Amsterdam 1973).
27. J. S. Langer, "Instabilities and Pattern Formation in Crystal Growth," *Rev. Mod. Phys.* **52**, 1 (1980).
28. M. Kahlweit, "Precipitation and Aging," in *Physical Chemistry, an Advanced Treatise*, Vol. 10, pp. 719–759, H. Eyring, D. Henderson, and W. Jost, Eds. (Academic Press, New York, 1970).
29. I. M. Lifshitz and V. V. Slyozov, "The Kinetics of Precipitation from Supersaturated Solid Solutions," *J. Phys. Chem. Sol.* **19**, 35 (1961).
30. C. Wagner, "Theorie der Alterung von Niederschlägen durch Umlösen," *Zeitschr. für Elektrochemie* **65**, 581 (1961).
31. P. W. Voorhees, "The Theory of Ostwald Ripening," *J. Stat. Phys.* **38**, 231 (1985).
32. B. Lewis and G. J. Rees, "Adatom Migration, Capture, and Decay among Competing Nuclei on a Substrate," *Phil. Mag.* **29**, 1253 (1974).
33. J. A. Venables and G. L. Price, "Nucleation of Thin Films," in *Epitaxial Growth*, Part B, pp. 381–436, J. W. Matthews, Ed. (Academic Press, New York, 1975).
34. W. K. Burton, N. Cabrera, and F. C. Frank, "The Growth of Crystals and the Equilibrium Structure of their Surfaces," *Phil. Trans. Roy. Soc.* (London) **A243**, 299 (1951).
35. J. Frenkel, "On the Surface Motion of Particles in Crystals and the Natural Roughness of Crystalline Faces," *J. Phys. USSR* **9**, 392 (1945).
36. M. Volmer, *Kinetik der Phasenbildung* (Th. Steinkopff Verlag, Dresden, 1939).
37. P. Bennema and G. H. Gilmer, "Kinetics of Crystal Growth," in *Crystal Growth: An Introduction*, pp. 263–327, P. Hartman, Ed. (North-Holland, Amsterdam, 1973).
38. A. A. Chernov, "The Spiral Growth of Crystals," *Usp. Fiz. Nauk.* **73**, 277 (1961). [Engl. transl. in *Sov. Phys. Usp.* **4**, 116 (1961).]
39. A. A. Chernov, *Modern Crystallography III: Crystal Growth* (Springer-Verlag, Berlin 1984).

40. R. Ghez and S. S. Iyer, "The Kinetics of Fast Steps on Crystal Surfaces and its Application to the Molecular Beam Epitaxy of Silicon," *IBM J. Res. & Dev.* **32**, 804 (1988).
41. G. H. Gilmer, "Growth on Imperfect Crystal Faces," *J. Cryst. Growth* **35**, 15 (1976).
42. G. H. Gilmer, "Computer Models of Crystal Growth," *Science* **208**, 355 (1980).

3

Diffusion Under External Forces

This short chapter is devoted to the relation between diffusion and mobility. Indeed, until now, the process of mass redistribution was attributed to *mixing* on a microscale. We now ask what effects force fields can have on this redistribution, as they affect particle flows through a drift or mobility term, in addition to the usual diffusion term.

3.1. The Anisotropic One-Dimensional Random Walk

A simple approach consists of relaxing one of the basic assumptions implicit in the random walk model of Chap. 1. There it was assumed that the jump frequencies were *isotropic*, which means that transition probabilities to the right and to the left had the same value $\frac{1}{2}\Gamma$. We now distinguish jumps to the right, with jump frequency Γ^+, from jumps to the left, with frequency Γ^-; these frequencies need not be equal. For the moment we ascribe no reason for this asymmetry. As in Sect. 1.1, it follows that the net number of particles that jump from site i to site $i + 1$ per unit time is

$$J_{i+1/2} = \Gamma^+ N_i - \Gamma^- N_{i+1}, \tag{3.1a}$$

and the net flux from $i - 1$ to i is

$$J_{i-1/2} = \Gamma^+ N_{i-1} - \Gamma^- N_i. \tag{3.1b}$$

Because of the anisotropy in jump frequencies, these fluxes, now, do not vanish when $N_i = N_{i+1}$. A constant distribution over sites is therefore no longer a necessary condition for equilibrium.

Again, the time rate of change of the particle distribution is evaluated by considering all possible transitions to and from the ith site:

$$\dot{N}_i = \Gamma^- N_{i+1} + \Gamma^+ N_{i-1} - (\Gamma^+ + \Gamma^-)N_i. \tag{3.2}$$

Hence, using Eqs. (3.1), we obtain the same expression (1.3)

$$\dot{N}_i = -(J_{i+1/2} - J_{i-1/2}) \tag{3.3}$$

in terms of fluxes. All the remarks (a–d) of Sect. 1.1 still hold, because we have done nothing except lift the assumption of isotropy of jump frequencies. In particular, Eq. (3.3) still has the form of a conservation law.

We pass to continuum expressions through the same interpolating scheme (1.5) and (1.6) as for the isotropic model. Expanding Eqs. (3.2) around the site x_i now yields

$$\frac{\partial \hat{N}}{\partial t} = \tfrac{1}{2}\Gamma a^2 \frac{\partial^2 \hat{N}}{\partial x^2} - \Delta a \frac{\partial \hat{N}}{\partial x} + \mathrm{O}(a^3), \tag{3.4}$$

where, for convenience, we have introduced the sum and difference

$$\Gamma = \Gamma^+ + \Gamma^- \qquad \text{and} \qquad \Delta = \Gamma^+ - \Gamma^- \tag{3.5}$$

of jump frequencies.

It is again delicate to analyze fluxes. Expanding Eqs. (3.1), we find that

$$J_{i\pm 1/2} = \Delta N|_{x_i} - \tfrac{1}{2}(\Gamma \mp \Delta)a \left.\frac{\partial \hat{N}}{\partial x}\right|_{x_i} + \mathrm{O}(a^2), \tag{3.6}$$

and, to first order in the jump distance a, these fluxes do *not* converge to the same expression. Since there seems to be nothing better to do, we *define* a continuous flux function

$$\hat{J}(x, t) = \Delta \hat{N} - \tfrac{1}{2}\Gamma a \frac{\partial \hat{N}}{\partial x} \tag{3.7}$$

as the arithmetic average of the right-hand side of Eq. (3.6). We note that Eqs. (3.4) and (3.7) both reduce to their homologs (1.7) and (1.9) when the jump frequencies are equal, i.e., when $\Delta = 0$. Then, as in Exercise 1.2, computing the difference $J_{i\pm 1/2} - \hat{J}(x_\theta, t)$ at intermediate points, we find that this difference is $O(a^2)$ when the discrete fluxes are defined at the midpoints $x_{i\pm 1/2} = (i \pm \frac{1}{2})a$. Consequently, the conservation law (3.3) has the continuum analog

$$\frac{\partial \hat{N}}{\partial t} = -a\frac{\partial \hat{J}}{\partial x} + O(a^3), \tag{3.8}$$

which, together with Eq. (3.7), is compatible with Eq. (3.4). These equations should be compared with their homologs (1.7)–(1.10). Note that the truncation error has increased by one order because of the asymmetry of the jump frequencies.

The required continuum equations emerge if, again, $C = \hat{N}/a$ is the average concentration per cell and if $D = \frac{1}{2}\Gamma a^2$ is the diffusivity. In addition, we define the *drift velocity* $v = \Delta a$, and Eqs. (3.7), (3.8), and (3.4) then become

$$J(x, t) = -D\frac{\partial C}{\partial x} + vC, \tag{3.9}$$

$$\frac{\partial C}{\partial t} = -\frac{\partial J}{\partial x}, \tag{3.10}$$

$$\frac{\partial C}{\partial t} = D\frac{\partial^2 C}{\partial x^2} - v\frac{\partial C}{\partial x}. \tag{3.11}$$

These equations, written in 1-d and for the case of constant coefficients, can be readily generalized. For example,

$$\mathbf{J} = -D\nabla C + \mathbf{v}C \tag{3.12}$$

is the flux vector in an isotropic body of any dimension.

We end this section by inquiring into the physical significance of the drift velocity. Since, by definition, v is proportional to the difference in jump frequencies, it is thus a measure of the *bias* in the system, a bias that subsists even if the particles are evenly distributed among sites. In a multicomponent fluid, one shows[1] that equations similar to Eqs. (3.9)–(3.12) describe its dynamics, as was implicit in the development of Sect. 1.8 [see Eqs. (1.45) and (1.48)]. A simple hopping model is then not applicable, but the drift velocity is interpreted simply as the fluid's mean velocity field. In general, the first term of Eq. (3.12) is called the diffusional flux, while the second, called the convective flux, measures mass transport due to an overall motion. Visualization follows if we think of a dye injected into a running stream.

In a solid there is, in general, no convection because crystal lattices do not move significantly.† Nonetheless, a potential force field $V(\mathbf{x}, t)$ can act on particles with a force $\mathbf{F} = -\nabla V$. These fields can be internal (e.g., stresses due to dislocation cores) or caused externally by applied fields (e.g., electric fields that act on charged particles, as in electromigration). Particles that move in a dense medium suffer collisions that tend to randomize their instantaneous velocities. The drift velocity $\mathbf{v}$ is a statistical average over these instantaneous velocities, biased by the force field. One can show[2] that there is often a linear relation

$$\mathbf{v} = B\mathbf{F} = -B\nabla V \tag{3.13}$$

between the drift velocity and the force field, where the coefficient B (necessarily positive) is called the *mobility*. Its dimensions are those of velocity per unit force, or $[B] = \text{s/g}$. We note that Eq. (3.13) is fundamentally different from Newton's law of motion, which relates force to acceleration. Here, because of collisions, the force is proportional to the velocity, as for an overdamped oscillator.

3.2. Diffusivity and Mobility Coefficients

Both coefficients D and B characterize the flow of particles. It is therefore not inconceivable that they be related. We shall develop two arguments. The first, owing to Mott (see Ref. 3), is based on equilibrium considerations (and is therefore misleading), while the second is dynamical (and is therefore somewhat more believable).

Mott's argument is very simple. The flux (3.12) must vanish identically at equilibrium. With Eq. (3.13), we get

$$-D\nabla C - BC\nabla V = 0. \tag{3.14a}$$

†This is not quite true when boundaries move, thermal expansion is taken seriously, or vacancy fluxes are significant (Kirkendall effect).

A scalar multiplication by an arbitrary displacement $d\mathbf{x}$ yields the differential equation $D\,dC + BC\,dV = 0$,† which has the solution

$$C \propto e^{-BV/D} \tag{3.14b}$$

if the diffusivity and mobility are constants. But, at equilibrium in a potential field V, classical particles obey a Boltzmann distribution $C \propto \exp(-V/kT)$. Comparing exponents, we get the desired relation

$$D = BkT, \tag{3.15}$$

first suggested by Einstein.[4] Note that the equilibrium concentration (3.14b) cannot be constant if the potential varies with position. The fundamental reason is that only the electrochemical potential (including fields) must be constant at equilibrium, and not the concentration, which, after all, is but a secondary variable from a basic thermodynamic point of view.[5]

Einstein's relation (3.15) was derived, arguably, under equilibrium conditions. Does it hold in general? The next calculation shows it to be true "near" equilibrium. First, we must ask why particles reside at lattice sites x_i, and why they can hop, occasionally, from site to site. Recall that an isolated particle in a local potential well can be in equilibrium: Any small enough displacement from the potential's minimum position causes a force that restores the particle to its original position. If, however, the potential has finite height, then a sufficiently energetic particle can overcome this barrier, and this particle moves away permanently from its initial position x_i. The site x_i is then metastable because it does not correspond to an absolute minimum of the particle's potential energy.

For an ensemble of particles, such as we are considering (N_i is really a statistical average of the number of particles at the site x_i), the role of potential energy is played by the Gibbs free energy G, and the Gibbs free energy per particle *is* the chemical potential. Referring to Fig. 3.1, we see, for example, that the potential energy is lower at site '2' than at site '1'. The transition $1 \rightarrow 2$, if successful, causes a negative change in energy, i.e., $\Delta G = G_2 - G_1 < 0$. In the language of thermodynamics, state (2) is more *stable* than state (1), and ΔG is the free-energy change in the "reaction" $1 \rightarrow 2$. The *rate* of this reaction

†Recall from vector analysis that $df = d\mathbf{x} \cdot \nabla f$ for any differentiable function $f(\mathbf{x})$.

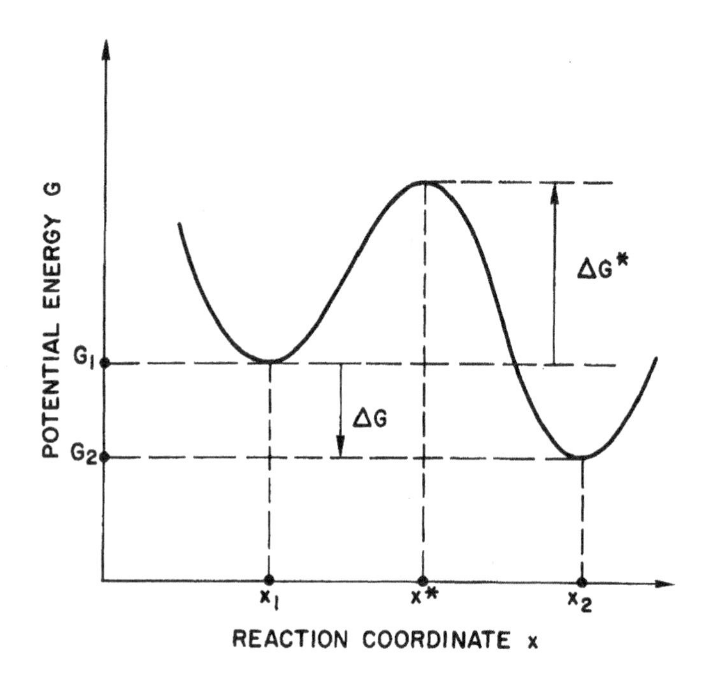

Fig. 3.1. Potential-energy diagram showing why the activation barrier ΔG^* differs from the energy change ΔG during reactions.

depends, however, on the potential barrier between the states $\Delta G^* > 0$, which is generally not calculable from thermodynamics alone. Particles at position (more generally, "state") x^* are said to be in an *activated* state, and the barrier ΔG^* is called the *activation energy*.[6] Energy must be imparted, often in the form of heat, to an ensemble of particles so that they can overcome this barrier. In statistical mechanics, one shows[6] that the probability P that a representative particle has an energy at least equal to the activation energy is[†]

$$P = e^{-\Delta G^*/kT}. \tag{3.16}$$

Equation (3.16) expresses this probability in terms of the ratio of the energy barrier to the thermal energy of particles.

[†]In this chapter all energies, including the activation energy, are expressed per particle. Expressions per mole only require the replacement of Boltzmann's constant k by the gas constant R.

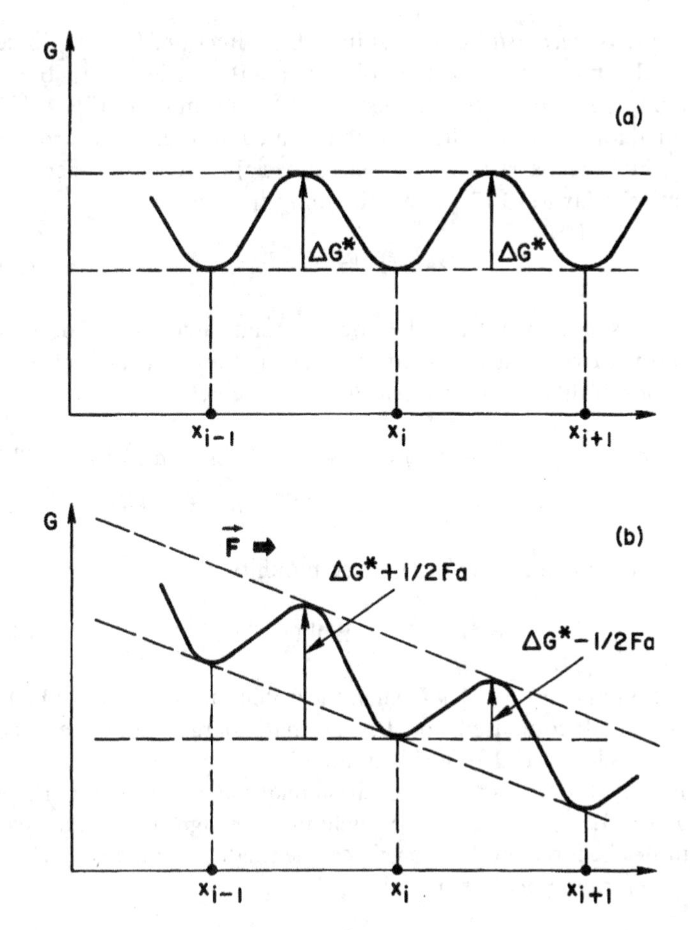

Fig. 3.2. (a) Periodic potential due to a lattice. (b) The bias introduced by the action of an external force **F**.

It is now very simple to derive Einstein's relation (3.15). In a crystal lattice, particles reside at sites x_i because they sit in periodic wells, as shown in Fig. 3.2a. These particles can hop from well to well if they have sufficient thermal energy. If ν is a characteristic vibration frequency in a given well (approximately the Debye frequency), then

$$\Gamma = 2\nu P = 2\nu e^{-\Delta G^*/kT} \tag{3.17a}$$

is the rate of *successful* attempts in either direction. If, now, forces $\mathbf{F}$ act on the particles (directed toward $x > 0$ in Fig. 3.2b), then the potential due to the lattice is decreased by an amount $V = -\int \mathbf{F}\, d\mathbf{x}$. For small forces, the maxima are still located at intermediate positions $x_{i\pm 1/2}$, but the barriers to the left and right are now different. An estimate similar to (3.17a) gives the jump frequencies

$$\Gamma^{\pm} = \nu e^{-(\Delta G^* \mp Fa/2)/kT}, \tag{3.17b}$$

where it is assumed that the force's magnitude does not change significantly over a lattice distance. Inserting these expressions into the definitions of diffusivity and drift velocity, we get

$$D = \tfrac{1}{2}a^2(\Gamma^+ + \Gamma^-) = a^2 \nu e^{-\Delta G^*/kT}\cosh(Fa/2kT), \tag{3.18a}$$

$$v = a(\Gamma^+ - \Gamma^-) = 2a\nu e^{-\Delta G^*/kT}\sinh(Fa/2kT), \tag{3.18b}$$

and, from the definition (3.13) of the mobility, the ratio

$$B/D = (2/Fa)\tanh(Fa/2kT). \tag{3.18c}$$

This ratio tends toward $1/kT$ when the potential drop due to the force field is small compared to the thermal energy,† and we recover Einstein's relation (3.15) in this limit.

Finally, the transcription of these relations to charged particles is instructive. Let q be the charge (including the sign) of the particles in question. The force on these particles due to an electric field is $\mathbf{F} = q\mathbf{E}$, and therefore the drift velocity is

$$\mathbf{v} = Bq\mathbf{E}. \tag{3.19a}$$

One calls

$$\mu = B|q| \tag{3.19b}$$

the *mobility coefficient* for electronic or ionic systems. It is defined here as a positive quantity so as to recover the semiconductor fluxes (1.40c) and (1.40d), and the symbol μ should not be confused with the same symbol used for chemical potentials. Note that its dimensions are

†Recall that $\tanh z = z + O(z^3)$.

$[\mu] = \text{Cbs/g} = \text{cm}^2/\text{V s}$. With these definitions, Eqs. (3.13) and (3.15) read†

$$\mathbf{v} = \text{sgn}(q)\mu\mathbf{E} \tag{3.19c}$$

and

$$D = \mu kT/|q|. \tag{3.19d}$$

In particular, the flux (3.12) for charged particles is

$$\mathbf{J}(x, t) = -D\nabla C + \text{sgn}(q)\mu C\,\mathbf{E}. \tag{3.20}$$

EXERCISE 3.1. *Concentration distribution in a slab:* Consider a dielectric, such as the SiO_2 layer of Fig. 2.2. Find the steady-state distribution of charged particles in that region, assuming that only one charged species is mobile, that the electric field is constant throughout the dielectric, and that the interface concentrations are known. Discuss for both positive and negative charges. □

The points we have broached are presently being actively investigated[5,7,8] as part of a study on the nature of fluctuations around, and the stability of, the steady state. Moreover, Eqs. (3.18) show that both D and B are functions of location if the external force F is not a constant, and are then said to be "state dependent." Conversely, a recent paper[9] shows how a given state-dependent diffusivity is equivalent to a bias in the potential energy and thus to a nonvanishing flux.

3.3. An Introduction to Double Layers

The last exercise was designed to illustrate that charged particles, even in 1-d steady state, are generally *not* distributed linearly. This is due to two competing influences on the flux (3.20): The diffusion term promotes mixing and thus leads to increased randomness or entropy, whereas the second term represents the system's attempt to lower its energy through the electric field. This field, in turn, depends on the *actual* distribution of charges, as well as on externally applied bias

†The *signum* function sgn(z) has values ± 1 according to the sign of its argument z.

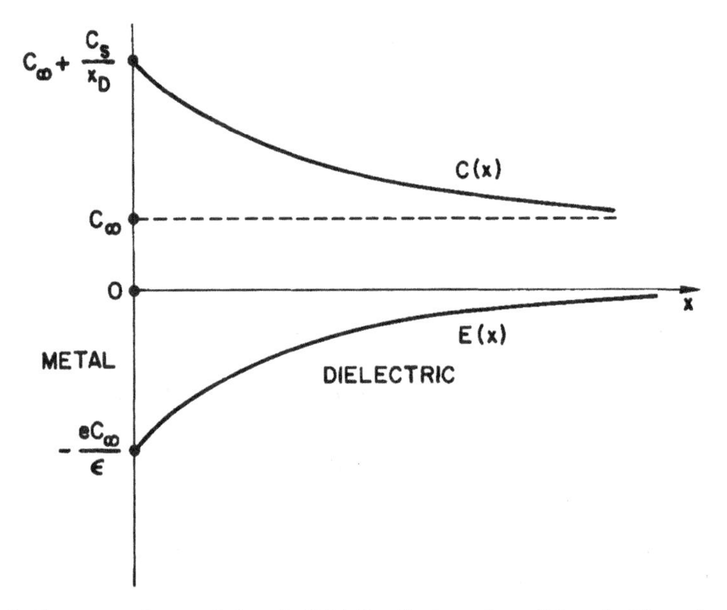

Fig. 3.3. Concentration and electric-field distributions in a dielectric. Note the charge separation within a Debye length of the interface.

voltages. Problems of this nature are thus nonlinear, and their solution depends heavily on the charges' behavior near surfaces. In general, we can expect charge imbalances close to phase boundaries, i.e., local charge neutrality breaks down. This is called a *double layer* or *space charge* because a net distributed charge must be compensated by a surface charge of opposite sign, thus giving rise to an electric dipole. Double layers are very common: Electrolytes in batteries, electrons and holes in semiconductors, and ions in dielectrics all share their features.

Here, we examine a particularly simple case that nevertheless, shows all the generic difficulties of double layers. We consider, as shown in Fig. 3.3, a semi-infinite, one-dimensional dielectric, of dielectric constant ϵ, that is terminated by a metal electrode. We assume, for example, that only one singly charged ionic species is mobile and that its charge e is say, positive. The ions' concentration distribution C in the dielectric is compensated by a fixed negative background at a concentration level C_∞. The ions are positive because they have discharged electrons at the electrode. Therefore, there is an induced negative charge on the metal whose surface concentration is C_s. We then wish to compute the steady-state distribution of these positive ions and their electric field or potential. In accordance with Sects. 2.1 and 2.2, the assumption of steady state in 1-d implies that the particle flux (3.20) is constant with respect to location x. If, further, we assume that the electrode is "blocking," meaning that J vanishes there, then it follows that

$$-D\frac{dC}{dx} + \mu CE = 0 \qquad \text{for all } x > 0, \tag{3.21a}$$

with $\mu/D = e/kT$ according to Einstein's relation (3.19d). The distributed charge density, however, is a source of an electric field, given by Gauss's law

$$\epsilon\frac{dE}{dx} = e(C - C_\infty), \tag{3.21b}$$

and the negative surface charge also gives rise to the field at the metal–dielectric phase boundary,

$$\epsilon E(0^+) = -eC_s. \tag{3.22a}$$

Finally, electrical neutrality must hold far from the electrode

$$C(\infty) = C_\infty. \tag{3.22b}$$

We must now solve the nonlinear system of two first-order differential equations (3.21) for the unknown functions $C(x)$ and $E(x)$, under the boundary conditions (3.22). Before doing so, however we note several consequences than can be drawn from these equations. First, the bulk neutrality condition (3.22b), together with Gauss's law (3.21b), implies that

$$E'(\infty) = 0. \tag{3.23a}$$

Next, we formally integrate Eq. (3.21b) with the boundary condition (3.22a) to give

$$\epsilon E(x) = -eC_s + e\int_0^x [C(\xi) - C_\infty]\,d\xi.$$

This equation (Gauss's) merely states that the field at any location x is proportional to the sum of charges contained in the region to the left of x. For large x this region contains all the charges, and, since global neutrality must hold, it follows that

$$E(\infty) = 0. \tag{3.23b}$$

Therefore, both the electric field and its gradient must vanish at infinity. The solution of Eqs. (3.21) follows by eliminating C in favor of the field

$$C = C_\infty + (\epsilon/e)E', \tag{3.24a}$$

which, inserted into Eq. (3.21a), yields the second-order differential equation:

$$E'' - \frac{e}{kT}E\left(E' + \frac{eC_\infty}{\epsilon}\right) = 0. \tag{3.24b}$$

Since this equation does not explicitly depend on x, the introduction of a new dependent variable $u = E'$, considered a function of E, transforms it into the first-order equation†

$$u\frac{du}{dE} - \frac{e}{kT}\left(u + \frac{eC_\infty}{\epsilon}\right)E = 0 \tag{3.25a}$$

whose variables separate. With the subsidiary conditions (3.23), we find that

$$\frac{\epsilon}{eC_\infty}E' - \ln\left(1 + \frac{\epsilon}{eC_\infty}E'\right) = \frac{\epsilon}{2kTC_\infty}E^2. \tag{3.25b}$$

This is as far as we can go analytically without further approximations because this equation gives E' implicitly. If, following Debye and Hückel (see Ref. 10), we assume that the field is "small" in the sense that

$$|(\epsilon/eC_\infty)E'| \ll 1, \tag{3.26}$$

†Use the identity $E'' = du/dx = (du/dE)(dE/dx) = u(du/dE)$.

then the logarithm can be expanded to second order in that quantity[†] to yield the equation

$$x_D E' = \pm E, \qquad (3.27a)$$

where the characteristic distance

$$x_D = \sqrt{\epsilon kT/e^2 C_\infty}, \qquad (3.27b)$$

called the Debye length, measures the extent of the double layer, as shown in what follows. Equation (3.27a) is readily solved under the boundary conditions (3.22a) and (3.23a) to give

$$E(x) = -\frac{eC_s}{\epsilon} e^{-x/x_D}, \qquad (3.28a)$$

and the concentration distribution follows from Eq. (3.24a) by differentiation

$$C(x) = C_\infty + (C_s/x_D)e^{-x/x_D}. \qquad (3.28b)$$

Thus, both C and E decay into the dielectric exponentially according to the Debye length.

EXERCISE 3.2. *When is the field small?:* First, carefully justify the steps that lead from Eqs. (3.24) to (3.28). Then, express the condition (3.26) in terms of material parameters, and interpret. Find the dipole moment. □

A few remarks are in order. First, it is clear that we could have obtained the same results by eliminating the field E rather than C. (How would you do this?) Next, a general solution of the first integral (3.25b) requires numerical methods. Further, the above calculation did not require the introduction of the electrostatic potential $\phi = -\int E\,dx$. Multidimensional problems, on the other hand, are more easily solved in terms of ϕ. We note, however, as with Mott's argument (3.14), that Eq. (3.21a) can be integrated in terms of the potential to give Boltzmann's distribution $C \propto e^{-e\phi/kT}$. Finally, several mobile charges

[†]Recall that $\ln(1 + z) = z - z^2 + O(z^3)$.

can be considered, such as electrons and holes—donors and acceptors in semiconductors.[11] Likewise, oxidation kinetics in metals[12] often depend on the motion of both ions and electrons. In all these cases, the surface charge plays a crucial part, and the identification and control of *surface states* due to dangling bonds or to adsorbed ions is of the utmost importance.[11,13–15]

EXERCISE 3.3. *The case of uncompensated bulk charge:* When the negative background C_∞ is negligible, the distributed charge is entirely compensated by the surface charge. Solve that corresponding problem (3.21) and (3.22) exactly, and define an appropriate Debye length. □

References

1. R. B. Bird, W. E. Stewart, and E. N. Lightfoot, *Transport Phenomena*, Chap. 18 (John Wiley, New York, 1960).
2. E. M. Lifshitz and L. P. Pitaevskii, *Physical Kinetics* (Vol. 10 of *Course on Theoretical Physics*), pp. 40–42 (Pergamon Press, Oxford, 1981).
3. N. F. Mott and R. W. Gurney, *Electronic Processes in Ionic Crystals*, 2nd ed., p. 63 (reprinted by Dover, New York, 1964).
4. A. Einstein, papers on brownian motion, collected in *Investigations on the Theory of the Brownian Movement* (reprinted by Dover, New York, 1956).
5. R. Landauer, "Stability and Relative Stability in Nonlinear Driven Systems," *Helv. Phys. Acta* **56**, 847 (1983).
6. A. S. Glasstone, K. J. Laidler, and H. Eyring, *The Theory of Rate Processes* (McGraw-Hill, New York, 1941).
7. R. Landauer, "The Ballast Resistor," *Phys. Rev.* **A15**, 2117 (1977).
8. R. Landauer, "Stability in the Dissipative Steady State," *Physics Today* **31**, 23 (Nov. 1978).
9. M. Büttiker, "Transport as a Consequence of State-Dependent Diffusion," *Z. für Physik* **B68**, 161 (1987).
10. J. Frenkel, *Kinetic Theory of Liquids*, pp. 36–40 (reprinted by Dover, New York, 1955).
11. E. Spenke, *Electronic Semiconductors* (McGraw-Hill, New York, 1958).
12. A. T. Fromhold, *Theory of Metal Oxidation*, Vols. I and II (North-Holland, Amsterdam, 1976 and 1980).
13. J. Bardeen, "Surface States and Rectification at a Metal Semi-conductor Contact," *Phys. Rev.* **71**, 717 (1947).
14. V. Heine, "Theory of Surface States," *Phys. Rev.* **A138**, 1689 (1965).
15. A. Many, Y. Goldstein, and N. B. Grover, *Semiconductor Surfaces* (North-Holland, Amsterdam, 1965).

4

Simple Time-Dependent Examples

Our ability to solve certain differential equations increases with our lexicon of available functions. Exponentials, for example, are basic solutions of ordinary linear differential equations with constant coefficients. Likewise, to solve time-dependent problems in diffusion theory requires the introduction of certain "special functions" that are closely related to gaussians.[1–3] In this chapter we follow a constructive path that issues from our earlier experience with random walks, through new physical examples, and finally to the family of error functions. We then show, in the next chapter, that these error functions are, in fact, part of a grander scheme—similarity—that also includes the classical method of separation of variables. This chapter ends with a consideration of problems in several space dimensions that can be expressed in terms of products of 1-d problems.

4.1. The Gaussian and One of Its Relatives

Let us return to the isotropic random walk and to its numerical solution presented in Sects. 1.4 and 1.5. It was suggested there that the reader plot and fit the output G_i^n of the program *EXPLICIT*, much as one would with experimental data. Figure 4.1 shows how $\ln G_i^n$ behaves when plotted against the square of the space index i. Except around the origin, we obtain essentially straight lines $\ln G_i^n \approx \alpha - \beta i^2$ with negative slopes. The absolute value β of these slopes appears to decrease with increasing time index n, as do the intercepts α. In fact, further data-fitting suggests that β is inversely proportional to n, but

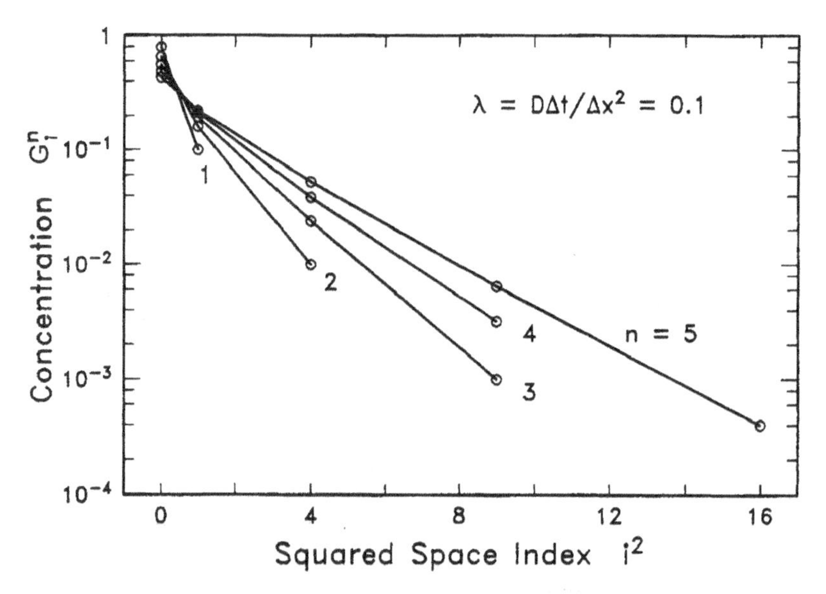

Fig. 4.1. Semilog plot of output from the explicit finite-difference scheme of Sect. 1.5.

that the intercepts α are slowly varying with n. More precisely, one finds that $\beta = (4\lambda n)^{-1}$, and we need not even find $\alpha(n)$. Exponentiation of the above logarithmic relation and remembering the definitions (1.24) and (1.21) of the modulus and grid points leads to

$$G_i^n = A(n)e^{-i^2/4n\lambda} = A(n)e^{-x_i^2/4Dt_n}, \tag{4.1a}$$

which is evidently a sample of the continuous function

$$G(x, t) = Ae^{-x^2/4Dt}. \tag{4.1b}$$

The coefficient of proportionality A is easily obtained, since we know that there was unit mass to begin with and that mass is conserved by diffusion in an infinite medium. Therefore we must have

$$\int_{-\infty}^{\infty} G(x, t)\, dx = 1 \tag{4.1c}$$

for all times. Inserting the expression (4.1b) into Eq. (4.1c) and recalling (Ref. 4, p. 371) a standard integral

$$\int_{-\infty}^{\infty} e^{-\xi^2} d\xi = \sqrt{\pi}, \tag{4.2}$$

we get the normalization constant $A = (4\pi Dt)^{-1/2}$ required in Eq. (4.1b). In sum, the continuum solution of the diffusion equation for a point source in an infinite medium is the gaussian

$$G(x, t) = \frac{1}{2\sqrt{\pi Dt}} e^{-x^2/4Dt}, \tag{4.3}$$

and Fig. 1.3 is its graph.

The resemblance between this solution (4.3) and the normal distribution of probability theory[5,6] is not fortuitous. To grasp this connection, recall that the explicit finite-difference equation (1.23),

$$G_i^{n+1} = (1 - 2\lambda)G_i^n + \lambda(G_{i+1}^n + G_{i-1}^n), \tag{4.4}$$

repeated here for convenience for the point source solution, samples the actual concentration distribution at grid points (x_i, t_n). But that equation can also be interpreted in terms of a random walk if the sampled concentration G_i^n represents the occupation probability of site i at the time step n, and if λ represents the *a priori* (conditional) probability of a jump (per unit time) either to the right or to the left. Indeed, there are three mutually exclusive events that lead to the occupation probability on the left-hand side of Eq. (4.4): a jump to the right from state $(i - 1, n)$, a jump to the left from state $(i + 1, n)$, and no jump at all from state (i, n). These transitions are represented by arrows in Fig. 1.6, and they correspond to the computational molecule of the finite-difference scheme. Thus, after n throws of a "three-sided coin," the probable positions of a particle that was initially at the origin, say, will be given by a trinomial distribution characterized by the three *a priori* probabilities λ, λ, and 1-2λ†. As the number of trials increases, one shows that this distribution converges, like the binomial

†In fact, the special case of marginal stability, $\lambda = \frac{1}{2}$, reduces Eq. (4.4) to a recursion that is very close to the one that defines Pascal's triangle. The *a priori* probability of "no jump" is then zero.

distribution, to the gaussian (4.3).[N1] In fact, the stability criterion (1.26) simply means that the *a priori* probabilities cannot be negative.

The gaussian (4.3) is a major building block for the construction of time-dependent solutions. It sometimes goes under the name of *fundamental solution* or *Green's function* (in an infinite domain). From its heuristic derivation, above, we have the following properties, which the reader can easily verify:

(a) It is a solution of the diffusion equation (1.14).
(b) It vanishes at infinity ($x \to \pm\infty$) for all positive values of time t.
(c) Its integral over all space is precisely 1.
(d) It tends to zero as t tends to zero,[N2] except at the origin $x = 0$, where it is singular. This, together with the last property, is characteristic of the delta function. Thus we have the representation $\lim_{t\to 0+} G(x, t) = \delta(x)$, which is nothing but the initial condition of the problem solved numerically in Sect. 1.5.†

These properties are also valid for higher dimensions, as shown in the exercise that follows.

EXERCISE 4.1. *Fundamental solution in d dimensions*: Show that

$$(4\pi Dt)^{-d/2}e^{-r^2/4Dt}$$

is a solution of the d-dimensional diffusion equation (1.33) if r is the length of the position vector. Verify also Properties (b–d). [Hint: It is easiest to verify Property (a) in rotationally symmetric coordinates (see Eq. 1.36), but the other proofs are more transparent in cartesian coordinates.] □

As a first example of the gaussian's use, we now give the continuum analog of the result (1.29b): An arbitrary initial distribution $\varphi(x)$ can be viewed as the superposition

$$\varphi(x) = \int_{-\infty}^{\infty} \delta(x - x')\varphi(x')\,dx' \tag{4.5a}$$

†We will have more to say on delta functions in Sect. 7.2. Its "selection" property (7.17d) will be used shortly.

of point sources. Thus, it evolves in time according to

$$C(x, t) = \int_{-\infty}^{\infty} G(x - x', t)\varphi(x')\, dx', \tag{4.5b}$$

where G is the gaussian (4.3). The reasoning is exactly the same as in the discrete case. It is also easy to verify directly that Eq. (4.5b) satisfies the diffusion equation and the boundary conditions at infinity. The verification of the initial condition is a consequence of the delta-function behavior of G.[†]

The elegance of an equation such as Eq. (4.5b) can be appreciated if we recognize that it produces, explicitly, the concentration distribution at all locations and all times through a simple integration over an arbitrarily given initial configuration. The boundary conditions are built in. The initial conditions might even be given numerically, as with experimentally determined data. For example, if diffusion occurs only in a half space $\{x > 0\}$ and if the flux vanishes at the boundary $x = 0$, then we find the surface concentration by setting $x = 0$ in Eq. (4.5b), *even before* performing the integral. For each set of boundary conditions there exists a Green's function representation similar to Eq. (4.5b). In Chaps. 7 and 8, we shall see that the Laplace convolution theorem allows us to find these representations, almost automatically.

EXERCISE 4.2. *A gaussian remains a gaussian:* An arbitrary initial gaussian $(Q/\sigma\sqrt{\pi})\exp[-(x - m)^2/\sigma^2]$ is entirely characterized by three constants: Q, m, σ. Show that it preserves its shape under diffusion in an infinite medium. Note how the mean, maximum, and variance (or dispersion) change with time. This problem is of some concern to those who practice ion implantation.[7] □

Consider now an initial distribution of the form

$$\varphi(x) = \begin{cases} -1 & \text{if } x < 0, \\ +1 & \text{if } x > 0. \end{cases} \tag{4.6a}$$

This nonphysical distribution (part of φ is negative) leads us quickly down the path of error functions. Inserting the function (4.6a) into

[†]A more formal proof, but by no means a rigorous one, follows from the change of variables $x' = x + 2\sqrt{(Dt)}\xi$ in Eq. (4.5b).

(4.5b) and changing variables, we get the expression

$$C(x, t) = \frac{2}{\sqrt{\pi}} \int_0^{x/2\sqrt{Dt}} e^{-\xi^2}\, d\xi. \tag{4.6b}$$

The integral on the right-hand side cannot be reduced to elementary functions. It is so common, however, that it is given a name meant to remind us of its origin in probability theory.

DEFINITION. *The error function is defined by the integral*†

$$\operatorname{erf}(z) = \frac{2}{\sqrt{\pi}} \int_0^{z} e^{-\xi^2}\, d\xi. \tag{4.7}$$

With this definition, the solution to our problem (4.6) is simply

$$C(x, t) = \operatorname{erf}(x/2\sqrt{Dt}). \tag{4.6c}$$

The error function has the following properties, which the reader should verify directly from the definition (4.7) to gain familiarity with this important function:

(a) It is an odd function, i.e., $\operatorname{erf}(-z) = -\operatorname{erf}(z)$, which implies that $\operatorname{erf}(0) = 0$.
(b) Its limit $\lim_{z\to\infty} \operatorname{erf}(z) = 1$, and with Property (a) its limit as $z \to -\infty$ is -1.
(c) The differentiation formula $(d/dz)\operatorname{erf}(z) = (2/\sqrt{\pi})\exp(-z^2)$.
(d) A Taylor expansion of the exponential and termwise integration yield the series representation

$$\operatorname{erf}(z) = \frac{2}{\sqrt{\pi}} \sum_{n=0}^{\infty} \frac{(-1)^n z^{2n+1}}{n!(2n+1)},$$

which is discussed and put to use in Appendix D.
(e) The function (4.6c) satisfies the diffusion equation under the boundary conditions

$$C(0, t) = 0 \quad \text{and} \quad C(\infty, t) = 1 \qquad \text{for } t > 0,$$

†To avoid confusion with x and t or any combination thereof, the variable z in this definition (and others to follow) represents any real variable. In fact, most of the following expressions in z can be extended to the complex domain.

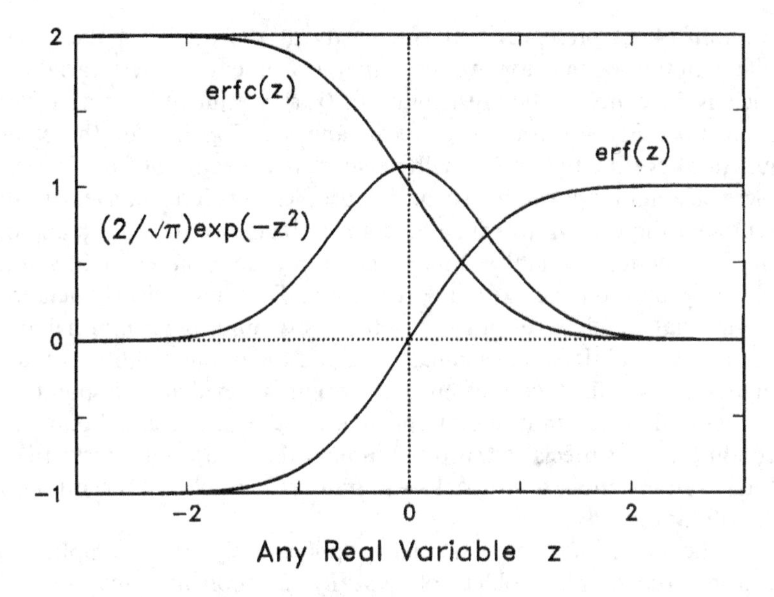

Fig. 4.2. Graph of the gaussian and error functions.

and the initial condition

$$C(x, 0) = 1 \qquad \text{for } x > 0.$$

What does all this mean? First, the parity of erf(z) follows from the gaussian's evenness. Both are represented in Fig. 4.2, and it is evident from the definition that erf(0) = 0. This is also clear from our method of construction, in which an odd function (4.6a) was allowed to relax by diffusion. Next, the crucial Property (e) follows, in fact, from the invariance properties C.1 in Sect. 1.3, because the error function (4.6c) is essentially an integral over space of the gaussian (4.3), which is itself a solution of the diffusion equation.† In fact, by completely different methods, we used symmetry arguments in Appendix B to construct this error function solution. Finally, the integration (4.5b) was performed through a point of discontinuity (the origin $x = 0$) of the initial distribution (4.6a). The "time-developed"

†The conditions of Exercise 1.7, namely $\partial G/\partial x|_{x=0}$, are met because G is an even function of x.

solution (4.6c) preserves that singularity at the origin of the (x, t)-plane. In that connection, we recall that any function of two variables, such as the concentration distribution $C(x, t)$, is equivalent to a surface (remember Fig. 1.3 for the gaussian and see Fig. B.3 for the error function). Rather than suffer with perspective drawings or from incomplete information given by isochronal profiles, it is often convenient to represent the behavior of such functions directly on the (x, t)-plane of the independent variables. Figure 4.3 shows how the error function (4.6c) behaves on the axes and at infinity. The figure clearly demonstrates that simple error function solutions require *constant* initial and boundary conditions on each half-axis, and the initial conditions must match the far-field conditions. The origin is evidently a point of discontinuity that must be excluded from the initial and boundary conditions. It is precisely this discontinuity that propagates forward in time—an infection, so to speak—to produce the actual concentration distribution.

The error function has many applications. For example, we consider the simple problem of *impurity diffusion in a semi-infinite region.* For definiteness, we assume that a dopant is initially distributed in a silicon wafer at a constant concentration level C_{∞}. At time $t = 0$, this wafer is placed in a reactive atmosphere inside a furnace. We

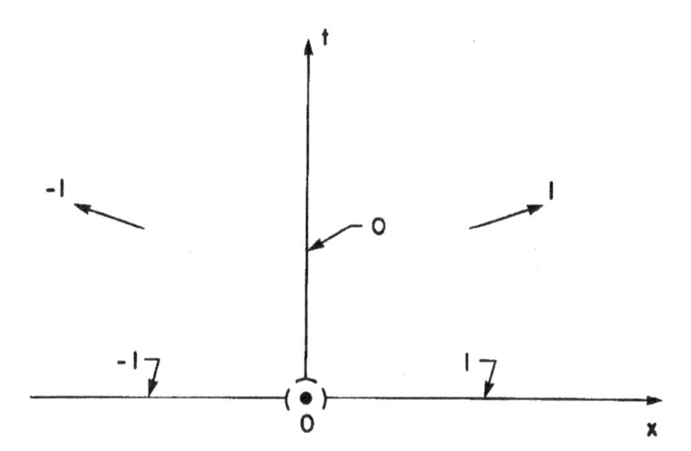

Fig. 4.3. Domain of the independent variables and particular values of the error-function solution of the diffusion equation. Note the singular behavior at the origin.

also assume, that the dopant reacts with this gas in such a way that local equilibrium is maintained at the wafer–gas surface, which is taken as the origin of coordinates. The dopant's concentration there will then be a constant $C_e < C_\infty$ that depends only on the reactant's temperature and pressure. We want to know at what rate the dopant escapes from the wafer. The solution is immediate if we scale the concentration properly. In fact, the function

$$u(x, t) = \frac{C(x, t) - C_e}{C_\infty - C_e} \tag{4.8a}$$

satisfies all the conditions of Property (e) above. Therefore,

$$C(x, t) = C_e + (C_\infty - C_e)\,\mathrm{erf}\big(x/2\sqrt{Dt}\big) \tag{4.8b}$$

is the required solution. It should be noted that we have used Property C.1 of Sect. 1.3 in the sense that an arbitrary constant is a solution of the diffusion equation.[N3]

Next, the rate at which dopant leaves the semiconductor is simply proportional to the surface flux J_0. Using Property (c), we easily get

$$J_0 = -DC_x|_{x=0} = -(C_\infty - C_e)\sqrt{D/\pi t}, \tag{4.8c}$$

which displays a characteristic $t^{-1/2}$ behavior. In addition, the minus sign reminds us that the flux is directed *into* the gas if $C_\infty > C_e$ and if the x-axis points into the solid. Finally, the solution (4.8b) is valid for any relative values of C_e and C_∞. Thus that solution can describe equally well the "indiffusion" into a semi-infinite body of an impurity from a constant external source. Problems of this type have no constant physical length scale, but we do note that the solution (4.8b) scales distances x with $\sqrt{(Dt)}$. This quantity, for reasons that will soon become clear, is called the *diffusion distance.*

The novice will perhaps be disturbed by the foregoing procedure. It may appear that we have pulled a solution out of a hat. The constructive method, first mentioned in the introduction to this chapter, consists essentially of taking known solutions of the diffusion equation, perhaps rescaled, and trying to match the given initial and boundary conditions. There are two possibilities: Either we are unsuccessful and we must look for other known solutions, or the conditions can all be met. Then, by virtue of the uniqueness theorem mentioned at the end of Sect. 1.3, we are sure to have found *the* solution to our problem, because there can be none other.

EXERCISE 4.3. *How much matter is gained or lost?:* The total mass lost by outdiffusion is clearly the integral of $C_\infty - C(x, t)$ over all $x > 0$. Perform that integral. One gets that same result (much more easily) by using Exercise 1.5. □

In spite of its simplicity, the error-function solution (4.8) has many physical applications. Perhaps the most famous was Kelvin's erroneous estimate (20–40 million years) of the age of the Earth; see Refs. 3 (pp. 85–87) and 8. It is justly famous, not because his was an underestimate by about two orders of magnitude (as the geologists well knew), not because he could be faulted for neglecting radioactivity (which had not yet been discovered), but because it clearly demonstrated that the sciences must be consistent as well as reproducible. The next two sections will relate to several other (more mundane) applications.

4.2. The Gaussian and Another of Its Relatives

To motivate the equally important *complementary error function* let us add and subtract the far-field concentration C_∞ from the solution (4.8b) of our previous impurity diffusion problem. Thus,

$$C(x, t) = C_\infty - (C_\infty - C_e)[1 - \mathrm{erf}(x/2\sqrt{Dt})]. \tag{4.9a}$$

The function in brackets occurs so frequently[N4] that one feels compelled to introduce the definition that follows.

DEFINITION. *The complementary error function is defined by*

$$\mathrm{erfc}(z) = 1 - \mathrm{erf}(z). \tag{4.10a}$$

With this definition, the standard integral (4.2), and the definition (4.7), we get the integral representation

$$\operatorname{erfc}(z) = \frac{2}{\sqrt{\pi}} \int_z^{\infty} e^{-\xi^2}\, d\xi. \tag{4.10b}$$

For example, the solution to the impurity diffusion problem can therefore be expressed as

$$C(x, t) = C_\infty - (C_\infty - C_e)\operatorname{erfc}(x/2\sqrt{Dt}), \tag{4.9b}$$

and the two representations of the same problem, Eqs. (4.8b) and (4.9b), differ merely in the choice of "baselines" C_e and C_∞, respectively.

The complementary error function also enjoys properties similar to the ordinary error function. They are easy to verify:

(a) It is an odd function around the point (0, 1), which implies that $\operatorname{erfc}(0) = 1$.
(b) Its limit for large positive z is $\lim_{z\to\infty} \operatorname{erfc}(z) = 0$, and, with Property (a), its limit as $z \to -\infty$ is 2. The symmetry property thus reads $\operatorname{erfc}(-z) = 2 - \operatorname{erfc}(z)$.
(c) The differentiation formula $(d/dz)\operatorname{erfc}(z) = -(2/\sqrt{\pi})\exp(-z^2)$.
(d) Its Taylor expansion is simply related to Property (d) of $\operatorname{erf}(z)$ through the definition (4.10a).
(e) The function $C(x, t) = \operatorname{erfc}[x/2\sqrt{(Dt)}]$ satisfies the diffusion equation under the boundary conditions

$$C(0, t) = 1 \quad \text{and} \quad C(\infty, t) = 0 \qquad \text{for } t > 0,$$

and the initial condition

$$C(x, 0) = 0 \qquad \text{for } x > 0.$$

These conditions are evidently complementary (and thus the name) to those satisfied by the ordinary error function. This function is also displayed in Fig. 4.2. It should be noted that it looks like a metallurgical junction, and we will soon see that this appearance is not misleading.

Another useful property of the complementary error function is the way in which it approaches zero for a large, *positive* argument. The decay to zero is quite fast; it is interesting analytically as well as being useful for numerical calculations. This leads us to the idea of an *asymptotic expansion*, a notion that generalizes the asymptotes of high-school geometry.

By repeated integration by parts we show formally in Appendix D that

$$\sqrt{\pi}\, z e^{z^2} \operatorname{erfc}(z) = 1 - \frac{1}{2z^2} + \frac{1 \times 3}{(2z^2)^2} - \cdots \tag{4.11a}$$

We seem to have generated a series in inverse powers of the argument z, and each term must thus be smaller than its predecessors as $z \to \infty$. But this series is not convergent, as is easily seen with the ratio test. In fact, for fixed z, the successive terms of this series will eventually increase. A series such as given by Eq. (4.11a) is useful, nevertheless, because the partial sums approach the function in question, with ever-increasing accuracy, as z increases. Using a tilde to symbolize an asymptotic relation, Eq. (4.11a) reads

$$\operatorname{erfc}(z) \sim \frac{e^{-z^2}}{z\sqrt{\pi}}[1 + O(z^{-2})] \qquad \text{for } z > 0, \tag{4.11b}$$

and in practice this is all we shall ever need.

Much more on asymptotic expansions can be found in Appendix D and in the references cited therein. Lin and Segel[9] give an excellent introductory account of the theory and its applications. For our purposes, however, Eqs. (4.11) show that the complementary error function behaves like a gaussian divided by the argument z to the first power, at least for large values of that argument. One says that it has a gaussian "tail." How large is "large" depends on what we mean by "small" values of the gaussian. For example, for $z = 2$ it follows from Eq. (4.11b) that the complementary error function is approximately 0.5% of its value (unity) at the origin. Such estimates allow operational definitions of "diffusion distances," "penetration lengths," or "boundary-layer thicknesses"—all equivalent terminology. For example, because our previous solutions (4.8b) and (4.9c) contain distance and time only in the particular combination $z = x/2\sqrt{(Dt)}$, it follows that the fields have decayed to less than a percent of their boundary values at distances of the order of $4\sqrt{(Dt)}$.

EXERCISE 4.4. *The "exterior" problem for a sphere:* Consider diffusion in the region exterior to a sphere of radius R (a precipitate or bubble, say). Assume spherical symmetry and use your previous knowledge to justify the trial solution

$$C(r, t) = A + \frac{B}{r}\operatorname{erfc}\left[(r - R)/2\sqrt{Dt}\right].$$

Determine the constants if the initial and boundary conditions are $C(r, 0) = C(\infty, t) = C_\infty$ and $C(R, t) = C_e$. Draw a (r, t)-diagram. Calculate the flux J_R at the sphere's surface. Is there a steady state? □

Very well, but how do we *compute* error functions? Of course, one can call on an appropriate subroutine or evaluate the integrals (4.7) and (4.10b) numerically. Table 4.1 displays, however, an excellent rational function approximation that is easily programmed. In fact, the

Table 4.1. Rational Function Approximation of the Complementary Error Function

$$\operatorname{erfc}(x) \approx e^{-x^2}\frac{P(x)}{Q(x)},$$

where

$$P(x) = \sum_{n=1}^{7} P_n x^n \quad \text{and} \quad Q(x) = \sum_{n=1}^{8} Q_n x^n.$$

The coefficients of these polynomials are:

$P_1 = 440.4137358247522000$	$Q_1 = 440.41373582475220$
$P_2 = 625.6865357696830000$	$Q_2 = 1122.64022017704700$
$P_3 = 448.3171659914834000$	$Q_3 = 1274.66726675766200$
$P_4 = 191.8401405879669000$	$Q_4 = 838.81036547840640$
$P_5 = 50.9916976282532000$	$Q_5 = 347.12292881178430$
$P_6 = 7.9239298287415450$	$Q_6 = 90.87271120353703$
$P_7 = 0.5641999455982575$	$Q_7 = 14.04533730546114$
	$Q_8 = 1.00000000000000$

curves in Fig. 4.2 were evaluated with this function. It gives 15 correct decimal places in the range $0 \leqslant x \leqslant 10$. For larger positive argument, we need only use the asymptotic expansion (4.11b), and for negative argument we use the symmetry property (b).

EXERCISE 4.5. *A few integrals:* Gaussians and error functions are widely used to express certain integrals. For real λ, show that:

1. $$\frac{2}{\sqrt{\pi}}\int_0^\infty e^{-\xi^2}e^{-2\lambda\xi}\,d\xi = e^{\lambda^2}\operatorname{erfc}(\lambda).$$

   [Hint: Complete the square.]

2. $$\frac{2}{\sqrt{\pi}}\int_0^\infty e^{-\xi^2}\cos(2\lambda\xi)\,d\xi = e^{-\lambda^2}.$$

   [Hint: Extend the integral over the range $(-\infty, \infty)$ and reduce to the previous problem.]

3. $$\frac{2}{\sqrt{\pi}}\int_0^\infty e^{-\xi^2}e^{-(\lambda/\xi)^2}\,d\xi = e^{-2|\lambda|}.$$

   [Hint: Change variables $\xi \to \eta = \lambda/\xi$ and form a differential equation.]

4. $$\int_0^x \operatorname{erf}(\xi)\,d\xi = x\operatorname{erf}(x) - \frac{1}{\sqrt{\pi}}(1 - e^{-x^2}).$$

   [Hint: Integrate by parts.] ☐

4.3. Two Applications of Error Functions to One-Dimensional Phases

We have just learned that error functions are particular solutions of the diffusion equation—particular in the mathematical sense that they satisfy given constant initial and boundary conditions *on the field*, as opposed to conditions on its derivatives. The impurity diffusion

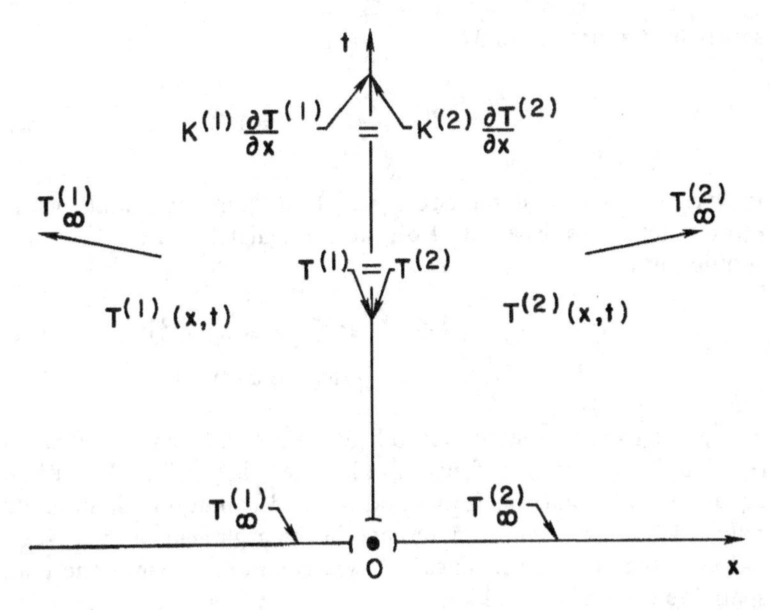

Fig. 4.4. Domains of the independent variables and the initial and boundary conditions for the two-phase thermal redistribution problem.

problem [Eqs. (4.8) and (4.9)] and Stokes's problem[N4] were both of this type. In this section, we apply these functions to the solution of problems that are just slightly more complicated than those just mentioned.

Our first task will be to compute the *time-dependent temperature field in a 1-d, infinite, two-phase system*. Specifically, we consider two large and different chunks of material, labeled 1 and 2, each initially at constant temperature $T_\infty^{(1)}$ and $T_\infty^{(2)}$. We now bring these materials together along a common plane phase boundary and let heat flow from the hotter body to the colder.† We ask how the temperature field varies with position and time, and, in particular, what is the temperature T_0 at the interface $x = 0$ joining these two phases.

The initial and boundary conditions for this problem are shown in Fig. 4.4, which is similar to Fig. 4.3, and the temperature field must

†Here we assume that these initial temperatures do *not* bracket the melting point or other phase transition points at which there is a release of latent heat.

satisfy heat equations (1.37)

$$\frac{\partial T^{(\alpha)}}{\partial t} = \kappa^{(\alpha)} \frac{\partial^2 T^{(\alpha)}}{\partial x^2} \qquad (\alpha = 1, 2) \tag{4.12}$$

in phase 1 $\{x < 0\}$ and in phase 2 $\{x > 0\}$. Because the complementary error function vanishes at $+\infty$, we immediately write the general solution in the form

$$T^{(1)}(x, t) = T_\infty^{(1)} + A^{(1)} \operatorname{erfc}[-x/2(\kappa^{(1)} t)^{1/2}], \tag{4.13a}$$

$$T^{(2)}(x, t) = T_\infty^{(2)} + A^{(2)} \operatorname{erfc}[x/2(\kappa^{(2)} t)^{1/2}], \tag{4.13b}$$

where the A's are constants yet to be determined. Because of Property (e) of the last section, the functions (4.13) certainly satisfy the diffusion equations (4.12) and the initial and far-field conditions. It should be noted that we have put a minus sign in the argument of the first erfc (4.13a), which is allowed since the heat equation is symmetric under spatial reflection.

The constants $A^{(1)}$ and $A^{(2)}$ are easily determined from the requirements for temperature and flux continuity at the phase boundary,[N5]

$$T^{(1)}|_{x=0^-} = T^{(2)}|_{x=0^+}, \qquad K^{(1)} \frac{\partial T^{(1)}}{\partial x}\bigg|_{x=0^-} = K^{(2)} \frac{\partial T^{(2)}}{\partial x}\bigg|_{x=0^+}, \tag{4.14a}$$

which hold for all t. With Properties (a) and (c) of the last section, we get a linear system of two equations, with the solution

$$A^{(1)} = \frac{T_\infty^{(2)} - T_\infty^{(1)}}{1 + (K^{(1)}/K^{(2)})(\kappa^{(2)}/\kappa^{(1)})^{1/2}}, \tag{4.14b}$$

$$A^{(2)} = \frac{T_\infty^{(1)} - T_\infty^{(2)}}{1 + (K^{(2)}/K^{(1)})(\kappa^{(1)}/\kappa^{(2)})^{1/2}}, \tag{4.14c}$$

Hence, we obtain the full temperature distributions (4.13) and, in particular, the interface temperature

$$T_0 = \frac{K^{(1)}\sqrt{\kappa^{(2)}}T_\infty^{(1)} + K^{(2)}\sqrt{\kappa^{(1)}}T_\infty^{(2)}}{K^{(1)}\sqrt{\kappa^{(2)}} + K^{(2)}\sqrt{\kappa^{(1)}}}. \tag{4.15}$$

The interface temperature is thus a weighted mean of the initial temperatures in the two phases before contact. For semi-infinite phases, it is a *constant*, independent of time, and it is *not* a free parameter under the experimentalist's control. This is in line with the phase rule for isobaric, c-component, two-phase systems†

$$f = c - 1 \qquad (\varphi = 2, p = \text{const.}), \tag{4.16}$$

where f is the number of degrees of freedom for such a system in equilibrium. In our case, the interface between the two different phases is assumed to be at equilibrium,[N5] and it follows that there are no controllable parameters for one-component systems. For multicomponent systems, on the other hand, local equilibrium at the interface implies that we should really consider the redistribution of all species, as well as of energy.[N6] It should be emphasized, however, that the interface temperature T_0 is a constant only for times such that the two adjacent phases can be considered semi-infinite. For finite phases and for long enough times, T_0's time dependence is controlled by the boundary conditions at the external phase boundaries.

Figure 4.5 shows the temperature distribution at various "times" $\sqrt{(\kappa t)}$, for the special case of equal thermal diffusivities and a ratio $K^{(2)}/K^{(1)} = 4$ of the thermal conductivities. The initial temperatures are taken to be 90° and 10°C for phases (1) and (2), respectively, but any other values could be used by properly scaling these graphical results. The figure shows continuous and monotonically decreasing temperature profiles. At the interface there is an abrupt change in slope that corresponds to the conductivity change (see Exercise 1.6). The units for distance are determined by those of κ because, again, there are no natural, constant distance scales. As with all our previous examples, $\sqrt{(\kappa t)}$ is thus the penetration distance, which can be verified from Fig. 4.5.

EXERCISE 4.6. *Mass redistribution in a 1-d two-phase system:* Consider an isothermal, isobaric system of two large phases such as Si/SiO_2. An impurity diffuses in both phases, and its constant, initial levels $C_\infty^{(1)}$ and $C_\infty^{(2)}$ are known. Assume that the interface "partitions" the impurity, i.e., the ratio $C^{(2)}/C^{(1)}$ at $x = 0$ is a known constant k (which can depend on temperature). Find the concentration distribution in both

†See Eq. (C.3) in Appendix C.

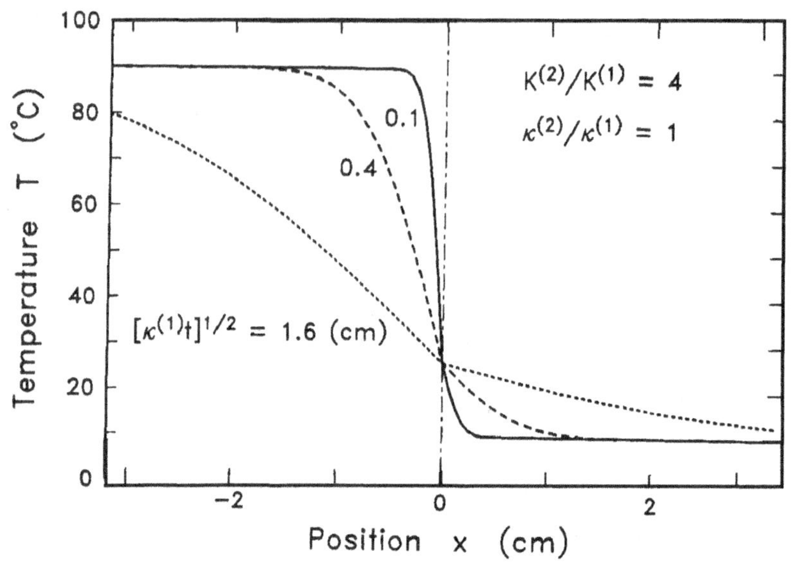

Fig. 4.5. Temperature distribution in the two phases for various times, measured in units of diffusion length.

phases and interpret the partitioning. Is the interface at "local thermodynamic equilibrium"? □

A second interesting problem whose exact solution is expressed in terms of error functions is a simple 1-d *Stefan problem*, that is to say, the question of interface motion between two phases that support diffusion processes. For definiteness, we consider the two-phase, binary system of Sects. 2.4 and 2.5, and assume that diffusion of component B can occur only in the liquid phase 2. According to the phase rule (4.16), there is but one degree of freedom at the interface if it is at local equilibrium. This is precisely what the phase diagrams in Figs. 2.5 and 2.6 tell us: The equilibrium compositions of a (large) solid and liquid are uniquely determined at a given temperature. We now take a large solid of composition C_s, and put it in contact with a prepared semi-infinite liquid of *arbitrary* constant composition C_∞. What are we to expect? First, at the instant of contact, the surface concentration in the liquid must assume its equilibrium value C_e compatible with the phase diagram. Then, because generally $C_e \neq C_\infty$, we expect mass

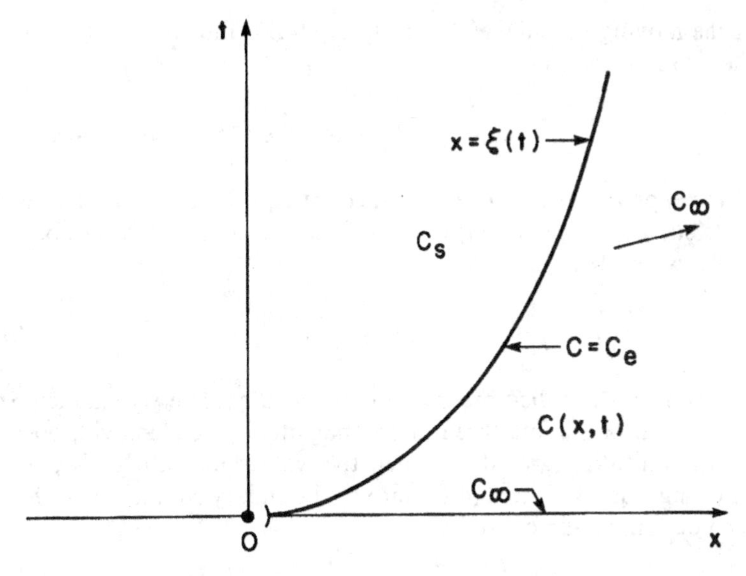

Fig. 4.6. Domains of the independent variables and the initial and boundary conditions for the one-phase Stefan problem.

redistribution in the liquid, according to the diffusion equation (1.14), and consequent interface motion $\xi(t)$.

The general solution to this problem is again

$$C(x, t) = C_\infty + A\,\mathrm{erfc}\big(x/2\sqrt{Dt}\big), \qquad x > \xi(t), \tag{4.17}$$

because of the initial and far-field conditions shown in Fig. 4.6. In contrast to the first problem in this section, the interface can move, as was the case for the spherical precipitate of Sect. 2.5.[N7] There are now two unknowns: the constant A in Eq. (4.17) and the interface's schedule $\xi(t)$. But we have, not so coincidentally, two equations: the specification of local equilibrium,

$$C(\xi, t) = C_e, \tag{4.18a}$$

and of local mass conservation,

$$\dot{\xi}(C_s - C_e) = DC_x|_{x=\xi(t)}, \tag{4.18b}$$

at the moving boundary.[†] Inserting Eq. (4.17) into Eqs. (4.18), we get the two requirements:

$$C_e = C_\infty + A\,\mathrm{erfc}\left(\xi/2\sqrt{Dt}\right) \quad \text{and} \quad \dot{\xi} \propto t^{-1/2}\exp(-\xi^2/4Dt).$$

These expressions cannot be satisfied unless $\xi(t) \propto t^{1/2}$. Therefore, we have succeeded in finding the interface's motion—it is parabolic—which we write as

$$\xi(t) = 2\lambda\sqrt{Dt}. \tag{4.19}$$

The constant λ, called the *growth constant*, completely characterizes the motion of the interface in both magnitude and sign, with positive values indicating growth and negative values indicating dissolution. Inserting Eqs. (4.17) and (4.19) into the boundary conditions (4.18), we then determine the constant

$$A = -\frac{C_\infty - C_e}{\mathrm{erfc}(\lambda)}, \tag{4.20a}$$

and we get the transcendental equation

$$\sqrt{\pi}\,\lambda e^{\lambda^2}\,\mathrm{erfc}(\lambda) = \frac{C_\infty - C_e}{C_s - C_e} \equiv \sigma \tag{4.20b}$$

for λ. The right-hand side of Eq. (4.20b) is precisely the (absolute) supersaturation σ that appeared in Eqs. (2.25) for the spherical-precipitate problem. In both cases, σ represents the maximum driving force for the diffusion process. A given supersaturation determines λ according to Eq. (4.20b), hence the constant A through Eq. (4.20a), and thus the full concentration distribution (4.17).

Figure 4.7 shows a plot of the function on the left-hand side of Eq. (4.20b). Because of the asymptotic relations (4.11), that function has an upper bound for large positive values of λ, which corresponds to $\sigma \to 1$. On the other hand, a Taylor expansion for small λ's shows that $\lambda \approx \sigma/\sqrt{\pi}$ when the supersaturation is small (see the next exercise). We note that there are no real solutions of Eq. (4.20b) for

[†]This last is the Stefan condition (1.20) if we neglect diffusion in the solid.

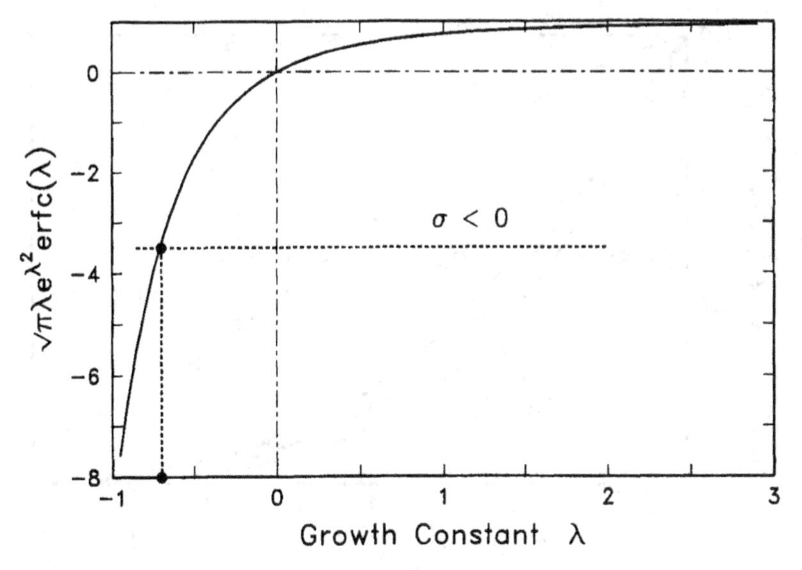

Fig. 4.7. The graph, as a function of λ, of the function on the left-hand side of Eq. (4.20b). This equation has a unique solution for each value of the absolute supersaturation σ (drawn negative on the figure).

supersaturations greater than unity, but that case is excluded by the inequalities (2.19). The relation (4.20b) can be inverted graphically or numerically for $\lambda(\sigma)$, and Fig. 4.8 shows the concentration distribution that is obtained for various values of the supersaturation. The distribution (4.17) is defined on the right-hand domain of Fig. 4.6, i.e., for $x \geqslant \xi(t)$. Anticipating somewhat the notion of similarity variable $\eta \equiv x/2\sqrt{(Dt)}$, explored in the next chapter, that condition reads simply $\eta \geqslant \lambda$. It should be noted that positive σ's imply growth and concentration profiles without inflection, whereas negative σ's mean dissolution and the appearance of an inflection point in the concentration. Can you interpret this?

EXERCISE 4.7. *Behavior of the function $\lambda(\sigma)$:* Using analytic properties of error functions, expand the left-hand side of Eq. (4.20b) for small and large values of λ, both positive and negative. Thus, invert that equation in these limits. Verify the monotonic behavior of the function in Fig. 4.7. □

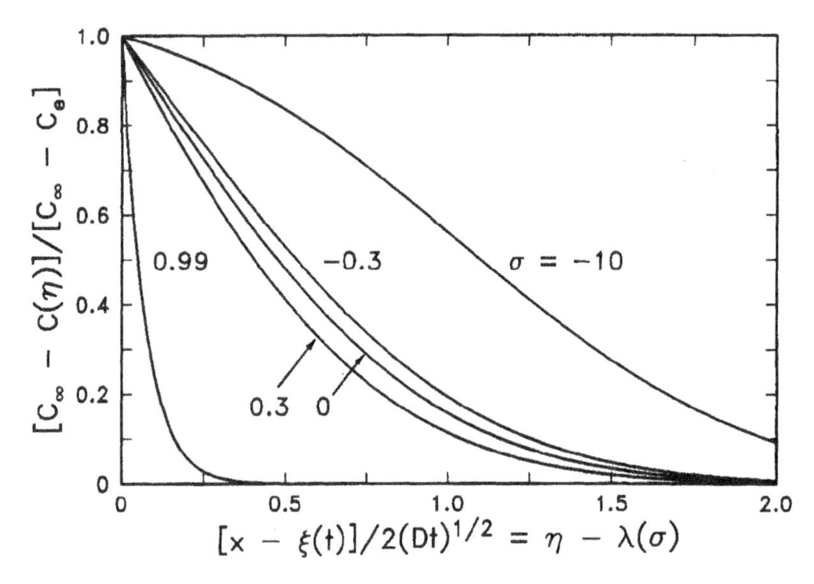

Fig. 4.8. Concentration distribution ahead of a moving interface for various supersaturations.

There is much more to say concerning Stefan problems. They constitute an active area of research, probably because in their full generality they are very difficult. These problems are always nonlinear, although this may not yet be evident (see, however, Sect. 5.2). Some of the earlier work can be found in Refs. 3 and 10, and Rubinstein's treatise[11] is a classic investigation. Suffice it to say that there are very few analytical results (we will examine another in Sect. 5.3), and today's investigations, especially for multidimensional problems, are largely numerical.[12–15]

4.4. The Gaussian Has Yet More Relatives

As mentioned at the beginning of the last section, all our examples so far have imposed given conditions on the field at physical boundaries. Such a specification is known as a *Dirichlet condition*. There are evidently many other realistic types of boundary conditions. For example, we consider the following *heating problem for a half-space*:

A large solid, initially at the constant temperature T_∞, is subjected to intense irradiation along its plane boundary, for example, by a laser or an electron beam.[N8] By continuity (1.19), the imposed energy flux F must equal the thermal flux in the solid at that boundary:

$$F = -KT_x|_{x=0}. \tag{4.21a}$$

Since, as is easily verified, the flux $J = -KT_x$ satisfies the *same* heat equation (1.37) at all interior points of the solid as does the temperature,† and since it satisfies the initial condition $J(x, 0) = 0$, it follows that

$$J(x, t) = F \operatorname{erfc}\left(x/2\sqrt{\kappa t}\right) \tag{4.21b}$$

gives the flux distribution. The temperature distribution is then obtained by integrating this relation over space. Thus,

$$\begin{aligned} T(x, t) &= (F/K)\int_x^\infty \operatorname{erfc}\left(x'/2\sqrt{\kappa t}\right) dx' + f(t) \\ &= (F/K)2\sqrt{\kappa t}\int_{x/2\sqrt{\kappa t}}^\infty \operatorname{erfc}(\eta)\, d\eta + T_\infty, \end{aligned} \tag{4.21c}$$

where the arbitrary function of time $f(t)$ must reduce to the constant T_∞ because of the far-field condition on T. Consequently, the temperature distribution produced by a constant flux is given by the integral of an error function.‡ Is this motivation enough to introduce the *iterated error functions*? In view of their wide application (laser processing being but a simple example), we now turn to this family of functions.

Before doing so, however, we must introduce another useful special function, the *gamma function* and some of its properties.[1,2,4] We recall the standard integral (4.2) and write it in the new variable $\alpha = \xi^2$:

$$\sqrt{\pi} = \int_0^\infty \alpha^{-1/2} e^{-\alpha}\, d\alpha.$$

This integral is no more expressible in terms of elementary functions than is the integral (4.2) from whence it came. When in doubt, we

†This was already mentioned in Sect. 1.3.

‡The condition of Exercise (1.7) is verified at infinity.

give it a name, $\Gamma(\frac{1}{2})$, and "embed" this result in the definition that follows.

DEFINITION. *The gamma function is defined by the integral*

$$\Gamma(z) = \int_0^\infty \alpha^{z-1} e^{-\alpha}\, d\alpha. \tag{4.22}$$

This integral converges for positive values of z, and one shows that it diverges logarithmically when $z \to 0$. Integrating (4.22) by parts yields the recursion

$$\Gamma(z+1) = z\Gamma(z), \tag{4.23a}$$

and we have $\Gamma(1) = 1$ directly from the definition (4.22). Therefore, $\Gamma(2) = 1, \Gamma(3) = 2, \ldots$, and generally, when z is a positive integer n, the gamma function interpolates the factorial

$$\Gamma(n+1) = n! \tag{4.23b}$$

of an integer. For this reason, Γ is sometimes called the "factorial function."

The recursion (4.23a) can be used in yet other ways. For example, with $\Gamma(\frac{1}{2}) = \sqrt{\pi}$, we have $\Gamma(\frac{3}{2}) = \Gamma(1 + \frac{1}{2}) = \frac{1}{2}\Gamma(\frac{1}{2}) = \sqrt{(\pi)}/2$, and the recursion can be carried forward to any half-integer value $n + \frac{1}{2}$. Similarly, we use Eq. (4.23a) to continue the definition (4.22) to negative values of z.† It suffices to replace z by $z - 1$ in Eq. (4.23a) to get

$$\Gamma(z-1) = \frac{\Gamma(z)}{z-1}. \tag{4.23c}$$

Hence, for example, $\Gamma(-\frac{1}{2}) = -2\sqrt{\pi}$. These results are generalized in the exercise that follows.

EXERCISE 4.8. *Gamma function for half-integer arguments:* Show that

$$\Gamma(n + \tfrac{1}{2}) = \frac{1 \times 3 \times \cdots \times (2n-1)}{2^n}\sqrt{\pi},$$

†Remember that the integral (4.22) converges only for positive values.

$$\Gamma(-n - \tfrac{1}{2}) = \frac{(-2)^{n+1}}{1 \times 3 \times \cdots \times (2n + 1)} \sqrt{\pi},$$

for any nonnegative integer n. Show, also, that the gamma function diverges for any negative integer: it has "poles" on the negative real axis. □

The gamma function enjoys many other interesting properties. Here, we mention only three that will prove useful. The first is a *duplication equation*, similar to what holds for trigonometric functions:

$$\sqrt{\pi}\,\Gamma(2z) = 2^{2z-1}\Gamma(z)\Gamma(z + \tfrac{1}{2}). \tag{4.24a}$$

The second is its asymptotic behavior for large positive z,

$$\Gamma(z + 1) \sim \sqrt{2\pi z}\,(z/e)^z\,[1 + O(z^{-1})], \tag{4.24b}$$

also known as *Stirling's formula*. For proofs of these two properties (4.24) see Refs. 2 and 4. The third property is the gamma function's ability to express some common integrals that involve exponentials. For example, by simple changes of variables and the definition (4.22), we easily get

$$\int_0^\infty \xi^{z-1} e^{-a\xi}\, d\xi = a^{-z}\Gamma(z), \tag{4.25a}$$

$$\int_0^\infty \xi^{2z-1} e^{-a\xi^2}\, d\xi = \tfrac{1}{2} a^{-z}\Gamma(z), \tag{4.25b}$$

for any positive values of the parameters z and a. In other words, arbitrary moments of exponentials and gaussians are expressible in terms of gamma functions. For arguments other than integers or half-integers, the evaluation of the gamma function, like the error function, depends either on tables and rational-function approximations[1] or on the numerical evaluation of the integral (4.22).

We are now ready to attack the integrals of error functions.

DEFINITION. *The iterated error functions are defined by the recursion*

$$\mathrm{i}^n \operatorname{erfc}(z) = \int_z^\infty \mathrm{i}^{n-1} \operatorname{erfc}(\xi)\, d\xi \tag{4.26a}$$

for any nonnegative integer n, with the initial condition

$$\mathrm{i}^{-1} \operatorname{erfc}(z) = \frac{2}{\sqrt{\pi}} e^{-z^2} \qquad \text{(the gaussian).} \tag{4.26b}$$

With this definition, the complementary error function is simply the zeroth member

$$\mathrm{i}^0 \operatorname{erfc}(z) = \operatorname{erfc}(z) \tag{4.26c}$$

of the sequence (4.26a), and the solution to our previous heating problem (4.21c) is simply

$$T(x, t) = T_\infty + (F/K) 2\sqrt{\kappa t}\, \mathrm{i}^1 \operatorname{erfc}\left(x/2\sqrt{\kappa t}\right). \tag{4.21d}$$

Consequently, the surface temperature is

$$T(0, t) = T_\infty + (F/K) 2\sqrt{\kappa t}\, \mathrm{i}^1 \operatorname{erfc}(0). \tag{4.21e}$$

and it increases as the square root of time.

But aren't these merely empty definitions? And how do we actually compute these functions? The following properties[1,3] show that iterated error functions are here to stay and that they are easy to evaluate.

(*a*) *Differentiation formula:* From the definition (4.26a) we immediately get

$$\frac{d}{dz} \mathrm{i}^n \operatorname{erfc}(z) = -\mathrm{i}^{n-1} \operatorname{erfc}(z) \tag{4.27}$$

(*b*) *Expression as a single integral:* According to the definition (4.26a), the nth iterated error function is the integral of the $(n-1)$th, and so forth, until we reach the gaussian (4.26b). We then have a sequence of nested integrals for which there is a theorem of calculus (see Sect. 7.5). Indeed, by repeated differentiation or by induction we verify that

$$\mathrm{i}^n \operatorname{erfc}(z) = \frac{2}{\sqrt{\pi}} \int_z^\infty \frac{(\xi - z)^n}{n!} e^{-\xi^2} \, d\xi. \tag{4.28}$$

(c) *Value at the origin:* From this last result we compute

$$\begin{aligned} \mathrm{i}^n \operatorname{erfc}(0) &= \frac{2}{\sqrt{\pi}} \int_0^\infty \frac{\xi^n}{n!} e^{-\xi^2} \, d\xi = \frac{1}{\sqrt{\pi}} \frac{1}{\Gamma(n+1)} \Gamma\left(\frac{n+1}{2}\right) \\ &= \frac{1}{2^n \Gamma(1 + n/2)}, \end{aligned} \tag{4.29}$$

where the second equality results from Eqs. (4.25b) and (4.23b), and where the duplication equation (4.24a) is used for the third equality; and, from Exercise 4.8, we know how to evaluate these gamma functions for integer n.

(d) *Recurrence relation:* The iterated error functions obey an important recursion,

$$2n\mathrm{i}^n \operatorname{erfc}(z) = -2z\mathrm{i}^{n-1} \operatorname{erfc}(z) + \mathrm{i}^{n-2} \operatorname{erfc}(z), \tag{4.30a}$$

which relates any member of that family of functions to its two immediate predecessors. For example, because of Eqs. (4.26b) and (4.26c), we compute in succession

$$2\mathrm{i}^1 \operatorname{erfc}(z) = -2z \operatorname{erfc}(z) + \frac{2}{\sqrt{\pi}} e^{-z^2}, \tag{4.30b}$$

$$4\mathrm{i}^2 \operatorname{erfc}(z) = (1 + 2z^2) \operatorname{erfc}(z) - \frac{2}{\sqrt{\pi}} z e^{-z^2}. \tag{4.30c}$$

This allows us to evaluate these two functions because we know how to compute both the gaussian and the complementary error function. The proof of recursion (4.30a) is not hard: Just change variables $\xi \to \alpha = \xi - z$ in Eq. (4.28) and differentiate this result with respect to z, bearing the property (4.27) in mind. Figure 4.9

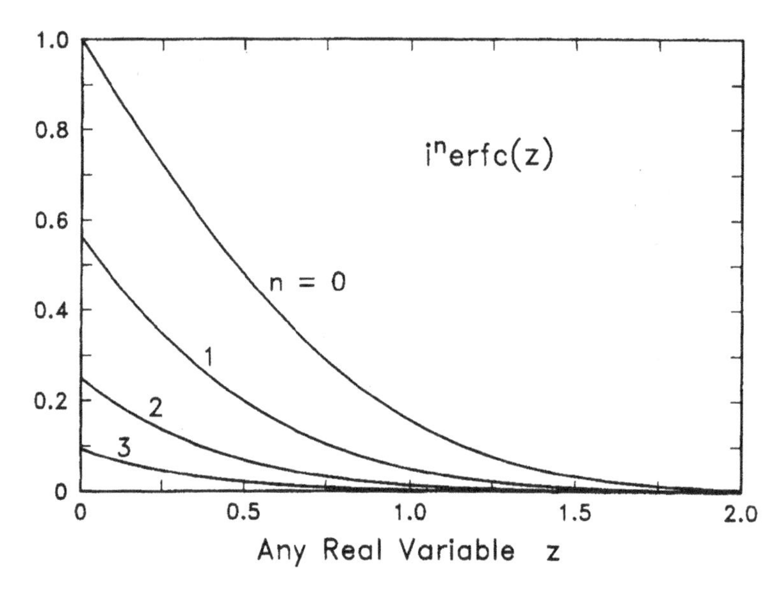

Fig. 4.9. The graph for positive argument of the first four iterated error functions.

is a graph, for a positive argument, of the first four iterated error functions. They all decrease monotonically with increasing argument, and they approach zero increasingly fast with the order n [see Eqs. (4.33) below].

(*e*) *Differential equation:* Expressing the functions of lower order in Eq. (4.30a) as derivatives of the nth iterate, we find that $\mathrm{i}^n\,\mathrm{erfc}(z)$ is a solution of the linear second-order differential equation (but with a variable coefficient of y')

$$y'' + 2zy' - 2ny = 0 \qquad \text{for } n \geqslant -1, \tag{4.31}$$

where primes mean differentiation with respect to z.

EXERCISE 4.9. *Another solution of this differential equation:* Show that $\mathrm{i}^n\,\mathrm{erfc}(-z)$ is also a solution of the differential equation (4.31). Computing the wronskian of the two functions $\mathrm{i}^n\,\mathrm{erfc}(\pm z)$, show that they are linearly independent solutions of that equation if n is non-negative. □

(*f*) *Taylor series:* Applying Taylor's theorem, together with the results (4.27) and (4.29), we get the expansion around the origin

$$\begin{aligned} \mathrm{i}^n \operatorname{erfc}(z) &= \sum_{k=0}^{\infty} \frac{z^k}{k!} \frac{d^k}{dz^k} \mathrm{i}^n \operatorname{erfc}(z)|_{z=0} \\ &= \sum_{k=0}^{\infty} \frac{(-1)^k z^k}{k! 2^{n-k} \Gamma[1 + (n-k)/2]}. \end{aligned} \tag{4.32}$$

(*g*) *Asymptotic expansions:* Under fairly general conditions, one shows that asymptotic expansions can be integrated term by term. For example, integrating Eq. (4.11b) once by parts, we get

$$\mathrm{i}^1 \operatorname{erfc}(z) \sim \frac{e^{-z^2}}{2z^2\sqrt{\pi}}[1 + O(z^{-2})], \qquad z > 0, \tag{4.33a}$$

and, more generally, one can show that

$$\mathrm{i}^n \operatorname{erfc}(z) \sim \frac{2}{\sqrt{\pi}} \frac{e^{-z^2}}{(2z)^{n+1}} [1 + O(z^{-2})] \tag{4.33b}$$

for large *positive* argument z.

EXERCISE 4.10. *Asymptotic behavior as* $z \to -\infty$*:* With the recursion (4.30a), show that

$$\mathrm{i}^1 \operatorname{erfc}(z) \sim -2z \quad \text{and} \quad \mathrm{i}^2 \operatorname{erfc}(z) \sim z^2 + \tfrac{1}{2}$$

for a large *negative* argument (plus exponentially small terms). Can you generalize? □

(*h*) *Solutions of the diffusion equation?* By direct differentiation and substitution, we verify that $\mathrm{i}^n \operatorname{erfc}[x/2\sqrt{(Dt)}]$ is *not* a solution of the diffusion equation. On the other hand, the function

$$C(x, t) \propto t^{n/2}\, \mathrm{i}^n \operatorname{erfc}(x/2\sqrt{Dt}), \qquad n \geqslant -1, \tag{4.34}$$

is such a solution, as can be easily verified. In fact, the solution (4.21d) of the previous thermal problem (4.21), of the impurity problem

(4.9), and the gaussian (4.3) (the response to a point source) are precisely of this form. In these cases, we note that the surface fields (at $x = 0$) change proportionally to $t^{1/2}$, t^0, and $t^{-1/2}$, respectively. In other words, the family of iterated error functions includes all our previous examples, and the functions (4.34) are therefore a natural extension of gaussian and error-function solutions to the diffusion equation. They describe diffusion when the surface concentration changes as an integer or half-integer power of time. We will take advantage of this property in the next section.

EXERCISE 4.11. *The gaussian generates the family of solutions (4.34):* Consider a gaussian

$$G(x, t) = \frac{1}{\sqrt{\pi D t}} e^{-x^2/4Dt}$$

that is "twice as large" as the gaussian (4.3), i.e., it is normalized in the interval [0, ∞]. Show that G generates the family (4.34) by repeated integration over space. □

4.5. Crystal Growth Under Conditions of Constant Cooling Rate

Our next application is of interest to those who grow crystals.[16-18] This problem may also be of interest to those who solve diffusion problems, because it introduces several new features: the possibility of decomposing a given problem into simpler parts (again, according to Property C.1 of Sect. 1.3) and the question of natural time scales.

Consider the physical situation that we encountered in Sects. 2.4 and 2.5 and in Sect. 4.3 above, namely, a binary liquid mixture brought into contact with a solid composed of the same elements. Here, we assume that the crystal is large in extent, and we focus our attention on a particular crystal face. Diffusion in the liquid, perpendicular to that face, is then essentially one-dimensional. Further, as we noted, the liquid can be arbitrarily prepared: its composition C_∞ in the minor component B is in the hands of the experimentalist. But he can do more than just join these two phases and let the concentration relax toward equilibrium. He can also impose controlled temperature variations on his solid–liquid system, e.g., by slowly ramping the power of his furnaces. This creates an extra driving force for diffusion because the equilibrium concentration shifts with temperature, as permitted by the phase rule (4.16): For binary, isobaric, two-phase systems, there is a single degree of freedom, which can be used to advantage, for example, by ramping the temperature according to the schedule

$$T(t) = T_0 - \alpha t, \tag{4.35a}$$

where T_0 is the initial temperature and α is the cooling rate.† Since the liquidus (see Fig. 2.6) is described by some $C_e(T)$ relation, it follows that the equilibrium concentration changes with time according to

$$\begin{aligned} C_e[T(t)] &\approx C_e(T_0) + (T - T_0)\left.\frac{dC_e}{dT}\right|_{T=T_0} + O[(T-T_0)^2] \\ &= C_0(1 - \alpha t/mC_0) + O(t^2). \end{aligned} \tag{4.35b}$$

In this second equality we have used Eq. (4.35a), and we define C_0 and m as the equilibrium concentration $C_e(T_0)$ and liquidus slope $dT/dC_e|_{T_0}$ at the initial temperature, respectively.

EXERCISE 4.12. *Temperature ramps and exponential behavior:* In many cases, the phase diagram $C_e(T)$ is well represented by an Arrhenius expression

$$C_e(T) = Ke^{-Q/RT},$$

where K and Q are constants and R is the gas constant. Insert the schedule (4.35a) and show that $C_e(t) \approx C_0 e^{-t/\tau}$. Determine the constants C_0 and τ. How "good" is this approximation? Compare with Eq. (4.35b). □

An exact solution of this Stefan problem is unavailable, but if we assume that the crystal–liquid interface moves slowly enough, then, as

†Thermal diffusivities are generally much larger than mass diffusivities in liquids and solids. Therefore, temperature equilibration occurs much more quickly than mass redistribution, and we can assume that the temperature is spatially constant. The temperature ramp rate α in Eq. (4.35a) is defined here as positive for *cooling*. Negative α's correspond to heating schedules.

in the oxidation problem of Sect. 2.3 and the spherical-precipitate problem of Sect. 2.5, we can solve the diffusion problem *as if* that interface were at rest. We then use the Stefan condition (1.20) to determine its motion. The diffusion problem is quite simple: The concentration in the liquid must be equal, initially and at infinity, to its prepared value C_∞, whereas, if the interface is again assumed to be at local thermodynamic equilibrium, then the surface concentration must be equal to that determined by Eq. (4.35b). Mathematically, this means

$$C(x, 0) = C(\infty, t) = C_\infty, \tag{4.36a}$$

$$C(0, t) = C_0(1 - \alpha t/mC_0), \tag{4.36b}$$

if the interface position essentially coincides with the spatial frame's origin. The extra "twist" here is the surface concentration's *given* (linear) time dependence. Equation (4.36b) deserves yet another remark. Although the temperature schedule (4.35a) has a characteristic time constant $T_0/|\alpha|$, the system responds with a time constant

$$\tau = mC_0/|\alpha|, \tag{4.35c}$$

and these need not be at all equal.

The solution to this diffusion problem follows easily if we recognize that the concentration can be viewed as the superposition $C = C_1 - C_2$ of two partial concentrations, each of which satisfies the diffusion equation under the conditions

$$C_1(x, 0) = C_1(\infty, t) = C_\infty, \tag{4.37a}$$

$$C_1(0, t) = C_0, \tag{4.37b}$$

and

$$C_2(x, 0) = C_2(\infty, t) = 0, \tag{4.38a}$$

$$C_2(0, t) = \mathrm{sgn}(\alpha) C_0 t/\tau. \tag{4.38b}$$

The solution for C_1 has been derived a number of times in this chapter. Since its initial and boundary conditions (4.37) are constant, we have

$$C_1(x, t) = C_\infty - (C_\infty - C_0)\,\mathrm{erfc}\big(x/2\sqrt{Dt}\big). \tag{4.39}$$

On the other hand, the partial concentration C_2 must vanish initially and at infinity, and it must be proportional to the first power of time at the interface. Looking back at our lexicon of available solutions (4.34) in unbounded media, we see that the second iterated error function fits our conditions, and we tentatively write $C_2 = At\,\mathrm{i}^2\,\mathrm{erfc}[x/2\sqrt{(Dt)}]$, where the constant A follows by matching this solution at the boundary with condition (4.38b). This solution satisfies the other conditions (4.38a) by virtue of the asymptotic expansion (4.33b). But $\mathrm{i}^2\,\mathrm{erfc}(0) = \frac{1}{4}$, in accordance with Eqs. (4.29) and (4.23b), so that $A = \mathrm{sgn}(\alpha)4C_0/\tau$, and we get

$$C_2(x, t) = \mathrm{sgn}(\alpha)(4C_0 t/\tau)\mathrm{i}^2\,\mathrm{erfc}\left(x/2\sqrt{Dt}\right). \tag{4.40}$$

To calculate the growth rate and hence the interface's schedule, we need only remember that the Stefan condition (1.20) must hold.† Therefore, using the differentiation equation (4.27), we get

$$\begin{aligned}\dot{\xi}[C_s - C(0, t)] &= D\left.\frac{\partial C}{\partial x}\right|_{x=0} = D\left(\left.\frac{\partial C_1}{\partial x}\right|_{x=0} - \left.\frac{\partial C_2}{\partial x}\right|_{x=0}\right)\\ &= \frac{1}{\sqrt{\pi}}(C_\infty - C_0)(D/t)^{1/2} + \mathrm{sgn}(\alpha)(2C_0/\tau)(Dt/\pi)^{1/2}. \end{aligned} \tag{4.41a}$$

With Eq. (4.36b), this is a quadrature for the interface schedule $\xi(t)$. The integration is notably simpler if we only consider times short with respect to the relaxation time τ. Then the surface concentration remains essentially at its initial value C_0, and we get the dimensionless form

$$\frac{\xi(t)}{\sqrt{D\tau}} \approx \frac{2}{\sqrt{\pi}}\frac{C_\infty - C_0}{C_s - C_0}(t/\tau)^{1/2} + \mathrm{sgn}(\alpha)\frac{4}{3\sqrt{\pi}}\frac{C_0}{C_s - C_0}(t/\tau)^{3/2}. \tag{4.41b}$$

This expression is the sum of two terms that correspond to the two driving forces of our system. The first, parabolic and proportional to $t^{1/2}$, is caused by the initial (absolute) supersaturation σ in the liquid. It is no different from the many other diffusion-limited examples we have seen. On the other hand, the second term is proportional to $t^{3/2}$;

†It is again assumed that diffusion in the solid phase is negligible.

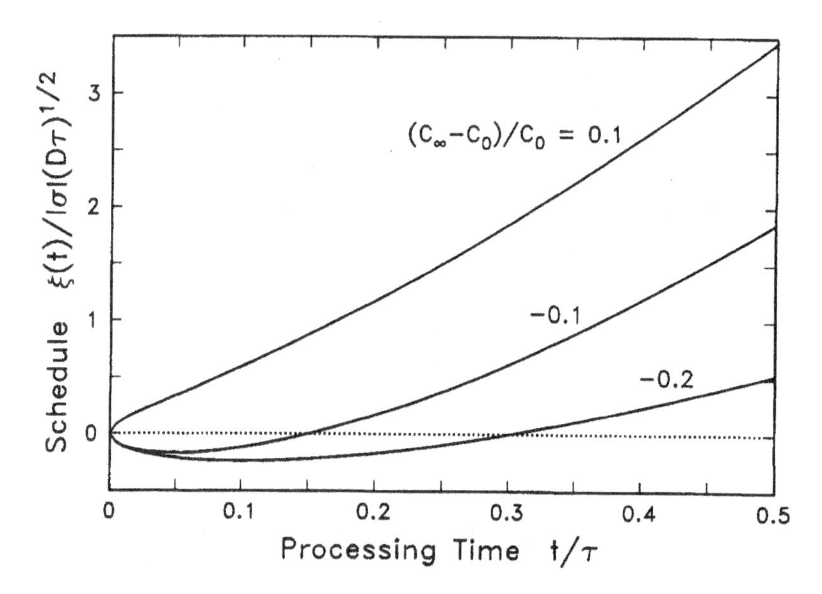

Fig. 4.10. Interface schedule for a crystallizing system under two distinct driving forces: initial supersaturation and cooling. Note the period of dissolution when these forces are opposed.

it is due to the *additional* supersaturation caused by temperature ramping. In other words, this second driving force is caused by agencies *exterior* to our system. Both terms can have either sign, and thus regimes of crossover from growth to dissolution can occur.

Figure 4.10 shows just such a case. The function (4.41b) is plotted for the case of cooling, $\alpha > 0$, and for various values of the relative supersaturation $\sigma_r = (C_\infty - C_0)/C_0$. It is evident that when the system is initially undersaturated, some time must elapse before the crystal can overcome this condition and begin to grow. This is sometimes observed and is called, somewhat in desperation, an "incubation time." The crossover from dissolution to growth occurs at a time

$$t = -\mathrm{sgn}(\alpha) 3\sigma_r \tau/2 \qquad (4.41c)$$

if the ramp rate and the relative supersaturation have opposite signs. On the other hand, the crystal always grows (or dissolves) monotonically when α and σ_r have the same sign.

The problem that we have just solved suggests an immediate generalization. If, for example, the surface concentration is given as a *known* power series in time, then, by virtue of the basic solutions (4.34), we expect a power-series solution in the *even* iterated error functions. In fact, we shall see later in Eq. (7.36) that there exists a closed-form representation of the diffusion equation's solution in a half-space if the surface concentration is an *arbitrarily given* function of time. Likewise, the last chapter will address the general problem of thermal stimulation, i.e., situations where temperature programming can be used to advantage to study diffusion and desorption. The procedure of this section is amplified in the exercise that follows.

EXERCISE 4.13. *Solutions in a half-space for variable Dirichlet conditions:* Assume that the surface concentration is a known function of time and that it can be expanded as a Taylor series in $\sqrt{t}$:

$$C_0(t) = \sum_{n=0}^{\infty} a_n t^{n/2}.$$

In addition, assume constant initial and far-field conditions such as those given by Eqs. (4.36a). Find the solution of the diffusion equation. What can you say about the surface flux? Can you generalize when the series for C_0 also contains a finite number of terms with negative powers $n \leqslant -1$? (This is called a Laurent series.) Can you solve similar problems in which the applied surface flux F, rather than C_0, is a known Laurent series in $\sqrt{t}$? □

4.6. Product Solutions for Higher-Dimensional Problems

We have just seen that linear combinations (i.e., essentially *sums*) of the diffusion equation's solutions can help solve problems when the boundary conditions themselves are linear combinations of powers of time. We now ask whether *products* of solutions can be put to advantage. For simplicity, we illustrate the procedure for two dimensions and in cartesian coordinates (x, y). Let $C_1(x, t)$ and $C_2(y, t)$ be any solutions of the 1-d diffusion equations

$$\frac{\partial C_1}{\partial t} = D\frac{\partial^2 C_1}{\partial x^2} \tag{4.42a}$$

and

$$\frac{\partial C_2}{\partial t} = D \frac{\partial^2 C_2}{\partial y^2} \tag{4.42b}$$

respectively. For the moment, we specify neither the domains[†] of the independent variables (x, t) and (y, t) nor the initial and boundary conditions. Nevertheless, the product

$$C(x, y, t) \equiv C_1(x, t) C_2(y, t) \tag{4.43}$$

satisfies the 2-d equation

$$\frac{\partial C}{\partial t} = D \left(\frac{\partial^2 C}{\partial x^2} + \frac{\partial^2 C}{\partial y^2} \right) \tag{4.44}$$

over some domain of the variables (x, y, t), because its substitution into Eq. (4.44) yields

$$C_2 \frac{\partial C_1}{\partial t} + C_1 \frac{\partial C_2}{\partial t} = D \left(C_2 \frac{\partial^2 C_1}{\partial x^2} + C_1 \frac{\partial^2 C_2}{\partial y^2} \right), \tag{4.45}$$

and this is an identity by virtue of Eqs. (4.42).

For example, the functions

$$C_1 = \operatorname{erfc}\left(x/\sqrt{Dt}\right) \tag{4.46a}$$

and

$$C_1 = \operatorname{erfc}\left(y/\sqrt{Dt}\right) \tag{4.46b}$$

are solutions of Eqs. (4.42) over the domains

$$\{x > 0, t > 0\} \tag{4.47a}$$

and

$$\{y > 0, t > 0\} \tag{4.47b}$$

[†]Topologically, a *domain* is any open and connected set. Here, we deal with ordered pairs or triples (enclosed in parentheses) of real numbers. Curly braces containing inequalities of the independent variables define domains. The reader is advised to sketch, as we go along, the indicated domains and the initial and boundary conditions.

respectively, and they satisfy the initial and boundary conditions

$$C_1(x, 0) = C_1(\infty, t) = 0, \qquad C_1(0, t) = 1, \tag{4.48a}$$

$$C_2(x, 0) = C_2(\infty, t) = 0, \qquad C_2(0, t) = 1, \tag{4.48b}$$

Consequently, their product $C = \text{erfc}[x/\surd(Dt)]\ \text{erfc}[y/\surd(Dt)]$, defined over the domain $\{x > 0, y > 0, t > 0\}$, satisfies Eq. (4.44), the initial condition $C(x, y, 0) = 0$, and the boundary conditions at infinity $C(x, \infty, t) = C(\infty, y, t) = 0$, all potentially reasonable requirements. But the boundary conditions at $x = 0$ and $y = 0$ would be most peculiar because, by construction, we have $C(0, y, t) = \text{erfc}[y/(Dt)]$ and $C(x, 0, t) = \text{erfc}[x/\surd(Dt)]$,. which are unlikely conditions in practice. This product solution is therefore *not* a suitable candidate, but it does indicate that we require *homogeneous* conditions at finite boundaries.† For example, the product

$$C(x, y, t) = \text{erf}(x/\sqrt{Dt})\,\text{erf}(y/\sqrt{Dt}), \tag{4.49a}$$

defined on the same "octant," satisfies the initial and boundary conditions

$$C(x, y, 0) = C(\infty, y, t) = C(x, \infty, t) = 1, \tag{4.49b}$$

$$C(0, y, t) = C(x, 0, t) = 0. \tag{4.49c}$$

These correspond to a concentration distribution about a rectangular corner. The initial concentration is everywhere unity, and the edges of the corner are held at zero. Evidently, we can multiply this solution by arbitrary constants and add other arbitrary constants to produce new solutions, where, for example, the concentration on the corner's edges could be a given constant.

EXERCISE 4.14. *The case of arbitrary initial conditions:* Consider solutions $C_1(x, t)$ and $C_2(y, t)$ of Eqs. (4.42) that satisfy *any* homogeneous conditions on finite boundaries and that go to the same constant at infinity if the domain is unbounded. Let $\varphi_1(x)$ and $\varphi_2(y)$ be their initial conditions. Show that the function C, defined as the product

†A homogeneous boundary condition is any relation of the form $f(C, \nabla C) = 0$ that vanishes with both C and its gradient. For example, $\alpha C + \beta \partial C/\partial x = 0$ is homogeneous.

(4.43), is a solution of Eq. (4.44) over the octant $\{x > 0, y > 0, t > 0\}$ if it satisfies the same homogeneous boundary conditions and if its initial condition $\varphi(x, y)$ is precisely $\varphi_1\varphi_2$. □

It would appear that this method of construction requires considerable skill. It is explained in more detail on pp. 33–35 of Carslaw and Jaeger's excellent survey[3] of known solutions, and it is extensively used in their Chaps. 5–7. Today, one tends to handle multidimensional problems through repeated Fourier transforms. Appendix E presents a particular case of these transforms.

We end this section with a few remarks about the method of *separation of variables*. As is well known, solutions of partial differential equations are often sought in the form of products. What relation, if any, is there with the previous method? To illustrate, take again the case of Eq. (4.44), and assume that its solutions C can be expressed as products of functions involving fewer variables. For example, assume that

$$C(x, y, t) = X(x, t)Y(y, t), \tag{4.50}$$

where X and Y are functions of the indicated variables. Introducing this product into Eq. (4.44), we also get the form (4.45), which we write

$$\frac{1}{X}\left(\frac{\partial X}{\partial t} - D\frac{\partial^2 X}{\partial x^2}\right) = -\frac{1}{Y}\left(\frac{\partial Y}{\partial t} - D\frac{\partial^2 Y}{\partial y^2}\right). \tag{4.51}$$

Now, however, the functions X and Y are *not* necessarily solutions of Eqs. (4.42). Consequently, Eq. (4.51) is not an identity. Nevertheless, because the left- and right-hand sides of this Eq. (4.51) are functions of different sets of variables, it follows that each side must be separately equal to the same arbitrary function λ of the common variable t. (Can you prove this?) Consequently, we have separated the 2-d equation (4.44) into the two 1-d equations,

$$\frac{\partial X}{\partial t} = D\frac{\partial^2 X}{\partial x^2} + \lambda(t)X, \tag{4.52a}$$

$$\frac{\partial Y}{\partial t} = D\frac{\partial^2 Y}{\partial y^2} - \lambda(t)Y, \tag{4.52b}$$

and the arbitrary function λ is yet to be determined from the initial and boundary conditions on C. Equations (4.52) are diffusion equations with sources.

Notes

1. This is, in fact, a special case of the "central limit theorem" which pervades probability theory, see Refs. 5 and 6.
2. This convergence to zero is faster than that of any power of t. This is called an *essential singularity*, and all the derivatives $\partial^k G/\partial t^k$ vanish as $t \to 0^+$. If an observer is sitting a distance x_0 away from the origin, he would "feel" a concentration rise $G(x_0, t)$ at any time following the initial impulse. Plotting this function (it is infinitely "flat" close to the origin) will convince the reader why "diffusion signals" appear to propagate with finite speed.
3. The scaling (4.8a) might not be evident to a beginner, who might prefer to write the solution as a "trial solution" $C(x, t) = A + B\,\mathrm{erf}[x/2\sqrt{(Dt)}]$, and then determine the constants A and B by using the initial and boundary conditions.
4. For example, error functions provide the solution of "Stokes's first problem," which was mentioned in Sect. 1.7. Indeed, Eqs. (1.38) show that the velocity distribution $v(x, t)$ must satisfy the diffusion equation $v_t = \nu v_{xx}$. We assume that the flat plate impulsively moves in its plane at a constant velocity V. The fluid is initially at rest, but, because it "sticks" at material boundaries, its tangential velocity there must equal to that of the plate: $v(0, t) = V$. The function $1 - v/V$ thus satisfies all the requirements of Property (e) of Sect. 4.1. The solution is therefore $v(x, t) = V\{1 - \mathrm{erf}[x/2\sqrt{(\nu t)}]\}$.
5. Flux continuity follows from energy conservation at *stationary* boundaries [see Eq. (1.19b)]. Temperature continuity, on the other hand, is really an additional assumption that can be understood as a condition of local thermal equilibrium [see Eq. (2.15a) and Chap. 6]. There are cases where it may not hold.
6. In many cases one assumes "partial equilibrium" for those species whose diffusivities are so low that they cannot redistribute significantly. Thus, each phase *separately* is in chemical equilibrium, although together they are not.
7. We disallowed phase transitions in our previous thermal problem Eqs. (4.12)–(4.14) and impurity diffusion problem Eq. (4.8), and there could be no interface motion. One can treat freezing and melting problems in the same way as the current constitutional problem in a binary.[3,10] Indeed, the freezing and melting of polar ice caps was Stefan's original concern.
8. We assume that the source is not powerful enough to cause ablation: There are no moving boundaries in this problem. We also assume that the energy

flux F is constant in time (more precisely, a step function). General time-dependent cases $F(t)$ are handled in Chap. 8, where we also explore the boundary condition (4.21a) in somewhat greater detail. The specification of the gradient at a boundary, rather than of the field, is called a *Neumann condition*.

References

1. M. Abramowitz and I. A. Stegun, *Handbook of Mathematical Functions* (reprinted by Dover, New York, 1965).
2. N. N. Lebedev, *Special Functions and their Applications*, R. A. Silverman, Ed. (reprinted by Dover, New York, 1972).
3. H. S. Carslaw and J. C. Jaeger, *Conduction of Heat in Solids*, 2nd ed. (Oxford University Press, London, 1959).
4. D. V. Widder, *Advanced Calculus*, 2nd ed. (Prentice-Hall, Englewood Cliffs, NJ, 1961). [Reprinted by Dover, New York.]
5. B. Gnedenko, *The Theory of Probability* (Mir Publishers, Moscow, 1976).
6. R. V. Hogg and A. T. Craig, *Introduction to Mathematical Statistics*, 3rd ed. (Macmillan, New York, 1970).
7. R. Ghez, G. S. Oehrlein, T. O. Sedgwick, F. F. Morehead, and Y. H. Lee, "Exact Description and Data-Fitting of Ion-Implanted Dopant Profile Evolution during Annealing," *Appl. Phys. Lett.* **45**, 881 (1984).
8. L. Badash, "The Age-of-the-Earth Debate," *Scientific American* **261**, 78 (August 1989).
9. C. C. Lin and L. A. Segel, *Mathematics Applied to Deterministic Problems in the Natural Sciences* (MacMillan, New York, 1974).
10. J. Crank, *The Mathematics of Diffusion*, 2nd ed., pp. 286–325 (Oxford University Press, London, 1975).
11. L. I. Rubinstein, *The Stefan Problem* (Translations of Mathematical Monographs, Vol. 27, by American Mathematical Society, Providence, RI, 1971).
12. J. M. Ockendon and W. R. Hodgkins, Eds., *Moving Boundary Problems in Heat Flow and Diffusion* (Oxford University Press, London, 1975).
13. G. H. Meyer, "An Alternating Direction Method for Multi-dimensional Parabolic Free Surface Problems," *Int. J. Numer. Methods Engng.* **11**, 741 (1977).
14. R. S. Gupta and D. Kumar, "Variable Time Step Methods for One-dimensional Stefan Problem with Mixed Boundary Conditions," *Int. J. Heat Mass Transf.* **24**, 251 (1981).
15. J. J. Derby and R. A. Brown, "A Fully-implicit Method for Simulation of the One-dimensional Solidification of a Binary Alloy," *Chem. Engng. Sci.* **41**, 37 (1986).

16. M. B. Small and J. F. Barnes, "The Distribution of Solvent in an Unstirred Melt under the Conditions of Crystal Growth by Liquid Epitaxy and its Effect on the Rate of Growth," *J. Cryst. Growth* **5**, 9 (1969).
17. M. B. Small and R. Ghez, "Conditions for Constant Growth Rate by LPE from a Cooling, Static Solution," *J. Cryst. Growth* **43**, 512 (1978). [There is a small misprint in Eq. (5) of this paper; its right-hand side should read $(4/3\sqrt{\pi})(t/\tau)^{3/2}$.]
18. M. B. Small, E. A. Giess, and R. Ghez, "Liquid-Phase Epitaxy," in *Handbook of Crystal Growth*, D. T. J. Hurle, Ed., Vol. 3a, pp. 223–253 (Elsevier, Amsterdam, 1994).

5

An Introduction to Similarity

With iterated error functions we have already greatly expanded our vocabulary of solutions of the diffusion equation. The question is, however, whether or not any useful further expansion should be expected. The study of physical symmetries provides a partial answer, and that study is called *similarity*. In essence, we shall find that the linear diffusion equation in one dimension has two types of solutions: superpositions either of iterated error functions or of trigonometric functions, otherwise known as Fourier series. The strength of similarity methods lies in their applicability to nonlinear problems as well. This chapter continues with more physical examples, including exact solutions for the kinetics of certain first-order phase transformations and of diffusion problems where the diffusivity depends on the field. In contrast, Chaps. 7 and 8 describe deductive methods for the solution of diffusion problems.

5.1. Similarity Solutions

In geometry, two figures are said to be similar if they are congruent after a uniform stretching of space. In turn, congruent figures are those that can be superposed after a rigid-body motion (rotations and translations, and, perhaps, reflections in space). The key notion for geometric similarity is our ability to change the length scale and, so doing, to introduce an equivalence between certain geometric figures. Theorems about right triangles, say, hold for *any* member of this class of figures, rather than for this or that particular triangle.

Similarity is also a useful tool for investigating classes of solutions of differential equations. In our case, the diffusion equation, we now ask if there exists a real *scale function* $T(t)$ of the dependent variable C and another, $L(t)$, of the independent variable x such that, for any two times t_1 and t_2, there are two locations x_1 and x_2 where the solution appears to be the same. In other words, when distance x is properly scaled, the ratios

$$\frac{C(x_1, t_1)}{T(t_1)} = \frac{C(x_2, t_2)}{T(t_2)} \tag{5.1a}$$

must have the same numerical value for all pairs of values of time t_1 and t_2. This is evidently possible if the concentration has the form

$$C(x, t) = T(t)F(\eta), \qquad \text{with } \eta = x/L(t). \tag{5.1b}$$

In that case the solution is the product of a pure function T of time and a function F that depends, at most, on a so-called *similarity variable* η. Such a solution (5.1b) is called a *similarity solution*. The search for these solutions is an important aspect of hydrodynamics[1,2] because of the Navier–Stokes equations' complexity: they are highly nonlinear. That search is also instructive for the diffusion equation, even though that equation is linear when the diffusivity is independent of the field.[N1]

Before proceeding with the formal development, let us note that we have already encountered similarity in at least two guises. The error-function solutions (4.8), (4.9), (4.13), and (4.17) of particular physical problems were all similarity solutions. In these cases, the concentration scale $T(t)$ was a constant, and the length scale was simply $L = 2\sqrt{(Dt)}$. Likewise, the class of solutions (4.34)—and the related physical problems (4.3), (4.21), and (4.40)—has the form (5.1b) when $T(t)$ is a power of $\sqrt{t}$. Our other encounter with similarity was through Exercises 1.8 and 1.19, and we meet it, perhaps more explicitly, in Appendix B. There, we find that the diffusion equation is invariant under the change of variables

$$\bar{x} = \alpha x + x_0 \qquad \text{and} \qquad \bar{t} = \alpha^2 t + t_0. \tag{5.2}$$

It follows that $\bar{x}/\sqrt{\bar{t}} = x/\sqrt{t}$ when the translations x_0 and t_0 are zero.

In other words, the scale function $L(t) \propto \sqrt{t}$ could have been predicted for the physical problems just mentioned. There are other possible scale functions, however, as we will see.

A second prediction follows from the transformation (5.2) by putting $\alpha = 1$ and the origin of time at $\bar{t} = 0$. We are then left with pure translations of space $\bar{x} = x + x_0$ and $\bar{t} = t$, which means that solutions can be periodic in space with period x_0. What is the meaning of periodicity, and what class of functions must be considered? We now turn to the consequences of similarity (5.1).

We wish to find the scales T and L such that the function (5.1b) is a solution of the diffusion equation (1.14). To do so we simply compute the partial derivatives

$$\frac{\partial C}{\partial t} = \dot{T}F + T\left(\frac{-x\dot{L}}{L^2}\right)F', \tag{5.3a}$$

$$\frac{\partial C}{\partial x} = \frac{T}{L}F' \quad \text{and} \quad \frac{\partial^2 C}{\partial x^2} = \frac{T}{L^2}F'', \tag{5.3b}$$

where the dots and primes indicate ordinary derivatives with respect to t (for T and L) and to η (for F), respectively. Inserting the partial derivatives (5.3) into the diffusion equation, we get

$$DF'' + (L\dot{L})\eta F' - (\dot{T}L^2/T)F = 0, \tag{5.3c}$$

an expression we would like to be as simple as possible. Viewed as a differential equation for the function F, Eq. (5.3c) reduces to an *ordinary* differential equation in $F(\eta)$ if the coefficients in parentheses are independent of time. Thus, we impose the additional conditions

$$L\dot{L} = 2D\alpha \tag{5.4a}$$

and

$$\dot{T}L^2/T = D\beta, \tag{5.4b}$$

where α and β are constants, and the factor D is thrown in for convenience. Equation (5.3c) then becomes simply

$$F'' + 2\alpha\eta F' - \beta F = 0. \tag{5.4c}$$

The condition (5.4a) is easily integrated with respect to t:

$$L^2(t) = L_0^2 + 4\alpha Dt, \tag{5.5a}$$

where L_0 is some initial value $L(0)$. Inserting this function into the second condition (5.4b), we get

$$T(t) = \gamma(L_0^2 + 4\alpha Dt)^{\beta/4\alpha}. \tag{5.5b}$$

Having found the functional form of the scales, these are now discussed in terms of the four integration constants: α, β, L_0, and γ. By "discussion" is meant, as in mechanics, the search for *physical solutions*, i.e., those solutions that are real and bounded except, perhaps, at isolated points in the (x, t)-plane.

We first observe that the multiplicative constant γ can be chosen arbitrarily because the (linear) diffusion equation is homogeneous in C, and thus in T. Then, α must be real and nonnegative, or else there would be times such that L, given by Eq. (5.5a), becomes imaginary. Accordingly, we distinguish the two cases: $\alpha = 0$ and $\alpha > 0$.

Case $\alpha = 0$. In this case L_0 must be nonzero or there would be no length scale (5.5a) at all. We choose the value $|L_0|^{-\beta/2\alpha}$ for the arbitrary constant γ, so that Eq. (5.5b) yields

$$T(t) = \lim_{\alpha \to 0} (1 + 4\alpha Dt/L_0^2)^{\beta/4\alpha} = e^{\beta Dt/L_0^2}. \tag{5.6a}$$

On the other hand, Eq. (5.4c) becomes simply

$$F'' - \beta F = 0 \tag{5.6b}$$

in the similarity variable[N2]

$$\eta = x/L = x/(\pm |L_0|). \tag{5.6c}$$

As just noted, we seek bounded solutions, in particular, those that decay for long times. It follows that the exponent in Eq. (5.6a) must be negative,† and thus we write $\beta = -\lambda^2$. The solutions of Eq. (5.6b)

†Strictly speaking, the exponent must have negative *real* part. Some physical problems, such as the diurnal temperature variations of the earth, are periodic in time. These problems require a complex exponent, yet with negative real part.

are then trigonometric functions, and, collecting our partial results, we have

$$C(x, t) \propto e^{-\lambda^2 Dt/L_0^2} e^{\pm i\lambda x/|L_0|}. \tag{5.7}$$

But this is merely the solution that is usually obtained under the heading of method of "separation of variables," which is illustrated in Sects. 4.6 and 5.3 and Appendix E. This method, usually introduced without much justification, now appears as a special case of similarity.[N3] The constant length scale L_0 is generally determined by the size of a given physical system, and we anticipate that solutions of the form (5.7) apply mainly to problems in *bounded* regions.

Case $\alpha > 0$. In this case we can put $L_0 = 0$ by an appropriate choice of the arbitrary origin of time in Eq. (5.5a). Then, for convenience, we choose $\gamma = \alpha^{-\beta/4\alpha}$ for the arbitrary integration constant in Eq. (5.5b). It follows that the scale functions are now

$$L(t) = \pm 2(\alpha D t)^{1/2} \tag{5.8a}$$

and

$$T(t) = (4Dt)^{\beta/4\alpha}. \tag{5.8b}$$

Next, we note that the transformation $\bar{\eta} = \eta\sqrt{\alpha} = x/2\sqrt{(Dt)}$ allows us to express the differential equation (5.4c) as

$$\frac{d^2F}{d\bar{\eta}^2} + 2\bar{\eta}\frac{dF}{d\bar{\eta}} - \frac{\beta}{\alpha}F = 0. \tag{5.8c}$$

We now state, without proof (see Ref. 3, p. 293), that the only bounded solutions of this differential equation are those for which the constant β/α is an even integer $2n$. But then, what do we have but the differential equation (4.31) for the iterated error functions.[N4] Collecting our results (5.8), we obtain the similarity solutions (4.34)

$$C(x, t) \propto (4Dt)^{n/2}\, \mathrm{i}^n \operatorname{erfc}\left(\pm x/2\sqrt{Dt}\right) \qquad \text{for all integers } n \geqslant -1 \tag{5.9}$$

that were derived earlier by construction, and *they are the only ones* when $\alpha > 0$. Since here there is no constant length scale, but rather a diffusion length $L(t) \propto \sqrt{t}$ that increases indefinitely with time, it follows that these solutions apply mainly (but not exclusively) to problems in *unbounded* regions.

In sum, similarity imposes strict conditions on the possible physical solutions of the diffusion equation. They must have either the form (5.7) or the form (5.9), possibly augmented by translations of the independent variables. Both forms result from the solution of ordinary differential equations, and these are applied to specific physical problems in the next sections.[N5]

We end this section on a note of hope for the *nonlinear* diffusion equation (1.15). Indeed, if the diffusivity varies with concentration, then sometimes there also exist similarity solutions that stem from the solution of ordinary differential equations. Some of these cases will be taken up later in this chapter. A careful study, however, requires the theory of Lie groups. This is beyond our scope, and we merely point the reader to some of the relevant literature.[4–6] Nonetheless, like Molière's *Monsieur Jourdain*, in this section we have, in fact, just analyzed—prosaically—two particular subgroups of the diffusion equation's group of physical symmetries.

5.2. Boltzmann's Transformation

It is perhaps of more than historical interest that the question of similarity for diffusion in unbounded phases was first raised by Boltzmann.[7] With remarkable intuition, he introduced the notion of similarity variable $\eta = x/2\sqrt{(Dt)}$ for the solution of a problem in nonlinear diffusion. A transformation of the diffusion equation only in terms of that variable is called *Boltzmann's transformation*, which is one of the more fruitful ideas for the solution of certain nonlinear problems. For example, it is instructive to rework our previous Stefan problem Eqs. (4.17)–(4.20) in this light. This problem is nonlinear, as will be shown presently, even for cases of constant diffusivity.[N1]

We first recall that we "guessed" the similarity solution (4.17) in the laboratory frame (x, t).[N6] In that frame, the solid–liquid interface moves with a schedule $\xi(t)$ given by Eq. (4.19). Rather than guess, we now change coordinates to an accelerated frame that "immobilizes" the boundary[8]:

$$\bar{x} = x - \xi(t) \qquad \text{and} \qquad \bar{t} = t. \tag{5.10a}$$

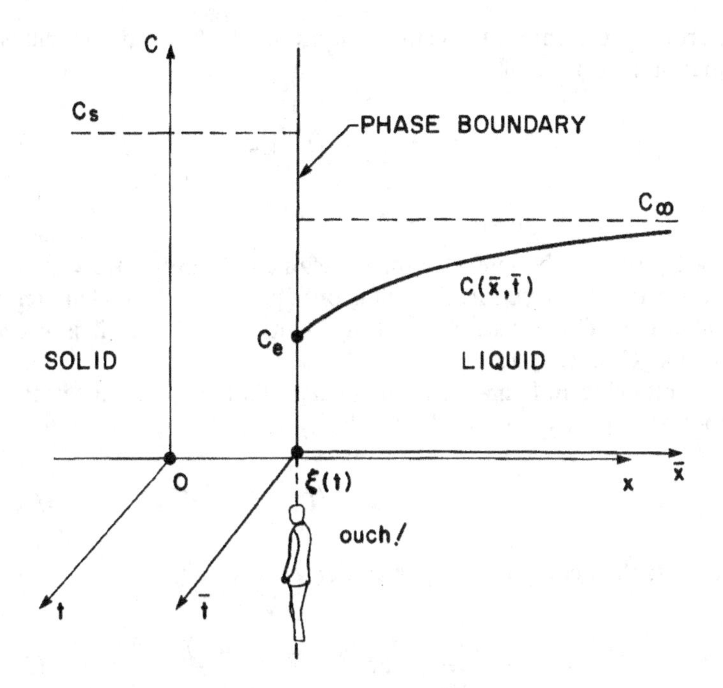

Fig. 5.1. Schematic representations of a reference frame that moves with the phase boundary.

In other words, the unfortunate observer of Fig. 5.1, rigidly skewered on the $(\bar{x}, \bar{t})$-frame, would always measure $\bar{x} = 0$ for the interface's position. Moreover, this frame is accelerated because ξ can be an *a priori arbitrary* function of time. Equations (5.10a) thus constitute an extension of galilean transformations that were included in the affine transformations of Exercises 1.8 and 1.19. As in these exercises and in Appendix B, we must compute partial derivatives with respect to the new variables:

$$\frac{\partial C}{\partial x} = \frac{\partial C}{\partial \bar{x}}\frac{\partial \bar{x}}{\partial x} + \frac{\partial C}{\partial \bar{t}}\frac{\partial \bar{t}}{\partial x} = \frac{\partial C}{\partial \bar{x}}, \tag{5.10b}$$

$$\frac{\partial C}{\partial t} = \frac{\partial C}{\partial \bar{x}}\frac{\partial \bar{x}}{\partial t} + \frac{\partial C}{\partial \bar{t}}\frac{\partial \bar{t}}{\partial t} = \frac{\partial C}{\partial \bar{x}}(-\dot{\xi}) + \frac{\partial C}{\partial \bar{t}}. \tag{5.10c}$$

Inserting these into the Stefan condition (4.18b) and the diffusion equation (1.14), we obtain

$$\dot{\xi}(C_s - C_e) = DC_{\bar{x}}|_{\bar{x}=0}, \tag{5.11a}$$

$$C_{\bar{t}} = DC_{\bar{x}\bar{x}} + \dot{\xi}C_{\bar{x}}. \tag{5.11b}$$

Clearly, we have bought a simpler Stefan condition† at the expense of a convective term $\dot{\xi}C_{\bar{x}}$ in the diffusion equation (5.11b). This term is nonlinear in C because, eliminating $\dot{\xi}$ with Eq. (5.11a), it is proportional to $C_{\bar{x}}(0, \bar{t})C_{\bar{x}}(\bar{x}, \bar{t})$.

Following Boltzmann, let us *assume* that the concentration distribution depends only on the similarity variable

$$\bar{\eta} = \bar{x}/2\sqrt{D\bar{t}}. \tag{5.12a}$$

We must then compute the partial derivatives

$$\frac{\partial C}{\partial \bar{x}} = \frac{dC}{d\bar{\eta}} \frac{1}{2\sqrt{D\bar{t}}} \quad \text{and} \quad \frac{\partial C}{\partial \bar{t}} = \frac{dC}{d\bar{\eta}}\left(-\frac{\bar{\eta}}{2\bar{t}}\right). \tag{5.12b}$$

Inserting these into Eqs. (5.11) and clearing fractions, we get

$$2\dot{\xi}\sqrt{\bar{t}/D}\,(C_s - C_e) = C'|_{\bar{\eta}=0}, \tag{5.13a}$$

$$C'' + 2(\bar{\eta} + \dot{\xi}\sqrt{\bar{t}/D})C' = 0, \tag{5.13b}$$

where primes denote ordinary derivatives with respect to $\bar{\eta}$. These equations (5.13) would be ordinary differential expressions if the common factor $\dot{\xi}\sqrt{(\bar{t}/D)}$ were independent of time. Therefore, in the spirit of the general similarity equation (5.3c), we assume that this factor is a constant λ. Integrating that relation, we recover Eq. (4.19) for the interface schedule, and Eq. (5.13b) becomes simply

$$C'' + 2(\bar{\eta} + \lambda)C' = 0. \tag{5.13c}$$

†Simpler because the boundary evaluations occur on a *fixed* line $\bar{x} = 0$ of the $(\bar{x}, \bar{t})$-plane.

This differential equation is easily twice integrated to yield the general solution

$$C(\bar{\eta}) = B + A\,\mathrm{erfc}(\bar{\eta} + \lambda), \tag{5.14}$$

which depends on two arbitrary constants. These, as usual, are determined by the initial and boundary conditions, which warrants a short digression.

We return to the moving $(\bar{x}, \bar{t})$-frame in which the initial and boundary conditions are

$$C(\bar{x}, 0) = C(\infty, \bar{t}) = C_\infty \qquad \text{and} \qquad C(0, \bar{t}) = C_e. \tag{5.15a}$$

These *three* conditions, shown again in Fig. 5.1, collapse into the *two* conditions

$$C(\bar{\eta} \to \infty) = C_\infty \qquad \text{and} \qquad C(\bar{\eta} \to 0) = C_e \tag{5.15b}$$

when expressed in terms of the similarity variable (5.12a).[N7]

To solve this mini-mystery of the disappearing condition, we first note that all our previous examples in unbounded regions were such that *the initial condition coincided with the far-field condition*, as is evident from Figs. 4.3, 4.4, and 4.6. This is precisely the effect of a similarity variable such as (5.12a): $\bar{\eta}$ goes to infinity when either $\bar{x} \to \infty$ (if $\bar{t} \neq 0$) or $\bar{t} \to 0$ (if $\bar{x} \neq 0$), and the error function's behavior (5.14) must be the same in either case. Consequently, it should be intuitively evident that the origin of coordinates plays a peculiar role. As noted in Chap. 4, the existence of a driving force (or supersaturation, because $C_\infty \neq C_e$) is closely related to the discontinuous behavior of the conditions (5.15a) at the origin. This leads to the question of singularities of the diffusion equation.

For simplicity, we consider a one-phase diffusion problem with no moving boundaries. The cases of several phases, e.g., the temperature-redistribution problem Eqs. (4.12)–(4.16) and the previous Stefan problem can be discussed in a similar way. We assume that *arbitrary* initial and (Dirichlet) boundary conditions are prescribed

$$C(x, 0) = \varphi(x) \qquad \text{and} \qquad C(0, t) = C_0(t), \tag{5.16a}$$

as shown on the left-hand side of Fig. 5.2, and observe that the origin

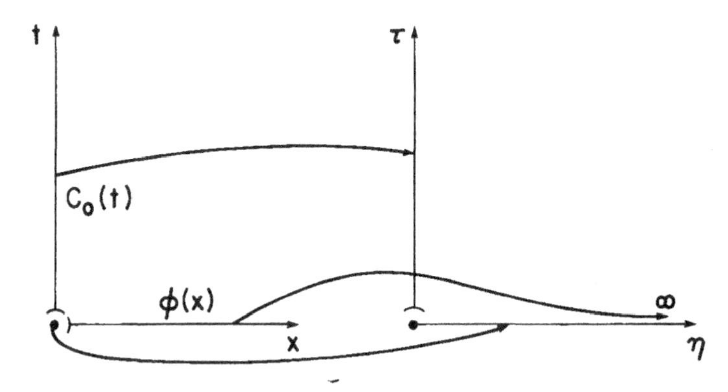

Fig. 5.2. Illustration of the coordinate transformation (5.17) and the corresponding lines over which the initial and boundary conditions must hold.

of coordinates is a point of discontinuity because, in general,

$$\lim_{x \to 0^+} \varphi(x) \neq \lim_{t \to 0^+} C_0(t). \tag{5.16b}$$

We then introduce the following change of coordinates:

$$\eta = x/2\sqrt{Dt} \qquad \text{and} \qquad \tau = \sqrt{t}, \tag{5.17}$$

which somewhat generalizes our earlier considerations of similarity (see Ref. 5, Vol. I, pp. 330–338). The jacobian of this transformation is easily computed, $\partial(\eta, \tau)/\partial(x, t) = 1/(4t\sqrt{D})$, and is singular when $t = 0$. Therefore, viewed as a mapping of the (x, t)-plane onto the (η, τ)-plane, we expect "one-to-oneness" to break down along the x-axis. Specifically, inspection of the transformation (5.17) leads to the following conclusions: (i) The whole positive x-axis (origin excluded) collapses onto the point at infinity of the η-axis; (ii) the t-axis (origin excluded) maps onto the τ-axis (origin excluded); (iii) the origin of the (x, t)-plane maps onto the whole η-axis (origin included, point at infinity excluded).† Consequently, the discontinuity (5.16b) is "smeared out" over the whole η-axis (which is helpful), but all the information on the

†The inquisitive reader should investigate what happens to other features of the (x, t)-plane, such as the negative x-axis, the lines $t = \text{const.}$, and the phase boundary's schedule (4.19).

initial condition $\varphi(x)$ must be contained in the single point at infinity, and this can happen *only if φ reduces to a constant.*

Next, we ask what form the diffusion equation takes in the new variables (5.17). A simple calculation, similar to that implied by Eqs. (5.12b), leads to

$$2\tau \frac{\partial C}{\partial \tau} = \frac{\partial^2 C}{\partial \eta^2} + 2\eta \frac{\partial C}{\partial \eta}, \tag{5.18}$$

and we still have a linear equation, albeit with variable coefficients. The main point, however, is that it reduces, when $\tau = 0$, to an ordinary differential equation

$$\frac{d^2 C}{d\eta^2} + 2\eta \frac{dC}{d\eta} = 0, \tag{5.19a}$$

analogous to Eq. (5.13c). Considered as a two-point boundary-value problem on the positive η-axis, its solution is clearly

$$C(\eta) = \varphi(0^+) + [C_0(0^+) - \varphi(0^+)] \operatorname{erfc}(\eta). \tag{5.19b}$$

This solution, identical to that of most of our previous examples, also shows its validity, for all time, if the initial condition φ and the boundary condition C_0 are independent of x and t, respectively. In other words, the first two conditions (5.15a) merge, and the solution (5.19b) "translates," unchanged, along the τ-axis if the initial and boundary conditions are constant.

To conclude this section, it should be emphasized that the simple error-function solution (5.19b) is inadequate to describe other situations. For example, we recall that Eq. (4.21d) described a physical circumstance for which $C_0(t) \propto \sqrt{t}$. Next, we note that such analytic expressions, valid on infinite regions, are often useful to "start" finite-difference schemes (see Refs. 9–12 and Appendix D) on bounded regions. These, we hinted in Sects. 1.4 and 1.5, often experience difficulties for the first few time steps, precisely because of discontinuities of the type (5.16b). Finally, we ought to ask if the variables (5.17) are the only possible choice. A partial answer will be given in Sects. 5.4 and 5.5, where we address some questions regarding the nonlinear diffusion equation.

EXERCISE 5.1. *Similarity in higher dimensions:* Express the rotationally symmetric diffusion equation (1.36) in terms of the similarity variable $\eta = r/2\sqrt{(Dt)}$. Find its general solutions. □

5.3. Two Applications of Similarity to Spheres

Our first application will be to *impurity diffusion in a sphere*, which is the analog for a bounded region of the impurity diffusion problem Eq. (4.8). Consider a sphere of radius R that contains a mobile species at an initial constant level C_∞. We place that sphere in a furnace, say, so that diffusion occurs at noticeable rate, and assume that the ambient causes the impurity's surface concentration to be pinned to its constant equilibrium value C_e. Thus, we must solve the diffusion equation in the *interior* of the sphere with the initial and boundary conditions

$$C(r, 0) = C_\infty, \qquad 0 < r < R, \tag{5.20a}$$

$$C(R, t) = C_e, \qquad t > 0, \tag{5.20b}$$

and we are left wondering what to do at the sphere's center, $r = 0$, where the diffusion equation (1.36) is singular.[†] In such cases, physical arguments suggest that C ought to be finite at the center.

The solution is easily found because, from Exercise 1.17 and the remark that introduces it, we know that the function $u = r(C + \text{const.})$ is a solution of the 1-d diffusion equation. Accordingly, we define a new unknown function u such that

$$C(r, t) = C_e + (C_\infty - C_e)\frac{u(r, t)}{r}, \tag{5.21a}$$

which satisfies the initial and boundary conditions

$$u(r, 0) = r, \qquad 0 < r < R, \tag{5.21b}$$

$$u(R, t) = 0, \qquad t > 0, \tag{5.21c}$$

$$\lim_{r \to 0} u/r < \infty. \tag{5.21d}$$

[†]Until now we have never worried about this possibility, because all previous calculations around spheres involved only the sphere's *exterior*. For example, see Exercise 4.4.

Because diffusion occurs in a bounded region $[0, R]$, we look to the general similarity solution (5.7) for guidance. The sphere's radius is a natural length scale L_0, and we write this solution in trigonometric form

$$u(r,t) \propto e^{-\lambda^2 Dt/R^2}[A\sin(\lambda r/R) + B\cos(\lambda r/R)]. \qquad (5.22a)$$

The coefficient B must vanish, or else we could not satisfy the condition (5.21d), and the other boundary condition (5.21c) requires that

$$\sin\lambda = 0, \qquad (5.22b)$$

whose roots are $\lambda_n = n\pi$, for any positive integer n.† The general solution to our problem is thus a superposition of "harmonics" (5.22a), and we write the Fourier series

$$u(r,t) = \sum_{n=1}^{\infty} A_n e^{-n^2\pi^2 Dt/R^2}\sin(n\pi r/R), \qquad (5.22c)$$

in which it remains only to determine the "Fourier amplitudes" A_n. This is easily done with the initial condition (5.21b), which now reads

$$r = \sum_{n=1}^{\infty} A_n \sin(n\pi r/R). \qquad (5.23a)$$

Multiplying this equation by $\sin(m\pi r/R)$, integrating over the interval $[0, R]$ and using the known "orthogonality"

$$\int_0^R \sin(n\pi r/R)\sin(m\pi r/R)\,dr = \tfrac{1}{2}R\delta_{n,m} \qquad (5.23b)$$

of the trigonometric functions, we find that

$$A_n = -\frac{2R(-1)^n}{n\pi}, \qquad (5.23c)$$

†The case $n = 0$ is excluded because Eq. (5.21d) would be indeterminate; negative integers n are unnecessary because they do not label linearly independent "eigenfunctions."

and thus the complete solution

$$C(r, t) = C_e + 2(C_\infty - C_e) \times \sum_{n=1}^{\infty} \frac{(-1)^{n+1} e^{-(n\pi\sqrt{Dt}/R)^2} \sin(n\pi r/R)}{n\pi r/R}. \tag{5.24}$$

It is instructive to compute the rate of impurity outdiffusion

$$J_0 = -D\frac{\partial C}{\partial r}\bigg|_{r=R} = 2D(C_\infty - C_e)R^{-1} \sum_{n=1}^{\infty} e^{-(n\pi\sqrt{Dt}/R)^2}, \tag{5.25a}$$

where we have merely differentiated the concentration distribution (5.24) with respect to r. This function is displayed (full line) in Fig. 5.3, and it should be compared with the following approximate calculation (dotted line) for short times. The latter allows a comparison with the

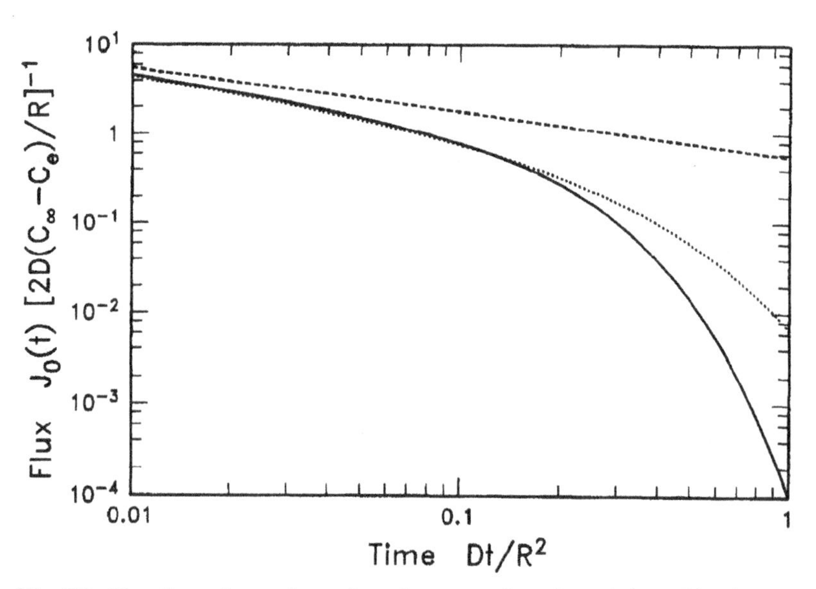

Fig. 5.3. Flux from the surface of a sphere as a function of time. The dotted line corresponds to the approximation (5.25b), and the dashed line to Eq. (4.8c).

earlier calculation (4.8c) for a half-space (dashed line). Indeed, if the diffusion length $\sqrt{(Dt)}$ is small compared to the sphere's radius R, then the impurity cannot distinguish a sphere from any other shape: All bodies look alike in the "tangent plane." In that case, the exponent of Eq. (5.25a) varies slowly with n, and, without any inquiry into rigor, we can replace the sum over n by an integral over the continuous variable $\xi = n\pi\sqrt{(Dt)}/R$. Thus, we get

$$J_0 \approx 2D(C_\infty - C_e)\frac{1}{\pi\sqrt{Dt}}\int_{\pi\sqrt{Dt}/R}^{\infty} e^{-\xi^2}\,d\xi$$

$$= (C_\infty - C_e)\sqrt{D/\pi t}\,\operatorname{erfc}\left(\pi\sqrt{Dt}/R\right), \tag{5.25b}$$

where the second equality results simply from the definition (4.10) of the complementary error function. For a small argument of the error function, this is nothing but Eq. (4.8c) with a change in sign, since the r-axis points out of the sphere. Approximations of this type for exponential sums are very useful numerically when their exponents vary slowly.

EXERCISE 5.2. *Short-time solution:* Verify that the distribution (5.24) for the spherical problem reduces to Eq. (4.8b) for the planar problem when the diffusion length is small. [Hint: Use the second integral derived in Exercise 4.5.] □

The *Frank–Zener theory of precipitate growth*[13,14] provides a second application of similarity solutions to spheres. This theory provides the kinetics of growth for isolated, rotationally symmetric phases. Here, we restrict our attention to spheres. We recall that the approximate treatment of this problem in Sect. 2.5 assumed the existence of a quasi-steady-state solution, but that the surface condition was allowed to vary in accordance with local equilibrium, including capillarity effects. The present theory develops an exact time-dependent solution, but the boundary conditions are drastically simplified. Specifically, this theory neglects the surface tension γ. Thus, we must solve the same equations (2.20) and (2.21) as before, except that the surface concentration C_R is assumed to be truly a constant C_e.

This is also a Stefan problem, and, taking a cue from the analogous planar case in Sects. 4.3 and 5.2, we write a trial solution

$$C(r, t) = C_\infty + A\frac{u(r, t)}{r}, \qquad r > R(t), \tag{5.26a}$$

where, as was observed previously, u must be a solution of the 1-d diffusion equation. We must now find the concentration distribution in the unbounded region *exterior* to the sphere. The similarity solutions (5.9), with r substituted for x, are now obvious candidates. But what value of n will do?† Since everything must be expressed in terms of the similarity variable $\eta = r/2\sqrt{(Dt)}$, by inspection of Eq. (5.26a) it follows that $n = 1$, and the concentration distribution must be of the form

$$C(r, t) = C_\infty + A\frac{\mathrm{i}^1\,\mathrm{erfc}(\eta)}{\eta}, \qquad \eta > R(t)/2\sqrt{Dt}. \tag{5.26b}$$

We have, again, two unknowns: the constant A above, and the surface's schedule $R(t)$. These are determined through the boundary conditions (2.21b) and (2.21c)

$$C|_{r=R(t)} = C_e \tag{5.26c}$$

and

$$\dot{R}(C_s - C_e) = D\frac{\partial C}{\partial r}\bigg|_{r=R(t)}, \tag{5.26d}$$

repeated here for convenience. As in Sect. 4.3, the only way to satisfy both of these conditions is again to set

$$R(t) = 2\lambda\sqrt{Dt}, \tag{5.27a}$$

i.e., a parabolic schedule. Inserting it into Eqs. (5.26c) and (5.26d), using the similarity form (5.26b), the Eq. (4.27), and the recursion (4.30b), we find the two unknowns

$$A = -\lambda\frac{C_\infty - C_e}{\mathrm{i}^1\,\mathrm{erfc}(\lambda)}, \tag{5.27b}$$

†The letter n, now used to label error functions, must be carefully distinguished from its previous use as a label for eigenfunctions.

$$2\sqrt{\pi}\lambda^2 e^{\lambda^2} \mathrm{i}^1\mathrm{erfc}(\lambda) = \sigma, \tag{5.27c}$$

where σ is again the absolute supersaturation. This problem is formally solved because, given a driving force σ, we solve the transcendental equation (5.27c) for the growth constant λ; hence we get the schedule (5.27a), the constant A from Eq. (5.27b), and then the full concentration distribution (5.26b).[†]

At this point we must step back and ask what we have gained over the simple solution presented in Sect. 2.5. We have, in fact, gained an exact solution to a moving-boundary problem that generalizes our previous estimate. To see this, we assume that the initial radius R_0 in Eq. (2.25b) is very small. That estimate for the schedule is then $R = \sqrt{(2\sigma Dt)}$. But this is precisely the limit of Eq. (5.27a) when the supersaturation is small. Indeed, using Eq. (4.32), we find from Eq. (5.27c) that $2\lambda^2 \approx \sigma$ for small supersaturations. Further, for each value of σ in the (semi-open) interval $[0, 1)$ we can solve Eq. (5.27c) numerically for λ. Figure 5.4 shows how the function $\lambda(\sigma)$ deviates from its small-σ values as $\sigma \to 1$. The restriction on possible values of σ is explored in the exercise that follows.

EXERCISE 5.3. *Analytic results for the Frank–Zener theory:* Using the expansion (4.32), show that a quasi-steady state is obtained for small σ. In particular, examine the concentration distribution. Next, with the expansion (4.33a) and Exercise 4.8, show that there are no real solutions for $\sigma \geqslant 1$ and $\sigma < 0$. [It is recommended to sketch the function on the left-hand side of Eq. (5.27c).] Interpretation? □

No matter what the initial conditions might be, the Frank–Zener calculation certainly represents an asymptotic solution for spherical precipitate growth as $t \to \infty$. On the other hand, we have lost several physical features of the precipitation problem. The most glaring are the similarity solution's inability to handle capillarity (and thus critical nucleation considerations) and its related requirement of a vanishing initial radius. Thus, in this problem, a sphere can only grow. Analyses of the effects of finite initial size, capillarity, and surface kinetics are

[†]The original papers by Frank and Zener derive these results by solving the diffusion equation written in the similarity variable η, much as we had done in Eqs. (5.12) and (5.13). In fact, these authors analyze rotationally symmetric cases in any number d of dimensions (see Exercise 5.1, above).

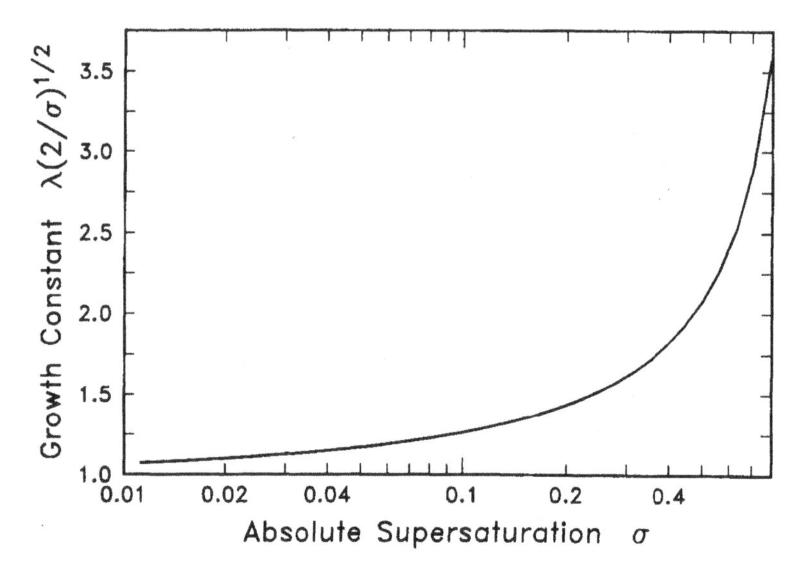

Fig. 5.4. The deviation of the growth constant from its expression for small supersaturations for the Frank–Zener problem.

mostly numerical. Reference 15 is one of many studies of this topic. Finally, as with our earlier treatment, it is implicitly assumed that growth is shape-preserving, e.g., spheres become spheres, and we have again evaded the issue of morphological stability during this diffusion process.

5.4. Variable Diffusivity and Boltzmann's Transformation

A world in which all transport coefficients are constant would be very poor indeed. In connection with the random-walk problem that was presented at the very beginning, we emphasized that any variation of the transition frequencies along sites would lead to a variable diffusivity. In fact, Exercise 1.4 provided a simple model for such an occurrence, and large force fields are bound to affect D, as is clear from Eqs. (3.18). Another way to variable D's is through nonequilibrium thermodynamics,[16] which takes the view that phenomenological relations, such as Fick's first law (1.12) and (1.31), must relate fluxes to gradients of chemical potential, rather than to concentration gradients.

After all, constancy of μ, rather than of C, is required at equilibrium. Thus, in accordance with Eq. (1.48), we write

$$\mathbf{J} = -L\nabla\mu, \tag{5.28a}$$

which defines the phenomenological coefficient L. If, further, the chemical potential of a solution can be expressed as in Eq. (2.17b), namely,

$$\mu = \mu^0(p, T) + kT\ln(\gamma C), \tag{5.28b}$$

where γ is the concentration-dependent "activity coefficient,"[N8] then a comparison of Eqs. (5.28) and (1.31) shows that

$$\mathbf{J} = -D\nabla C,$$

with

$$D = \frac{LkT}{C}\left(1 + \frac{\partial \ln \gamma}{\partial \ln C}\right). \tag{5.28c}$$

Therefore, no matter what the concentration dependence of the coefficient L might be, the diffusivity D is certainly variable, as first shown by Darken,[17] because the activity coefficient is generally a function of the fields p, T, and C.[N9]

The world of variable diffusivity is considerably more complicated to handle analytically, however, mainly because the diffusion equation (1.15),

$$\frac{\partial C}{\partial t} = \frac{\partial}{\partial x}\left(D\frac{\partial C}{\partial x}\right), \tag{5.29}$$

repeated here for convenience, is nonlinear. In essence, we lose the ability to superpose elementary solutions, either as sums or as integrals. Consequently, distinct methods are expected for each given $D(C)$ dependence and for every set of initial and boundary conditions. In other words, at most, we can hope to find *particular* solutions.

Let us first examine *nonlinear diffusion in a half-space under an arbitrary Dirichlet condition*

$$C(0, t) = C_0(t), \tag{5.30a}$$

together with an arbitrary initial condition

$$C(x, 0) = \varphi(x) \tag{5.30b}$$

that matches a (constant) far-field condition

$$C(\infty, t) = C_\infty = \lim_{x \to \infty} \varphi(x). \tag{5.30c}$$

This is the nonlinear analog of the problem investigated in Sect. 5.2. We now ask whether Boltzmann's transformation (5.12a) is at all applicable. Consider, then, a mild generalization of the transformation (5.17):

$$\tau = At^\alpha \qquad \text{and} \qquad \eta = x/(Bt^\beta), \tag{5.31a}$$

where α and β are *arbitrary* exponents, and A and B are constants that can be conveniently chosen.† Calculating the partial derivatives as in Sect. 5.2, we transform Eq. (5.29) into

$$B^2 \alpha \tau \frac{\partial C}{\partial \tau} = t^{1-2\beta} \frac{\partial}{\partial \eta}\left(D \frac{\partial C}{\partial \eta}\right) + B^2 \beta \eta \frac{\partial C}{\partial \eta}. \tag{5.31b}$$

For the right-hand side to depend exclusively on η, we require that the exponent of t vanish. Therefore, $\beta = \frac{1}{2}$, as with Boltzmann's transformation, but α ($\geqslant 0$) and A remain arbitrary because the left-hand side is homogeneous in τ. Choosing $B = 2$, we get

$$4\alpha\tau \frac{\partial C}{\partial \tau} = \frac{\partial}{\partial \eta}\left(D \frac{\partial C}{\partial \eta}\right) + 2\eta \frac{\partial C}{\partial \eta}, \tag{5.32}$$

with $\eta = x/2\sqrt{(Dt)}$, i.e., an equation very similar to Eq. (5.18), which was studied earlier. Moreover, the transformation (5.31a) has the same singularities and effect on the initial and boundary conditions as does the transformation (5.17). Therefore, if the surface concentration and the initial values are constant, and only under these circumstances, the system of equations (5.32) and (5.30) reduces to the solution of the

†The exponent α must be nonnegative, or else τ would blow up as $t \to 0$.

ordinary (but nonlinear), second-order differential equation for $C(\eta)$,

$$\frac{d}{d\eta}\left(D\frac{dC}{d\eta}\right) + 2\eta\frac{dC}{d\eta} = 0, \tag{5.33a}$$

under the conditions

$$C(0) = C_0 \qquad \text{and} \qquad C(\infty) = C_\infty. \tag{5.33b}$$

There are very few closed-form solutions of these equations (5.33). The search consists essentially of finding the necessary functional $D(C)$ relation compatible with a possible solution $C(\eta)$.[18,19] On the other hand, Eqs. (5.33) are the basis of a very powerful numerical procedure, the *Boltzmann–Matano scheme* (see Ref. 20), for finding the diffusivity's concentration dependence from experimental data. This, in fact, was Boltzmann's goal.

Interdiffusion, phase transformations, and reactions at phase boundaries are facts of life for the metallurgist. These questions are reviewed in Refs. 21 and 22 for metallic systems and silicides, respectively. In addition, device engineers are often engaged in the production of "heterostructures," a fancy name for a sequence of compositionally distinct, thin, semiconductor films. In that case, one is concerned that interdiffusion not alter these layers during subsequent processing steps.[23,24] To model these problems may require numerical methods because the phase boundaries are often neither stationary nor at local equilibrium. Furthermore, significant diffusion can occur over large portions of thin films. On the other hand, analytic modeling, when successful,[25–27] is based mainly on Boltzmann's transformation, which necessarily leads to equations similar to Eqs. (5.33).

5.5. Analytic Solutions for Variable Diffusivity

The search for analytic solutions of Eq. (5.29) requires further specialization. For example, there are many physical cases (e.g., diffusion in porous media or in heavily doped semiconductors) where D varies as some power ν of C.† Consequently, let us consider *power-law diffusion in a half-space*, such that no diffusant is present,

†Recall Exercises 2.2 and 2.3; ν need not be an integer.

initially or far from the surface. Thus, the diffusivity now has the form

$$D(C) = D_*(C/C_*)^\nu, \tag{5.34}$$

where C_* and D_* are some corresponding reference values. In addition, we have

$$C(x, 0) = C(\infty, t) = 0 \qquad \text{for } x > 0, t > 0, \tag{5.35a}$$

but we do *not* yet specify the boundary condition at the surface $x = 0$.

Here, we follow the treatment of Pattle,[28] as expanded by Boyer,[29] and we seek similarity solutions in the sense of Eqs. (5.1).[N10] To simplify their algebra, we will guess the form (power laws in t) of the scales $L(t)$ and $T(t)$. Thus, we assume that

$$C(x, t) = At^\alpha F(\eta), \tag{5.36a}$$

with $\eta = x/(Bt^\beta)$, where, again, the exponents α and β must be found, and the constants A and B will be chosen conveniently.

A few remarks concerning the exponents α, β, and ν are in order. First, it should be clear that α is quite arbitrary, but that β must be strictly positive, or else the presumptive form (5.36a) could not satisfy the conditions (5.35a). As we have seen in Sect. 5.2, these two conditions collapse into the single condition

$$F(\infty) = 0 \tag{5.35b}$$

on the unknown function $F(\eta)$. Next, the case of constant diffusivity corresponds to the exponent $\nu = 0$ in Eq. (5.34). Further, negative ν's are physically unreasonable, since they would imply an infinite diffusivity at zero concentration. Finally, cases of positive ν correspond to simultaneous vanishing of D and C. Thus, we expect a steep diffusion front when the concentration is low. From a mathematical point of view, vanishingly small diffusivities in equations of the type (5.29) mean that the term containing the highest derivative vanishes. Such circumstances, in the theory of differential equations, are called *singular*, and they are closely related to the existence of such fronts.

We proceed, as before, by computing the partial derivatives

$$\partial/\partial x = (Bt^\beta)^{-1}\, d/d\eta \qquad \text{and} \qquad \partial C/\partial t = At^{\alpha-1}(\alpha F - \beta\eta F'). \tag{5.36b}$$

Inserting these, together with Eq. (5.34), into Eq. (5.29), we get

$$(D_* A^\nu / C_*^\nu B^2) t^{\alpha\nu + 1 - 2\beta} (F^\nu F')' + \beta \eta F' - \alpha F = 0. \qquad (5.36c)$$

Let us now choose the leading coefficient so that this equation becomes as simple as possible an ordinary differential equation. Choosing

$$(D_* / C_*^\nu)(A^\nu / B^2) = 1, \qquad (5.37a)$$

and satisfying the condition

$$\alpha\nu + 1 - 2\beta = 0, \qquad (5.37b)$$

we get the ordinary differential equation

$$(F^\nu F')' + \beta \eta F' - \alpha F = 0 \qquad (5.38)$$

for the unknown function $F(\eta)$. Given a physical problem (i.e., for given ν), the conditions (5.37) relate the constant A to B and the exponent α to β. The yet unspecified boundary condition will provide another set of conditions. We now consider three different cases, only the last of which yields a solution in closed form.

Case 1: Constant Dirichlet Condition. Here, we impose a constant concentration

$$C|_{x=0} = C_0 = \text{const.} \qquad (5.39a)$$

at the free boundary. It follows from Eq. (5.36a) that

$$A t^\alpha F(0) = C_0. \qquad (5.39b)$$

Consequently, the exponent α must vanish for this equation to be always true. With the condition (5.37b), this implies

$$\alpha = 0, \qquad \beta = \tfrac{1}{2}, \qquad (5.40a)$$

and we are still free to choose A and B. For example, if we take $A = C_0$, then Eq. (5.39b) is a boundary condition,

$$F(0) = 1, \qquad (5.41)$$

on the unknown function F, and Eq. (3.35b) provides another. Furthermore, since a unit of concentration, C_0, is available, we can choose this to be the reference concentration C_*, and we denote the corresponding diffusivity by D_0. Their insertion into the condition (5.37a) determines these constants:

$$A = C_0, \qquad B = \sqrt{D_0}. \tag{5.40b}$$

Therefore, the differential equation (5.38) reduces to

$$(F^{\nu}F')' + \tfrac{1}{2}\eta F' = 0, \tag{5.42}$$

with $\eta = x/\sqrt{(D_0 t)}$, which is akin to Eq. (5.33a), discussed in connection with the Boltzmann transformation. Unfortunately, Eq. (5.42) is no more integrable in closed form than is Eq. (5.33a), but it is important to remember that we have reduced the diffusion equation to an ordinary differential equation in a similarity variable. This can be easily solved numerically under the two boundary conditions (5.35b) and (5.41). Readers familiar with hydrodynamics[1,2] will recognize an analogy (which runs deep) with the analysis of boundary layers.

Even if no analytic solution is available, valuable information can be gained directly from what we know about the exponents. For example, the surface flux $J_0 = -DC_x|_{x=0}$ has a time dependence, which we determine from Eqs. (5.36a) and (5.40):

$$J_0(t) = -F'(0)C_0\sqrt{D_0/t}, \tag{5.43}$$

and this expression is almost identical with Eq. (4.8c) for constant diffusivity.

Case 2: Constant Neumann Condition. This second case concerns constant flux conditions at the free surface. It is very similar to the first case, and it also does not provide an integrable differential equation. This is the object of the exercise that follows.

EXERCISE 5.4. *Case of constant Neumann condition:* Assume that the surface flux is a given constant J_0. Determine the exponents (α, β) and choose the constants (A, B) appropriately. How does the surface concentration vary with time? □

Evidently, the cases of variable, but given, surface concentration $C_0(t)$ or surface flux $J_0(t)$ are handled in an analogous manner, as long as these dependences are power laws in time.

Case 3: Instantaneous Point Source. This elementary case corresponds to the finite-difference calculation of Sect. 1.5 and to its continuum analog, the gaussian (4.3): A unit source is initially placed at the surface and is then allowed to relax by diffusion into the bulk. With the form (5.36), the total mass

$$M = \int_0^\infty C(x, t)\, dx = At^\alpha Bt^\beta \int_0^\infty F(\eta)\, d\eta \tag{5.44}$$

is conserved. Consequently, for M to be a constant M_0, independent of time, we require that $\alpha = -\beta$, and we can always choose $AB = M_0$. Using the previous conditions (5.37), we get

$$A = \left[\frac{C_*^\nu M_0^2}{D_*}\right]^{1/(\nu+2)}, \qquad B = \left[\frac{D_* M_0^\nu}{C_*^\nu}\right]^{1/(\nu+2)}, \tag{5.45a}$$

and

$$\alpha = -\beta = -1/(\nu + 2). \tag{5.45b}$$

Inserting these values into the differential equation (5.38), we can integrate once† to obtain

$$(\nu + 2)F^\nu F' + \eta F = \text{const.} \tag{5.46c}$$

Now we need the result of the next exercise.

EXERCISE 5.5. *A boundary condition for nonlinear diffusion:* If mass is conserved, show, using Eqs. (5.29) and (5.44), that the flux at the origin must vanish. Is this a general result or is it only true for power-law diffusivities? □

The integration constant vanishes because of this exercise, and the differential equation is integrable once again because it is separable:

$$F^\nu = \frac{1}{2}\frac{\nu}{\nu + 2}(\eta_0^2 - \eta^2), \tag{5.46a}$$

†Because $\eta F' + F = (\eta F)'$.

where η_0 is another integration constant. Next, the function F must be real, so that, taking the νth root, we get

$$F(\eta) = \begin{cases} [\nu/2(\nu + 2)]^{1/\nu}(\eta_0^2 - \eta^2)^{1/\nu} & \text{if } \eta \leqslant \eta_0, \\ 0 & \text{if } \eta > \eta_0. \end{cases} \tag{5.46b}$$

Finally, to determine η_0 we insert this function (5.46) into the mass conservation condition (5.44) to get

$$1 = \int_0^\infty F(\eta)\, d\eta = [\nu/2(\nu + 2)]^{1/\nu} \int_0^{\eta_0} (\eta_0^2 - \eta^2)^{1/\nu}\, d\eta. \tag{5.47a}$$

The integral in question is a special case of the so-called "beta function," itself expressible in terms of gamma functions. This will be clear from Eqs. (7.37), because, putting $\xi = \eta^2/\eta_0^2$, the integral is simply

$$\begin{aligned} \int_0^{\eta_0} (\eta_0^2 - \eta^2)^{1/\nu}\, d\eta &= \tfrac{1}{2}\eta_0^{(\nu+2)/\nu} \int_0^1 (1 - \xi)^{1/\nu}\xi^{-1/2}\, d\xi \\ &= \tfrac{1}{2}\eta_0^{(\nu+2)/\nu} B(1 + 1/\nu, 1/2) \\ &= \tfrac{1}{2}\eta_0^{(\nu+2)/\nu} \frac{\Gamma(1 + 1/\nu)\Gamma(1/2)}{\Gamma(3/2 + 1/\nu)}. \end{aligned} \tag{5.47b}$$

With the recursion (4.23a) for gamma functions, it follows from Eqs. (5.47) that

$$\eta_0 = \left[\frac{2(\nu + 2)^{\nu+1}}{\nu\pi^{\nu/2}}\right]^{1/(\nu+2)} \left[\frac{\Gamma(1/2 + 1/\nu)}{\Gamma(1/\nu)}\right]^{\nu/(\nu+2)}, \tag{5.48}$$

which depends only on ν.

This case is entirely solved, because the function F follows from (5.46) if we know the integration constant η_0, which we do from Eq. (5.48). Consequently, we know the full concentration distribution (5.36a) because Eqs. (5.45) determine all the coefficients.

Exercise 5.6. *Limiting cases for small and large exponents ν:* With Stirling's formula (4.24b), compute η_0 for small ν. Then, show that the concentration tends toward (twice) the gaussian (3.3). Why "twice"? Also compute these limits when $\nu \to \infty$. □

It is instructive to rewrite this solution (5.36a), (5.46) in the form

$$C(x, t) = C_0(t)\left[1 - \left(\frac{x}{L(t)}\right)^2\right]^{1/\nu} \qquad \text{for } x \leqslant L(t), \tag{5.49}$$

with the understanding that the concentration vanishes elsewhere. By virtue of Eqs. (5.45), the surface concentration varies with time according to

$$C_0(t) = At^{\alpha}F(0) = \left[\frac{\nu\eta_0^2}{2(\nu + 2)}\right]^{1/\nu}\left[\frac{C_*^{\nu}M_0^2}{D_*t}\right]^{1/(\nu+2)}, \tag{5.50a}$$

and the length scale is

$$L(t) = Bt^{\beta}\eta_0 = \eta_0\left[\frac{D_*tM_0^{\nu}}{C_*^{\nu}}\right]^{1/(\nu+2)}. \tag{5.50b}$$

This function is plotted in Fig. 5.5 for increasing values of the

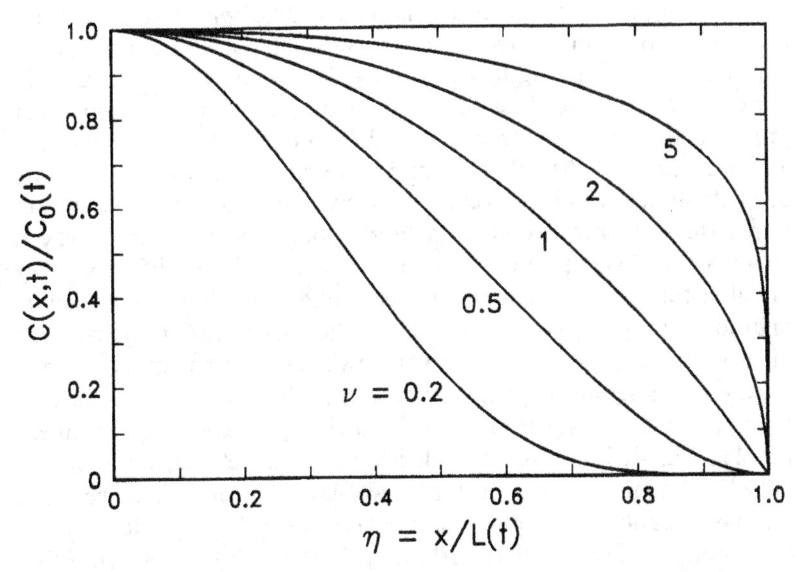

Fig. 5.5. Scaled concentration distribution for power-law diffusivity. Note the abrupt front that develops when ν is large.

diffusivity exponent ν, and several of its features deserve comment. First, it is clear that the concentration vanishes when $x > L(t)$. In other words, there exists a moving front, akin to a propagating wave, that separates disturbed regions from others that have not yet felt the influence of the initial point source. This is in marked contrast to the gaussian (4.3), its analog for cases of constant diffusivity, for which *all* regions of space are immediately affected (albeit infinitesimally, it is true, for distant points) by an initial point disturbance. Further, when the exponent ν is large, the concentration distribution is quite flat over large portions of its domain. When ν is small, Fig. 5.5 indicates that we approach a gaussian shape, as should be the case. This assertion is proved in Exercise 5.6 above. Next, the surface concentration (5.50a) decreases as $t^{-1/(\nu+2)}$, and its dependence on the total mass M_0 is quite different: linear in the gaussian case and to the $[2/(\nu + 2)]$th power when $\nu \neq 0$. The same differences appear for the length scale: independent of M_0 in the gaussian case, and to the $[\nu/(\nu + 2)]$th power otherwise. Again, both C_0 and $L(t)$ converge to gaussian behavior when $\nu \to 0$. Indeed, the reader is urged to consult the excellent treatise by Zel'dovich and Raizer,[30] particularly their Chap. 10, for a discussion of this problem by dimensional analysis.

In sum, we have shown that nonlinear diffusion presents features resembling traveling waves and that its characteristic parameters (concentration and length scales) may depend on the total mass in unexpected ways. Needless to say, there are very few other nonlinear problems that can be solved in closed form (Ref. 30 presents another: an initial "dipole" disturbance), and there is every reason to ask why our attention should not center exclusively on numerical solutions, rather than on these isolated analytical solutions. There is, however, an important asymptotic result in the case of linear diffusion: *Any* initial disturbance in an infinite or semi-infinite medium will evolve by diffusion into a gaussian if we observe the concentration distribution in faraway regions. This somewhat awkward statement is given a precise meaning and is proved in Appendix F. The beauty of asymptotic methods, however, resides in their application to nonlinear problems, as well. In a remarkable paper, Zel'dovich and Barenblatt[31] prove an analogous result for power-law diffusion: The solution we have found for a point source can also be regarded as the asymptotic behavior of an arbitrary initial disturbance. In other words, concentration "tails" for nonlinear diffusion can be fitted to a known analytic expression.

EXERCISE 5.7. *Moments and power-law diffusion:* Consider the moments (F.8) defined in Appendix F. Explore the consequences of their constancy for power-law diffusion in a half space. □

Notes

1. Even the linear diffusion equation (1.14) sometimes leads to a nonlinear problem because nonlinearities can arise from the boundary conditions. In Sect. 5.2 we show that almost all Stefan problems are intrinsically nonlinear, and Chap. 8 provides other examples of nonlinearities caused by boundary conditions.
2. The double sign is a consequence of the diffusion equation's symmetry with respect to spatial reflections.
3. The following calculations are typical of problems where variables separate. The method goes back to Fourier's monumental work. [See J. Fourier, *Théorie Analytique de la Chaleur* (1822) (also see the annotated edition of Fourier's early manuscript by I. Grattan-Guinness, *Joseph Fourier* 1768–1830, MIT Press, Cambridge, MA, 1972).] He discovered that trigonometric series can be used to represent diffusion phenomena, as well as periodic solutions of the wave equation.
4. The integer n can have either sign, so that, strictly speaking, the class of allowable functions is larger than (5.9): It includes Hermite polynomials multiplied by gaussians. These functions, however, are rarely used for diffusion, although they are essential for the analysis of the quantum-mechanical harmonic oscillator.
5. The reader may justifiably ask what connection, if any, exists between this chapter and Appendix B. In that appendix, we discussed general point transformations of the independent variables that leave the dependent variable unchanged. In this chapter, we consider a very particular transformation of the independent variables, but the dependent variable can be "stretched." A tasty application of similarity is offered by M. S. Klamkin in his justly famous and entertaining paper, "On Cooking a Roast," *SIAM Rev.* **3**, 167 (1961).
6. By "laboratory frame" we mean a spatial reference frame in which the diffusion equation (1.14) or (1.15) is valid, for example, a frame rigidly attached to the solid. This is not quite true for the case at hand if solid and liquid have different densities, for then the liquid has a convective motion induced by the interface's motion (e.g., see Ref. 11 in this chapter and Note 11 in Chap. 2).
7. Regardless of how that may ever happen, it is clear that we have solved our problem because these two boundary conditions (5.15b) are sufficient

to determine the constants A and B of the general solution (5.14). In fact, we easily get $B = C_\infty$, and A is again given by Eq. (4.20a). Further, insertion of this solution into the Stefan condition (5.13a) yields precisely the same Eq. (4.20b) for the growth constant λ. The reader should verify that the concentration distribution (5.14) has an inflection point at $\bar{\eta} = -\lambda$. Why?

8. Traditionally, the activity a is represented by the product γC, which defines the activity coefficient. Collisions in notation are unavoidable. Thus, γ represents either this coefficient or the surface tension, just as μ stands for either the chemical potential or the mobility coefficient, and L for either a phenomenological coefficient or a length scale.
9. There can be other reasons for variable diffusivity: In inhomogeneous media D can depend explicitly on position, and the action of external driving forces, such as temperature ramps (see Sect. 4.5), can cause D to depend on time. This last case will be considered further in Chap. 8.
10. The prehistory of this type of problem fades into World War II. It appears that solutions were first published by Zel'dovich and Kompane'ets in the USSR; see Ref. 30. Nevertheless, since these problems have bearing on detonations, it is likely that the solutions to follow were known to those skilled in the art of making bombs.

References

1. H. Schlichting, *Boundary-Layer Theory* 6th ed. (McGraw-Hill, New York, 1968).
2. G. Birkhoff, *Hydrodynamics: A Study in Logic, Fact and Similitude*, 2nd ed. (Princeton University Press, Princeton, NJ, 1960).
3. N. N. Lebedev, *Special Functions and their Applications* (reprinted by Dover, New York, 1972).
4. G. W. Bluman and J. D. Cole, *Similarity Methods for Differential Equations* (Springer-Verlag, Berlin, 1974).
5. W. F. Ames, *Nonlinear Partial Differential Equations in Engineering*, Vols. I and II (Academic Press, New York, 1965 and 1972).
6. L. Dresner, *Similarity Solutions of Nonlinear Partial Differential Equations* (Pitman Advanced Publishing Program, Boston, 1983).
7. L. Boltzmann, "Zur Integration der Diffusionsgleichung bei variabeln Diffusionscoefficienten," *Ann. der Phys.* **53**, 959 (1894).
8. G. Rosen, "Galilean Invariance and the General Covariance of Non-relativistic Laws," *Amer. J. Phys.* **40**, 683 (1972).
9. G. W. Evans II, E. Isaacson, and J. K. L. MacDonald, "Stefan-like Problems," *Quart. Appl. Math.* **8**, 312 (1950).

10. B. Boley, "A General Starting Solution for Melting and Solidifying Slabs," *Int. J. Engng. Sci.* **6**, 89 (1968).
11. M. B. Small and R. Ghez, "Growth and Dissolution Kinetics of III-V Heterostructures Formed by LPE," *J. Appl. Phys.* **50**, 5322 (1979).
12. R. Ghez, "Expansions in Time for the Solution of One-Dimensional Stefan Problems of Crystal Growth," *Int. J. Heat Mass Transfer* **23**, 425 (1980).
13. F. C. Frank, "Radially Symmetric Phase Growth Controlled by Diffusion," *Proc. Roy. Soc.* **201A**, 586 (1950).
14. C. Zener, "Theory of Growth of Spherical Precipitates from Solid Solution," *J. Appl. Phys.* **20**, 950 (1949).
15. R. J. Schaeffer and M. E. Glicksman, "Fully Time-Dependent Theory for the Growth of Spherical Crystal Nuclei," *J. Cryst. Growth* **5**, 44 (1969).
16. S. R. de Groot and P. Mazur, *Non-equilibrium Thermodynamics* (North-Holland, Amsterdam, 1962). [Reprinted by Dover, New York.]
17. L. S. Darken, "Diffusion, Mobility and their Interrelation through Free Energy in Binary Metallic Systems," *Trans. AIME* **175**, 184 (1948).
18. J. R. Philip, "General Method of Exact Solution of the Concentration-Dependent Diffusion Equation," *Australian J. Phys.* **13**, 1 (1960); see also *ibid*, p. 13.
19. B. Tuck, "Some Explicit Solutions to the Non-linear Diffusion Equation," *J. Phys.* **D9**, 1559 (1976).
20. B. Tuck, *Introduction to Diffusion in Semiconductors*, pp. 199–203 (Peter Peregrinus, Stevenage, 1974).
21. M. Hillert, "Diffusion and Interface Control of Reactions in Alloys," *Metall. Trans.* **6A**, 5 (1975).
22. F. M. d'Heurle and P. Gas, "Kinetics of Formation of Silicides: A Review," *J. Mater. Res.* **1**, 205 (1986).
23. L. L. Chang and A. Koma, "Interdiffusion between GaAs and AlAs," *Appl. Phys. Lett.* **29**, 138 (1976).
24. K. S. Seo, P. K. Bhattacharya, G. P. Kothiyal, and S. Hong, "Interdiffusion and Wavelength Modification in $In_{0.53}Ga_{0.47}As/In_{0.52}Al_{0.48}As$ Quantum Wells by Lamp Annealing," *Appl. Phys. Lett.* **49**, 966 (1986).
25. G. V. Kidson, "Some Aspects of the Growth of Diffusion Layers in Binary Systems," *J. Nucl. Mater.* **3**, 21 (1961).
26. C. Wagner, "The Evaluation of Data Obtained with Diffusion Couples of Binary Single-Phase and Multiphase Systems," *Acta Met.* **17**, 99 (1969).
27. D. S. Williams, R. A. Rapp, and J. P. Hirth, "Phase Suppression in the Transient Stages of Interdiffusion in Thin Films," *Thin Solid Films* **142**, 65 (1986).
28. R. E. Pattle, "Diffusion from an Instantaneous Point Source with a Concentration-Dependent Coefficient," *Quart. J. Mech. and Appl. Math.* **7**, 407 (1959).
29. R. H. Boyer, "On Some Solutions of a Non-linear Diffusion Equation," *J. Math. & Phys.* **40**, 41 (1961).

30. Ya. B. Zel'dovich and Yu. P. Raizer, *Physics of Shock Waves and High-Temperature Hydrodynamic Phenomena*, Vol. II (Academic Press, New York, 1967).
31. Ya. B. Zel'dovich and G. I. Barenblatt, "The Asymptotic Properties of Self-Modelling Solutions of the Nonstationary Gas Filtration Equations," *Dokl. Akad. Nauk SSSR* **118**, 4 (1958). [Engl. transl. in *Sov. Phys. Doklady* **3**, 44 (1958).]

6

Surface Rate Limitations and Segregation

The purpose in this chapter is to clarify the notion of "local equilibrium," namely, when can we assert that the concentration at a phase boundary is pinned to its equilibrium value, even as the core of an adjacent bulk phase undergoes nonequilibrium processes. To that end, we must first explain what is a "phase boundary," and what does "concentration at a phase boundary" really mean. These questions are of immense practical importance for such applications as the change in catalytic properties of metal surfaces over time and the change in doping character of semiconductors during annealing. They are all closely related to the question of *surface segregation* to free surfaces and to grain boundaries, which are topics of scientific interest in their own right.[1–3]

6.1. What *Are* Boundary Conditions?

For definiteness, consider a "semi-infinite" body.[N1] This means that we focus our attention on the half-space $\{x > 0\}$, and phenomena that occur in directions transverse to the x-axis play no significant role. This region does have a geometrical bounding surface at $x = 0$, and this boundary is our main concern.

We assume that the body contains a diffusing species that is continuously distributed. Its concentration $C(x, t)$ is related to its flux

$J(x, t)$ at all interior points according to the continuity equation (1.13),

$$\frac{\partial C}{\partial t} = -\frac{\partial J}{\partial x} \qquad \text{for } x > 0, t > 0, \tag{6.1}$$

repeated here for convenience. Thus the total mass of diffusant contained in this solid is

$$M(t) = \int_0^\infty C(x, t)\, dx. \tag{6.2}$$

As usual, differentiating this quantity and using Eq. (6.1) yields

$$\dot{M} = J(0^+, t), \tag{6.3}$$

a result that does not depend on the concentration values that one might assign at the boundary (see Exercise 1.5). Moreover, Fick's first and second laws (1.12) and (1.14) [or (1.15)] were not even used.† Of course, Fick's first law

$$J = -D\frac{\partial C}{\partial x} \qquad \text{for } x > 0, t > 0, \tag{6.4}$$

serves to define the flux in terms of the concentration distribution.

If the flux is assigned at the boundary (a Neumann condition), then, assuming continuity with interior points immediately adjacent to that boundary, Eq. (6.3) immediately indicates how mass changes in the course of time. We can then solve the full diffusion problem to find the concentration distribution. But what if the concentration is assigned at the boundary (a Dirichlet condition)? In that case we must first solve the diffusion problem under this condition, calculate the surface flux, and only then can we use Eq. (6.3) to calculate the total mass.

Mathematically, these problems appear well posed. Physically, however, they beg the question of what the concentration at a boundary might mean. Indeed, if "concentration" (equivalent to "mass or particle density") makes operational sense *for interior points*, it is not quite clear what element of volume we can choose *at a boundary point* to measure concentrations.

†This statement is not quite correct for an unbounded region because we need to prove that the flux at infinity vanishes; see Appendix F.

To grasp this problem, we should bear in mind that any experimental method that professes to measure concentration actually reckons the total mass or the total number of particles in a more or less well-defined sampling volume. This volume determines the resolution of the experimental method, and the mass it contains is the only quantity that can actually be observed. No method probes surface atoms exclusively, for the simple reason that phase boundaries are not geometrical 2-d surfaces existing in isolation of the adjacent bulk phases.[N2] This is also true for temperature measurements or, in fact, for measurements of *any* field.

The reader might justifiably feel that we waver between mathematical and physical (or operational) notions of concentration and flux. To further that perception, we note that if the observable mass contained in some volume is expressed by an integral similar to Eq. (6.2), then even that integral raises questions when viewed as a purely mathematical construct. For example, if integration is understood in the sense of Riemann, then the integrand may be modified arbitrarily at a countably infinite number of points.[4] It follows that the boundary value, $C(0, t)$, can be arbitrarily assigned without causing any change to the computed value of $M(t)$. Doesn't this deny the possibility of describing surface phenomena?

These questions become acute when we numerically estimate an integral of the form (6.2). The introduction of an equidistant mesh on the x-axis

$$x_i = ia \qquad i = 0, 1, 2, \ldots \tag{6.5a}$$

and a sample

$$C_i(t) \equiv C(x_i, t) \tag{6.5b}$$

of the concentration distribution allows several useful estimates. For example, referring to Fig. 6.1, if we estimate the integral by a sum of trapezoids $(x_i, C_i, C_{i+1}, x_{i+1})$ for which the truncation error is $O(a^3/12)$ (e.g., see Ref. 5), we have the expression

$$M(t) \approx \tfrac{1}{2}aC_0(t) + a \sum_{i=1}^{\infty} C_i(t), \tag{6.6}$$

which depends on the boundary value $C_0(t) \equiv C(x_0 = 0, t)$.

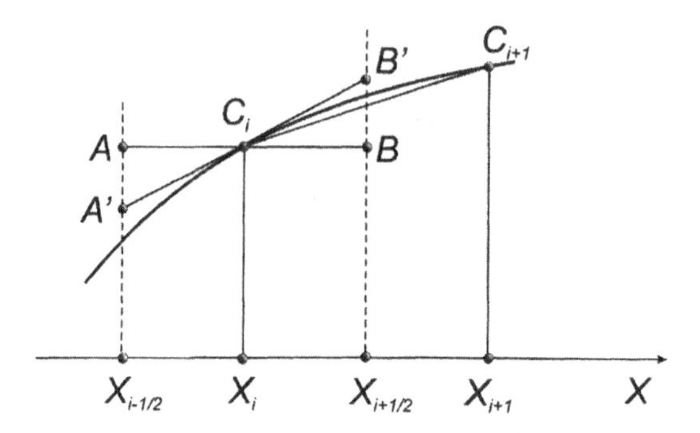

Fig. 6.1. Illustrating the trapezoidal and midpoint integration rules.

On the other hand, if we allow the mesh points (6.5a) to bisect "cells" $(x_{i-1/2}, x_{i+1/2})$, midpoint evaluation by rectangles $(x_{i-1/2}, A, B, x_{i+1/2})$ yields the slightly better estimate†:

$$M(t) \approx a \sum_{i=1}^{\infty} C_i(t). \tag{6.7}$$

Equation (6.7) is slightly better than (6.6) because "swiveling" the segment $\overline{AB}$ around its midpoint C_i, thus conserving the area under the resulting trapezoids, finally produces the tangent $\overline{A'B'}$ to the concentration distribution. This line is optimal in the sense that the area bounded by the distribution and any of its tangents is minimum when the tangent is evaluated exactly at the midpoint, as we have just done.[6,7] Note that Eq. (6.7) does *not* require the boundary value C_0.

We now recognize the *numbers* of particles, $N_i = aC_i$ (per unit of transverse area), that were first introduced in Sect. 1.1. It follows that the summation in Eqs. (6.6) and (6.7) is merely the sum of these numbers evaluated over all *interior* sites.[N3] But what are we to do with the first term, $\frac{1}{2}aC_0$, in Eq. (6.6)? Does it mean that the concentration C_0 extends over only "half" a cell, or should one fictitiously "extend the cell" by an amount $\frac{1}{2}a$ to the left of the point $x_0 = 0$?‡ Yes, we are forced to take phase boundaries more seriously.

†The truncation error is now of order $(a^3/24)$.

‡That is the essence of the "fictitious point method" described in Sect. 1.4 to evaluate gradients at a boundary.

6.2. Gibbs's Dividing Surface

Our gross powers of observation lead us to believe that boundaries between phases are sharp, that is to say, we expect discontinuous behavior of various observable fields at 2-d surfaces. But, if we are committed to a continuous description of natural phenomena,† then we might feel, intuitively perhaps, that a continuous variation of fields—albeit rapid—occurs at these boundaries. Figure 6.2 represents, schematically, what might happen at the interface between two phases at equilibrium. For example, the concentration distribution of some species, far from the boundary, then takes the *constant* values $C_\infty^{(1)}$ and $C_\infty^{(2)}$.‡ Indeed, Gibbs had proved, long ago,[8] that equilibrium demands the constancy of temperature, pressure, and chemical potentials in each phase if force fields are negligible. This implies discontinuities at geometric surfaces if these represent phase boundaries.

Nevertheless, Gibbs also recognized that the true concentration distribution in the neighborhood of a phase boundary must be rapidly varying over distances of the order of the range of atomic interactions. Call it $\tilde{C}(x)$. In other words, the actual boundaries, though sharp, cannot be infinitely abrupt.

†Even a quantum-mechanical description is continuous in the sense that one calculates the probability of a particle's presence at position **x**.

‡That "far-field" behavior would be, at most, slowly varying in the presence of force fields such as gravitation.

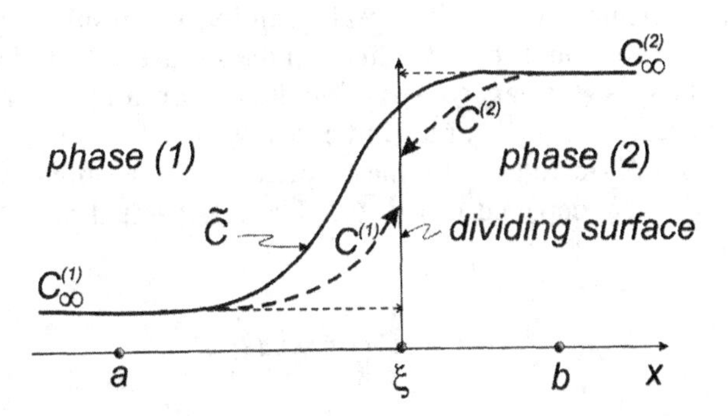

Fig. 6.2. The real distribution $\tilde{C}$, Gibbs's dividing surface, and the fictitious, bulk concentration distributions (dashed). The latter are terminated by arrowheads to suggest left and right limits.

In 1893, van der Waals[9] proposed a physical model to calculate $\tilde{C}$; it was rediscovered and greatly extended by Cahn and Hilliard[10] some 60 years later. In essence, Gibbs's fundamental relation between energy and entropy densities was modified to include concentration *gradients*, as well as concentration *values*. These ideas led to an outpouring of papers concerning the structure and dynamics of phase boundaries. In particular, they led to an improved understanding of "spinodal decomposition," a phenomenon that had long been recognized experimentally by metallurgists.

In more modern language, perhaps, Gibbs perceived that the description of phase boundaries requires two very different *length scales*. The first, large with respect to interatomic distances is operative far from the boundaries and was discussed in earlier chapters under the heading of "diffusion distance"; the second, of the order of interatomic distances, describes the spatial variation of various fields close to these boundaries. This reminds us of fluid dynamical boundary-layer theory, first introduced by Prandtl in 1904[11] and highly developed since then as a branch of "singular perturbation" problems[12,13] in applied mathematics.[†] Wishing to preserve the thermodynamic properties of bulk phases that he knew so well, but unable to estimate the rapid changes close to a phase boundary, Gibbs introduced the following useful construction: (i) Extrapolate the known bulk behavior of the concentrations right up to a geometrical surface that he calls the *dividing surface*, and whose position is as yet undecided. (ii) Calculate the total masses on each side of this surface by using these extrapolated values; these masses are fictitious. (iii) Subtract these masses from the system's total mass, which is a real quantity; the result is generally nonzero. (iv) Attribute this "residual" to the surface, which defines the surface mass. A similar construction holds for any other density function: energy, stress, charge, and entropy.

Let us see how to perform these operations for a simple geometry, a slab $a < x < b$, depicted in Fig. 6.2. The total, actual mass[N1] of this system is evidently

$$M = A \int_a^b \tilde{C}(x)\, dx, \tag{6.8}$$

[†]The analogy stops here, however, since the boundary-layer equations (which are parabolic) are generally simpler than the full, hyperbolic Navier–Stokes equations that govern behavior in the bulk of the fluid. Both are continuum descriptions, and the first is derived from the second in the limit of large Reynolds number, which is a *bulk* quantity.

where A is the area of the body transverse to the x-axis. This integral cannot be calculated if we do not know the real distribution $\tilde{C}$, but the total mass M is known. Choose a position $x = \xi$ for the dividing surface, which here is a plane. Since the equilibrium values $C_\infty^{(1)}$ and $C_\infty^{(2)}$ are constant, they are easily extrapolated toward that plane (dashed on the figure). The fictitious masses to the left and right of the dividing plane are then

$$\begin{aligned} M^{(1)} &= C_\infty^{(1)} A(\xi - a) = C_\infty^{(1)} V^{(1)}, \\ M^{(2)} &= C_\infty^{(2)} A(b - \xi) = C_\infty^{(2)} V^{(2)}, \end{aligned} \tag{6.9}$$

where $V^{(1)}$ and $V^{(2)}$ are the volumes thus constructed. The difference $M - M^{(1)} - M^{(2)}$ is generally nonzero. Call it $M^{(s)}$ and attribute it to the dividing plane. Dividing by its transverse area allows us to define the (excess) *surface concentration*:

$$C^{(s)} = M^{(s)}/A = (M - M^{(1)} - M^{(2)})/A. \tag{6.10}$$

Its dimensions are evidently "per unit area," $[C_s] = \text{cm}^{-2}$.

To summarize briefly, we subtract two fictitious quantities from the total (real) mass. That calculated difference is attributed to the dividing surface: it is also fictitious; it might even be negative! It is amazing that the whole body of classical surface thermodynamics, used for example in Sects. 2.4 and 2.5 and in Appendix C, is based on this representation. Of course, many questions remain open. For example, where should one place this dividing surface? How does $C^{(s)}$ depend on its position? Does each density field require a different dividing surface? We refer to classical texts (Refs. 17–20 in Chap. 2) for partial answers to these questions. They are often hotly debated, even today.

6.3. Nonequilibrium Behavior of Dividing Surfaces

Gibbs's theory, just sketched, concerns macroscopic systems at equilibrium. Can it be extended to diffusion, clearly a nonequilibrium process? The answer is affirmative if we are willing to examine his steps more closely.

In essence, Gibbs attempts to ignore the detailed structure of a phase boundary and to replace this "zone of ignorance" by what he does know, namely, the far-field behavior of the (real) concentration distribution and the requirement of total mass balance. For a system that is not at equilibrium, we can no longer extrapolate by requiring the constancy of concentration levels within certain regions. Rather, we must use whatever other information we may have about bulk nonequilibrium behavior and the local conservation of mass. The continuity equation (6.1) is an ideal candidate because it has this property and because it does hold far enough from the phase boundary. Having chosen a geometrical surface, we "continue" the concentration fields on either side, fictitiously perhaps, but in a way that is compatible with Eq. (6.1). Total mass must be conserved, however, which allows us to attribute physical quantities, such as surface concentration, to this dividing surface. We now explain this in more detail.

We consider once again the situation depicted in Fig. 6.2, but now allowing all quantities to depend on time as well as on distance. The dividing plane is at position $x = \xi(t)$, as yet undefined, and we demand that the fictitious concentration distributions $C^{(1)}(x, t)$ and $C^{(2)}(x, t)$ satisfy the continuity equation *right up to the dividing plane*:

$$\frac{\partial C^{(1)}}{\partial t} = -\frac{\partial J^{(1)}}{\partial x} \qquad \text{for } a < x < \xi(t),\ t > 0, \tag{6.11a}$$

$$\frac{\partial C^{(2)}}{\partial t} = -\frac{\partial J^{(2)}}{\partial x} \qquad \text{for } \xi(t) < x < b,\ t > 0. \tag{6.11b}$$

The system's (real) total mass is now the sum of three fictitious terms

$$M(t) = A\int_a^{\xi(t)} C^{(1)}(x, t)\,dx + A\int_{\xi(t)}^b C^{(2)}(x, t)\,dx + AC^{(s)}(t), \tag{6.12}$$

an expression that can be differentiated.

Why should the total mass ever change over time? As was explained in Sect. 1.8, there are two reasons for change: (i) the system contains sources or sinks; (ii) there are mass fluxes at its external boundaries. For simplicity, let us suppose that there are no sources of mass, meaning that there are no chemical reactions, homogeneous or heterogeneous, within the slab $a < x < b$. We also assume that the external boundaries are not "active" in the sense that they are not the seat of physical processes such as chemical reactions. We then have the very simple balance equation

$$\dot{M} = A(J_a - J_b), \tag{6.13a}$$

where these fluxes represent interactions with the environment. On the other hand, Eq. (6.12) can be differentiated using Leibniz's rule, and, inserting the continuity equations (6.11), we get†

$$\dot{M} = A[J^{(1)}(a^+, t) - J^{(1)}(\xi^-, t) + C^{(1)}(\xi^-, t)\dot{\xi} - J^{(2)}(b^-, t) + J^{(2)}(\xi^+, t) - C^{(2)}(\xi^+, t)\dot{\xi} + \dot{C}^{(s)}] \tag{6.13b}$$

Comparing these two equations (6.13), we get

$$\dot{C}^{(s)} = (J^{(1)} - C^{(1)}\dot{\xi})|_{\xi^-} - (J^{(2)} - C^{(2)}\dot{\xi})|_{\xi^+} \tag{6.14}$$

because the fluxes must balance at the external boundaries.

This mouthful invites several comments. First, Eq. (6.14) is an obvious generalization of the Stefan condition (1.20). It connects the time variation of the surface concentration $C^{(s)}$ to the discontinuity of the "molar flux" $J - C\dot{\xi}$ at the dividing plane. That it is a balance equation for the surface concentration is intuitively obvious: Any change of $C^{(s)}$ is caused by the limiting values of the bulk fluxes on either side of the dividing plane. Equation (6.14) can easily be generalized if the phase boundary supports heterogeneous reactions and if diffusion occurs along its plane. Second, if the phase boundary is not planar, then the surface concentration depends on curvilinear coordinates transverse to the x-axis of Fig. 6.2. A good dose of differential geometry is required to handle curved surfaces, especially if they are in motion. Third, and most important, we need constitutive relations to express the fluxes in terms of concentrations. Yet, even if we suppose that Fick's first law (6.4) holds in each bulk phase *right up to the dividing plane,* we still have insufficient information to solve the three coupled equations (6.11a), (6.11b), and (6.14). We are missing what amounts to conditions that connect $C^{(s)}$ to the limiting values of $C^{(1)}$ and $C^{(2)}$ as one approaches the surface. This, the author believes, is a fundamental problem in the irreversible thermodynamics of surfaces.

†These calculations should remind the reader of those performed to obtain Properties A in Sect. 1.3.

There exist three avenues in pursuit of these "missing" conditions. The first is to assume mere proportionality between $C^{(s)}$ and the limiting values of $C^{(1)}$ and $C^{(2)}$ as $x \to \xi^{\pm}$. This assumption, sometimes called the law of "proportional disturbances" in the description of grain-boundary diffusion,[14] is the basis for a description of equilibrium segregation to grain boundaries.[1] The physical reasons given for its validity are flimsy, and experiments indicate that it is not always satisfied.[15,16†] The second avenue consists of constructing models of the diffusive process, either continuous[17,18] or discrete descriptions[19,20] These have firmer physical underpinnings and will be examined in the next two sections.

The last avenue requires that we take Gibbs's picture more seriously to develop a theory of irreversible thermodynamics of surfaces. There are literally dozens of papers in this direction, most burdened by murky formalism to which the author has, alas, also contributed.[21] An interesting point of view is taken by a well-known school of rheologists[22] who have applied singular perturbation methods to the description of diffuse phase boundaries. Generally speaking, taking a cue from 3-d irreversible thermodynamics,[23] we must write the surface balance equations [similar to Eq. (6.14)] for all species, momentum, and energies (kinetic and potential). A balance equation for surface entropy finally yields an entropy source, which, like its 3-d counterpart, is a bilinear form in what can be identified as thermodynamic forces and fluxes. Among the latter, the normal mass flux, evaluated as one approaches the surface from one side, is "conjugate" to the *difference* between the chemical potential of the surface, itself, and its limiting value in the adjacent bulk. If a linear phenomenological relation is postulated, consistent with the second principle of thermodynamics, then the required coupling condition is obtained.[21]

To sharpen the argument, consider again diffusion in a half-space (see Sect. 6.1). It can be shown [see Eq. (8.33) in Chap. 8 or pp. 75–77 of Ref. 24] that the formal solution of the diffusion problem in a half-space under a constant initial and far-field condition C_∞ is

$$C(x, t) = C_\infty + \int_0^t J_0(s) \frac{e^{-x^2/4D(t-s)}}{\sqrt{\pi D(t-s)}} ds, \tag{6.15a}$$

†It resembles Henry's law, an expression of the equality of chemical potentials between *bulk* phases at equilibrium.

where J_0 is precisely the limiting value of the flux on the right-hand side of Eq. (6.3). If the bounding surface at $x = 0$ is immobile and is somehow endowed with surface mass, then the mass balance equation (6.14) reads simply

$$\frac{dC^{(s)}}{dt} = -J_0 = D\frac{\partial C}{\partial x}\bigg|_{x=0^+}. \tag{6.15b}$$

Here, in the spirit of Gibbs's construction, we have supposed that Fick's first law (6.4) holds right up to the boundary. But, try as we may, this system of two equations (6.15) cannot be solved unless we are given some relation between $C^{(s)}$ and the limiting values of the concentration or flux. For example, if "local thermodynamic equilibrium" is assumed, then a relation of the form

$$C^{(s)}(t) = \lambda C(0^+, t) \tag{6.16}$$

might hold, where the coefficient λ has dimensions of length. In that case, Eqs. (6.15) reduce to an integral equation for $C^{(s)}$, say. Indeed, the linear relation (6.16) can easily be generalized to nonlinear functional forms that describe adsorption isotherms such as Langmuir's.[25]

EXERCISE 6.1. *An integral equation:* Carry out the steps suggested above for the case of local equilibrium (6.16). Can you think of ways to solve this integral equation? Could you write it in terms of $C_0 \equiv C(0^+, t)$ or $J_0 \equiv J(0^+, t)$, instead? □

6.4. Grain Boundary Diffusion According to Whipple

The mathematical model proposed by Whipple[17] to analyze grain boundary diffusion, although published in 1954, has stood the test of time.[N4] It also ideally serves our purpose in the context of this chapter. Here, we allow a mild extension of his model.

Consider a single grain boundary that separates two different bulk phases. As shown in Fig. 6.3, suppose that it juts perpendicularly to a free surface into the bulk. A grain boundary is a region of atomic disorder, and therefore we expect the diffusivity $\tilde{D}$ of an impurity in that region to be much greater than the bulk diffusivities $D^{(1)}$ and $D^{(2)}$.

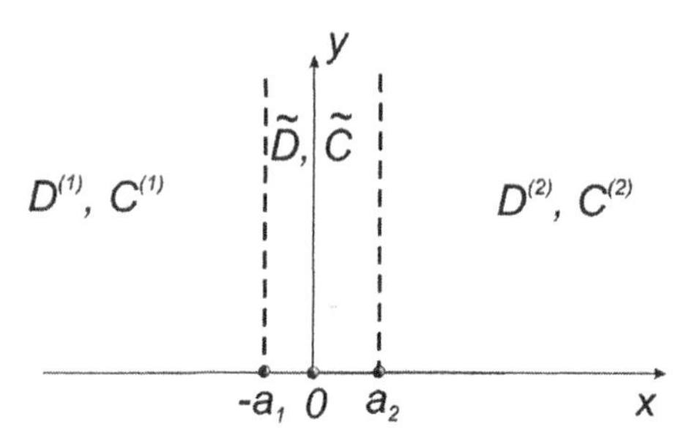

Fig. 6.3. Schematic of Whipple's model.

The width of this region is unknown, but Whipple suggests that it can be represented by two fiduciary planes, at $x = -a_1$ and $x = a_2$, respectively.† Further, he postulates that ordinary diffusive processes occur in the two bulk regions:

$$\frac{\partial C^{(1)}}{\partial t} = D^{(1)}\left(\frac{\partial^2 C^{(1)}}{\partial x^2} + \frac{\partial^2 C^{(1)}}{\partial y^2}\right) \qquad \text{for } -\infty < x < -a_1, \quad (6.17a)$$

$$\frac{\partial C^{(2)}}{\partial t} = D^{(2)}\left(\frac{\partial^2 C^{(2)}}{\partial x^2} + \frac{\partial^2 C^{(2)}}{\partial y^2}\right) \qquad \text{for } a_2 < x < \infty, \quad (6.17b)$$

and also in the disordered region

$$\frac{\partial \tilde{C}}{\partial t} = \tilde{D}\left(\frac{\partial^2 \tilde{C}}{\partial x^2} + \frac{\partial^2 \tilde{C}}{\partial y^2}\right) \qquad \text{for } -a_1 < x < a_2. \quad (6.18)$$

Because the planes that separate these phases are fictitious, he demands concentration and normal flux continuity at these stations:

$$C^{(1)} = \tilde{C} \quad \text{and} \quad D^{(1)}C_x^{(1)} = \tilde{D}\tilde{C}_x \qquad \text{at } x = -a_1, \quad (6.19a)$$

$$C^{(2)} = \tilde{C} \quad \text{and} \quad D^{(2)}C_x^{(2)} = \tilde{D}\tilde{C}_x \qquad \text{at } x = a_2. \quad (6.19b)$$

This diffusion problem can be solved if we prescribe initial and

†It is convenient to choose the origin of coordinates somewhere within the disordered region. The y-axis points into it, and the x-axis is chosen along the free surface. In all the next specifications of domains $y > 0$ and $t > 0$ are understood.

boundary conditions at the free surface. It is analogous to the problem of heat dissipation from a fin. The analytic solution is complicated, however, and we seek a simplified model that will also serve to illustrate Gibbs's construction.

To avoid unpleasant algebra, let us consider the case of a *symmetric* grain boundary between identical bulk phases. We then have $D^{(1)}=D^{(2)} \equiv D$, $a_1=a_2 \equiv a$, $C^{(1)}(-x, t)=C^{(2)}(x, t) \equiv C(x, t)$, and we only have to analyze processes over the right-hand side $\{x > 0\}$ of Fig. 6.3.

The reduction to a gibbsian surface model requires several steps. First, we assume an analytic form for the rapidly varying concentration $\tilde{C}$, at least in the direction perpendicular to the grain boundary. Because of x-symmetry, Whipple proposes that

$$\tilde{C}(x, y, t) = \alpha + \gamma x^2, \tag{6.20}$$

where the "coefficients" α and γ of this quadratic in x are yet unknown functions of (y, t). Next, we call $C^{(s)}$ the concentration averaged over the disordered region:

$$C^{(s)}(y, t) = 2\int_0^a \tilde{C}(x, y, t)\, dx, \tag{6.21}$$

where we have again taken advantage of the system's symmetry. From Eq. (6.20) it follows that

$$C^{(s)} = 2a\alpha + \frac{2a^3}{3}\gamma, \tag{6.22}$$

and we can apply that same averaging operation (6.21) to the diffusion equation (6.18). Thus

$$\frac{\partial C^{(s)}}{\partial t} = \tilde{D}\frac{\partial^2 C^{(s)}}{\partial y^2} + 2\tilde{D}\left.\frac{\partial \tilde{C}}{\partial x}\right|_{x=a}, \tag{6.23a}$$

and, with the flux continuity condition (6.19), we get the mass balance equation

$$\frac{\partial C^{(s)}}{\partial t} = \tilde{D}\frac{\partial^2 C^{(s)}}{\partial y^2} + 2D\left.\frac{\partial C}{\partial x}\right|_{x=a}, \tag{6.23b}$$

for the surface concentration $C^{(s)}$. It has exactly the form (6.14), in the limit $a \to 0$, if diffusion is also allowed in the plane of the immobile dividing surface.

It remains to calculate the unknown functions α and γ. This is easily done because conditions (6.19) at the boundary $x = a$ provide two equations for these two unknowns. We thus obtain

$$\alpha = C|_{x=a} - a^2\gamma \qquad \text{and} \qquad \gamma = \frac{D}{2a\tilde{D}} C_x|_{x=a}, \tag{6.24}$$

which we insert into expression (6.22) for the surface concentration. This last, rearranged, reads

$$D\frac{\partial C}{\partial x}\bigg|_{x=a} = \frac{3\tilde{D}}{a}(C|_{x=a} - C^{(s)}/2a). \tag{6.25}$$

If $\tilde{D} \gg D$, then the surface concentration $C^{(s)}$ remains roughly proportional to the boundary value $C|_{x=a}$. This is the law of proportional disturbances,[1,14] quoted earlier, and that Whipple used for his further calculations. We begin to understand that it is a law of "mass action," but in a dynamic sense because only D and $\tilde{D}$ are involved. However, if these diffusivities are of the same order of magnitude, then the two diffusion equations, (6.23b) and (6.17b) must be solved simultaneously, and Eq. (6.25) provides the extra coupling that we sought.[N5]

EXERCISE 6.2. *Anisotropic grain boundaries:* Assume that the two bulk phases are now distinct. Carry out the calculations described in Eqs. (6.20)–(6.25) by supposing that $C^{(s)}$ can be represented by a full quadratic, $\alpha + \beta x + \gamma x^2$, in each interval $[-a_1, 0]$ and $[0, a_2]$. Find a condition for the optimal position of this grain boundary, namely, find the best values of a_1 and a_2 under the condition that their sum (the grain boundary's width) is given. This fixes the position of the dividing surface. □

Averaging procedures over one of the system's space dimensions are very common. They originated in fluid dynamics where the momentum boundary-layer equations were averaged over the coordinate normal to the bounding surface, the so-called von Kármán–Pohlhausen procedure (see Ref. 12, Chap. 10). They are also widely used in chemical and mechanical engineering.[26]

6.5. Grain Boundary Diffusion According to Benoist and Martin

Concerned with the assumptions inherent in Whipple's model, Benoist and Martin developed a remarkable 2-d random walk model of grain boundary diffusion.[19] This model is particularly interesting for our purpose because it nicely generalizes the 1-d random walk that we put to good use in Chaps. 1 and 3.[N6]

We imagine a square lattice of integer points that overlays Fig. 6.3. Benoist and Martin allow four kinds of nearest-neighbor jumps: (i) Bulk jumps to four nearest-neighbors, with transition frequency Γ.† (ii) Jumps out of the plane at $x = 0$ representing the grain boundary, with frequency Γ_0. (iii) The reverse process, toward the plane, with frequency Γ_1.‡ (iv) Jumps in the plane of the grain boundary, with frequency $\tilde{\Gamma}$. They then solve the master equations (1.4) under appropriate initial and boundary conditions.

Because we are mainly interested in analyzing the boundary conditions that hold near phase boundaries, we consider a simplified version of this model, represented schematically in Fig. 6.4. We will not even have to solve the rate equations that follow. This three-parameter

†This frequency was called $\Gamma/2$ for the 1-d model in Sect. 1.1 or $\Gamma/4$ for a 2-d isotropic model in Appendix A.

‡These frequencies would have been called $\Gamma_{0,\pm 1}$ and $\Gamma_{\pm 1,0}$ in Sect. 1.1, or Γ^+ and Γ^- in Sect. 3.1.

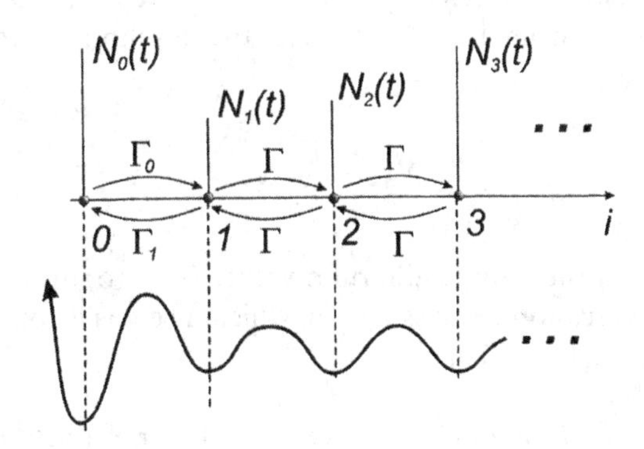

Fig. 6.4. Illustrating Benoist and Martin's model, its jump frequencies, and a schematic diagram of a potential distribution of particles near a phase boundary.

version is one-dimensional, and represents the dynamics of particle exchange between a plane at $i = 0$ and the nearest neighbors to the right $i \geqslant 1$. It is thus a good representation of segregation to a free surface or of diffusion around the grain boundary of a symmetric bicrystal in regions far from its free interface.

Let $N_i(t)$ be, once again, the number of particles that occupy site i at time t. Introducing particle fluxes between neighboring sites in the direction of increasing coordinate i leads to

$$J_{1/2} = \Gamma_0 N_0 - \Gamma_1 N_1, \tag{6.26a}$$

$$J_{i+1/2} = \Gamma(N_i - N_{i+1}) \qquad \text{for } i \geqslant 1, \tag{6.26b}$$

analogous to Eqs. (1.1), we easily obtain the "continuity equations"

$$\dot{N}_0 = -J_{1/2}, \tag{6.27a}$$

$$\dot{N}_i = J_{i-1/2} - J_{i+1/2} \qquad \text{for } i \geqslant 1. \tag{6.27b}$$

In the terminology of Remark (b) in Sect. 1.1, this model is inhomogeneous because the transition frequencies are not all equal. Indeed, the uniform and pleasing-looking Eqs. (6.27b) are misleading: The two fluxes on the right-hand side of the equation for $\dot{N}_1$ have different analytic forms. All is well for $i \geqslant 2$, however.

Before proceeding further, let us quickly examine the equilibrium state. By definition, all fluxes must then vanish. From the difference equations (6.26b) we obtain $N_1 = N_2 = \cdots$. We call N_e their common value. It follows from Eq. (6.26a) that the equilibrium value of the surface quantity N_0

$$N_{0,e} = \frac{\Gamma_1}{\Gamma_0} N_e \tag{6.28}$$

is different from the bulk equilibrium value. The proportional relation (6.28) is an expression of mass action, which, itself, is a consequence of detailed balance Eq. (6.26).

Exercise 6.3. *Langmuirian adsorption:* Modify Eq. (6.26a) in such a way that the "backward" component of the flux also depends on the number of *available* sites. Does one get Langmuir's adsorption isotherm at equilibrium? □

EXERCISE 6.4. *The steady state:* Examine the steady state, namely, when all $\dot{N}_i = 0$. Is it distinct from the equilibrium state? □

Otherwise, given an initial condition, $N_i(0) = N_i^0$, that is different from the equilibrium state, the system of equations (6.26) and (6.27) can be solved analytically.[19,20] Rather than discuss the *solution* of these equations, we examine the equations themselves to understand their behavior close to the surface in the continuum limit.

To motivate the next steps, perhaps, we define the system's total mass[N3]

$$M(t) = \sum_{i=0}^{\infty} N_i(t). \tag{6.29}$$

We take its time derivative and insert Eqs. (6.27). All fluxes cancel, two-by-two, and we are left with†

$$\dot{M} = 0. \tag{6.30}$$

This result makes sense (the total mass is a constant), but it seems to contradict its continuum analog (6.3). We note, however, that if by total mass we mean the sum (6.29), *excluding* the surface quantity N_0, then we do get the analogous result $\dot{M} = J_{1/2}$.

Our goal is the reduction of the discrete system (6.26) and (6.27) to a continuum form that is consistent with the nonequilibrium gibbsian construction detailed in Sect. 6.3. If we eliminate the fluxes, these equations read

$$\dot{N}_i = \Gamma(N_{i+1} + N_{i-1} - 2N_i) \qquad \text{for } i \geqslant 2, \tag{6.31a}$$

which is well known, but

$$\dot{N}_1 = \Gamma_0 N_0 - \Gamma_1 N_1 - \Gamma(N_1 - N_2), \tag{6.31b}$$

$$\dot{N}_0 = \Gamma_1 N_1 - \Gamma_0 N_0, \tag{6.31c}$$

for the first two sites. The rate equations (6.31a) have the standard form required to obtain the diffusion equation, but Eq. (6.31b) does not. Because we wish to extend that diffusion equation as far as

†Under the condition that the flux at infinity vanishes. This is a consequence of boundedness of solutions for large i.

possible toward the surface, we replace the functions $N_i(t)$ by fictitious, but equivalent, functions $N_i'(t)$ that obey the rate equations

$$\dot{N}_i' = \Gamma(N_{i+1}' + N_{i-1}' - 2N_i') \qquad \text{for } i \geqslant 1, \tag{6.32a}$$

$$\dot{N}_0' = \Gamma_1 N_1' - \Gamma_0 N_0'. \tag{6.32b}$$

These functions are "equivalent" in the sense that their sum must provide the *same* total mass M as would have been computed according to Eq. (6.29).[N7]

We are now in a position to effect the reduction to continuum form. As usual, we introduce an arbitrary interpolating function $\hat{N}(x, t)$ that collocates with all the discrete $N_i'(t)$ at lattice sites $x_i = ia$, and define the (fictitious) concentration distribution $C(x, t) = \hat{N}(x, t)/a$. The total mass is

$$M(t) = N_0' + \sum_{i=1}^{\infty} N_i', \tag{6.33a}$$

$$= N_0' + a \sum_{i=1}^{\infty} C(x_i, t), \tag{6.33b}$$

and that last sum is precisely the integral on the right-hand side of Eq. (6.2), if it is evaluated according to the midpoint rule (6.7). In other words, the "residual" $N_0'(t)$ is what we must attribute to the surface, up to a truncation error $O(a^3)$. Calling it $C^{(s)}(t)$, we have Gibbs's representation

$$M(t) \approx C^{(s)}(t) + \int_0^{\infty} C(x, t)\, dx, \tag{6.34}$$

of the real total mass as the sum of two fictitious terms: The first is attributed to the dividing surface at $x = 0$; the second is an integral over all interior points $x > 0$. It has exactly the form (6.12) if the mass M is interpreted "per unit transverse area."

This construction might appear arbitrary, but it is not because we must also satisfy Eqs. (6.32). As we know, Eqs. (6.32a) admit the continuum limit (6.1) and (6.4), at least for all interior points $x > 0$. Therefore, differentiating Eq. (6.34), and remembering Eq. (6.30), we get

$$\dot{C}^{(s)} = -J(0^+, t) = D\left.\frac{\partial C}{\partial x}\right|_{x=0^+}. \tag{6.35}$$

This is exactly the mass balance equation (6.14) for "one-sided" phases when the phase boundary is immobile.

The rest is formally easy, even as it is a logical minefield. We differentiate Eq. (6.33a), remembering Eq. (6.30), and sum all Eqs. (6.32a), to get

$$\dot{N}'_0 = -\sum_{i=1}^{\infty} \dot{N}'_i = -\Gamma(N'_0 - N'_1). \tag{6.36a}$$

This appears to contradict the rate equation (6.32b) since the three jump frequencies are distinct. But we must remember what the right-hand side of Eq (6.36a) really represents, to wit, the first discrete flux $-J'_{1/2}$. From Exercise 1.2, it has an optimal continuum representation $\Gamma a \partial \hat{N}/\partial x$ at the midpoint $x = \frac{1}{2}a$. Introducing concentrations and the bulk diffusivity $D = \Gamma a^2$, we obtain

$$\dot{C}^{(s)} = D \left.\frac{\partial C}{\partial x}\right|_{x=a/2}, \tag{6.36b}$$

in accordance with Eq. (6.35).

On the other hand, the rate equation (6.32b) represents surface-to-bulk interactions. Evaluating N'_1 at the midpoint $x = \frac{1}{2}a$, we can write

$$\dot{N}'_0 = \Gamma_1 \left[\hat{N} + \tfrac{1}{2}a \frac{\partial \hat{N}}{\partial x} + \mathrm{O}(a^2) \right]_{x=a/2} - \Gamma_0 N'_0. \tag{6.37a}$$

If we convert to concentrations, this reads

$$\dot{C}^{(s)} = \Gamma_1 \left[aC + \tfrac{1}{2}a^2 \frac{\partial C}{\partial x} + \mathrm{O}(a^3) \right]_{x=a/2} - \Gamma_0 C^{(s)}. \tag{6.37b}$$

Therefore, comparing the two expressions (6.36b) and (6.37b), we have

$$D \left.\frac{\partial C}{\partial x}\right|_{x=a/2} = \Gamma_1 aC|_{x=a/2} - \Gamma_0 C^{(s)} + \mathrm{O}(a^2). \tag{6.38}$$

In the limit as $a \to 0$, we get the relation

$$D \frac{\partial C}{\partial x}\bigg|_{x=0+} = kC|_{x=0+} - C^{(s)}/\tau \tag{6.39}$$

we have been seeking, where it is convenient to introduce the macroscopic rate constants $k = \Gamma_1 a$ and $\tau = 1/\Gamma_0$.

EXERCISE 6.5. *Rate constants:* What are the dimensions of the macroscopic rate constants, D, k, and τ? From their definition, D and k are obtained in the limit $a \to 0$, together with Γ and $\Gamma_1 \to \infty$. Justify why one may neglect the second term on the right-hand side of Eq. (6.37b). After all, it is second order in a, just like the left-hand side of that equation. □

Equation (6.39) has exactly the same form (6.25) that we found for Whipple's problem. As mentioned above, it can be justified, more generally, in the context of irreversible thermodynamics of surfaces. It is not really a boundary condition but rather a *relation* that couples three unknown functions: the gibbsian surface concentration $C^{(s)}(t)$ to the limiting values of the bulk concentration $C(0^+, t)$ and the normal component of the bulk flux $J(0^+, t) = -D\partial C/\partial x|_{x=0^+}$.

The manner in which we have just derived relation (6.39) also solves the problem we had set in Sect. 6.1: What *is* the "surface concentration"? Indeed, recalling this derivation, Eq. (6.39) is a consequence of Eq. (6.38) in the limit of a vanishingly small jump distance a. Clearly, by bulk concentration at the surface $C_0(t) \equiv C(0^+, t)$, we must mean the number of fictitious particles attributed to a very small "cell" $0 < x < a$ that is centered at $x = \frac{1}{2}a$. It is a fictitious quantity, just as is $C^{(s)}$, because both are constructed by means of an interpolation of the equivalent set of discrete functions N_i'. Likewise, Eq. (6.36a) shows that the limiting value of the flux $J(0^+, t)$ is obtained by optimal interpolation of $\Gamma(N_0' - N_1')$. *It is in this sense that we are to understand the symbols $C_0(t)$ and $J_0(t)$.* No matter how fictitious these functions, the masses expressed through sums of the form (6.33) are real, experimentally accessible quantities.

The physical interpretation of relation (6.39) is obvious. In the language of chemical kinetics, the net (normal) flux exchanged at a phase boundary (more precisely, at a Gibbs dividing surface) is the detailed balance of an adsorption (or "attachment") rate, characterized by a rate constant k, and a desorption (or "detachment") rate that is characterized by a lifetime τ. Relation (6.39) has been used to model certain extensions of the BCF theory of crystal growth.[27]

EXERCISE 6.6. *Two-sided 1-d random-walk:* Consider a phase boundary around $i = 0$ between two semi-infinite lattices, $i \leqslant -1$ and $i \geqslant 1$, characterized by different bulk transition frequencies $\Gamma^{(1)}$ and $\Gamma^{(2)}$, respectively. Write the rate equations at the phase boundary. (There are now four independent transition frequencies similar to Γ_0 and Γ_1.) Discuss this problem and find its continuum form. In particular, show that a relation of the form (6.39) must hold on *each* side of the boundary. Discuss the steady and equilibrium states. □

6.6. The Radiation Boundary Condition and a Typical Diffusion Problem

Relation (6.39) reminds us of a type of boundary condition that is often introduced without much justification to model nonequilibrium heat transfer at boundaries:

$$\pm K \frac{\partial T}{\partial x} + hT = f(t) \tag{6.40}$$

at a boundary, where h is a "heat transfer coefficient" and f is a known function. In other words, a linear combination of temperature and its gradient is imposed at the boundary. Equation (6.40) stems from Newton's "law of cooling" (see Ref. 24, pp. 18–21). For that reason it is sometimes called a *radiation boundary condition*.[N8] We shall now see that the radiation boundary condition is a consequence of the coupling relation (6.39).

For a semi-infinite body we have the mass conservation laws (6.1) and (6.35)

$$\frac{\partial C}{\partial t} = -\frac{\partial J}{\partial x} \qquad \text{for } x > 0,\ t > 0, \tag{6.41a}$$

$$\dot{C}^{(s)} = -J|_{x=0^+} \qquad \text{for } t > 0, \tag{6.41b}$$

and the constitutive relations (6.4) and (6.39)

$$J = -D\frac{\partial C}{\partial x} \qquad \text{for } x > 0,\ t > 0, \tag{6.42a}$$

$$-J|_{x=0^+} = kC|_{x=0^+} - C^{(s)}/\tau \qquad \text{for } t > 0, \tag{6.42b}$$

transcribed once again for convenience. These form a complete set of differential equations for the unknown functions $C(x, t)$, $J(x, t)$, and $C^{(s)}(t)$.[N9] Their solution, by Laplace transformation, for example, follows if we are given initial and far-field conditions that characterize some diffusion problem. These equations depend on three physical parameters: D, k, and τ. For the following theoretical discussion, it is advisable to retain both flux and concentration, rather than eliminate the former.

Let us now examine Eqs. (6.41) and (6.42) under equilibrium conditions. All fluxes must then vanish. If $J = 0$, then it follows from Eq. (6.42a) that C must be, at most, an arbitrary function of time. But that function can only be a constant, call it C_e by virtue of Eq. (6.41a). Likewise, because of Eq. (6.41b), $C^{(s)}$ must be a constant; call it $C_e^{(s)}$. Finally, these two constants must be related by Eq. (6.42b):

$$C_e^{(s)} = \tau k C_e, \tag{6.43}$$

similar to Eq. (6.28), which has the form of the law of "mass action" so familiar from chemical kinetics. This takes us to the simplification required to warrant the radiation boundary condition.

We assume that the surface concentration $C^{(s)}$ is always close to its equilibrium value (6.43). In that case, we throw away Eq. (6.41b), or else J_0 would always vanish, and we use the mass action condition (6.43) to rewrite Eq. (6.42b). Collecting our results, we have

$$\frac{\partial C}{\partial t} = D\frac{\partial^2 C}{\partial x^2} \qquad \text{for } x > 0,\ t > 0, \tag{6.44a}$$

$$D\left.\frac{\partial C}{\partial x}\right|_{x=0^+} = k(C|_{x=0^+} - C_e) \qquad \text{for } t > 0. \tag{6.44b}$$

Contrary to the physical model underlying Eqs. (6.41) and (6.42) this is only a two-parameter model: D and k. From the boundary condition

(6.44b) we see that the bulk concentration at the surface C_0 remains close to its equilibrium value when k is "large." The meaning of "local thermodynamic equilibrium" at a surface is now clear.

How large is "large"? From the form of Eqs. (6.44) it is easy to see that their solutions must depend on two dimensionless groups: $x/2\sqrt{(Dt)}$ and $k\sqrt{(t/D)}$. In other words, any diffusion problem based on these differential equations will tend toward a local equilibrium condition when $t \gg D/k^2$, and any surface kinetics, if described by k, can only be observed for "short" times on the scale of D/k^2.

These qualitative statements can be substantiated by the following typical diffusion problem: We solve Eqs. (6.44) in a half-space under constant initial and boundary conditions

$$C(x, 0) = C(\infty, t) = C_\infty. \tag{6.45}$$

The problem is very simple if we note that the function

$$u(x, t) = C(x, t) - \frac{D}{k}\frac{\partial C}{\partial x} \tag{6.46a}$$

satisfies all the conditions of the "impurity diffusion problem" Eq. (4.8). Consequently, we have

$$u(x, t) = C_\infty - (C_\infty - C_e)\,\mathrm{erfc}\left(x/2\sqrt{Dt}\right). \tag{6.46b}$$

Equation (6.46a), considered as an ordinary differential equation† whose inhomogeneous part is given precisely by the function (6.46b), is easily solved by the method of "variation of constants." This integration is not difficult because the differential equation is linear, with constant coefficients, and we find that

$$C(x, t) = C_\infty - (C_\infty - C_e)\left[\mathrm{erfc}\frac{x}{2\sqrt{Dt}} - e^{(kx + k^2 t)/D}\,\mathrm{erfc}\left(\frac{x}{2\sqrt{Dt}} + k\sqrt{\frac{t}{D}}\right)\right]. \tag{6.47a}$$

†We must be careful, however, just as in the "heating problem" Eq. (4.21), to treat the "arbitrary constant" as an arbitrary function of time.

Hence, for example, we have the concentration at the surface

$$C_0(t) = C_\infty - (C_\infty - C_e)[1 - e^{k^2t/D}\,\mathrm{erfc}(\sqrt{k^2t/D})], \quad (6.47b)$$

which, indeed, only depends on the argument k^2t/D. We can now tell exactly how close or how far we are from equilibrium at the surface, which is the object of the next two exercises.

EXERCISE 6.7. *The approach to equilibrium:* Using what you know about the error function's behavior for small and large arguments, discuss carefully the behavior of Eq. (6.47b) in those limits. Under what conditions does the solution (6.47a) approach Eq. (6.46b). □

EXERCISE 6.8. *Measurable quantities:* Concentration profiles are not always measurable, but quantities proportional to the total mass

$$M(t) = \int_0^\infty (C(x, t) - C_\infty)\,dx$$

often are. Calculate M (there is an easy way) for the diffusion problem (6.47), and discuss the limiting cases of short and long times. □

To summarize briefly the main points of this chapter, we showed that mass conservation (both global and local) allows an extension of Gibbs's construction to nonequilibrium situations. Then, on the basis of two segregation models, we showed that there must exist linear constitutive relations, on each side of the surface, that couple the surface concentration to the limiting values of bulk concentration and flux.† These relations can also be justified on the basis of nonequilibrium thermodynamics of surfaces. In fact, the BCF model of crystal growth (Sect. 2.6), the theory of double layers (Sect. 3.3), and thermally stimulated desorption (Sect. 8.3) are all illustrations of the gibbsian surface model. Therefore, the symbol C_s, used there for surface concentrations, should have been more properly called $C^{(s)}$—the surface excess mass density.

†This, of course, in addition to the usual Fick's first law that holds at all interior points of the bulk phases.

It is somewhat surprising that acceptance of these ideas is resisted. After all, no one doubts the "reality" of a density of surface charges in electrodynamics and in device structures. Further, no one would doubt that the stress tensor properly accounts for deformations, both at interior and at boundary points of a body. Perhaps we forget that we never measure mass and charge densities or stresses, but rather their integrals: total mass and total charge within a sampling volume or total force on that volume. We introduce these fictitious surface quantities because of mathematical manipulations (mainly Green's theorem) *that yield the right result*, "right" in the sense of equivalent circuits.[N7]

Notes

1. All the considerations of this chapter can be extended to other geometries and dimensionalities. Except in Sects. 6.2 and 6.3, M is the mass *per unit area* transverse to the direction of diffusion.
2. The adjective "bulk" is often used as a synonym for "3-d."
3. The reason for the approximate equality $M \approx \Sigma_i N_i$ in Eq. (6.7) stems from our point of view in Sects. 6.1–6.4, namely, that "reality" is represented by the continuum equations (6.1)–(6.4). That equality is exact in Eq. (6.29) because our "reality" is discrete in Sect. 6.5. In this context, D. Ruelle's incisive comments regarding "pieces of reality" are worth reading in his fascinating book *Chance and Chaos* (Princeton University Press, Princeton, 1991).
4. Through the grape-vine, it appears that Whipple (an applied mathematician at Harwell) was approached by A. D. Le Claire (an eminent authority on diffusion, also at Harwell) regarding this problem. Whipple produced his analysis in 24 hours, his paper a few days later, and has never bothered with grain boundaries since then.
5. The author performed a similar calculation in the context of heat flow in composite media. See Appendix A of "Laser Heating and Melting of Thin Films on Low-Conductivity Substrates" by R. A. Ghez and R. A. Laff, *J. Appl. Phys.* **46**, 2103 (1975).
6. This is certainly not the first application of a 2-d random walk model. Its steady state was analyzed in the context of the BCF theory of crystal growth by R. Gevers, "The Advance of Mono-Molecular Steps on the Surface of a Growing Kossel Crystal as a Random Walk Problem," *Physica* **12**, 832 1956), and by D. E. Temkin, "A Kossel Model of the Kinetics of Growth and Evaporation Normal to a $\{100\}$ Face," *Kristallografiya* **14**, 417 (1969). [Engl. transl. in *Sov. Phys. Crystallogr.* **14**, 344 (1969).]

7. It should be remembered that the total mass in some volume is the only observable. The adjective "equivalent" is used in the same sense as for electrical circuits.
8. It is called a "boundary condition of the *third* kind" in the Russian literature, quite reasonably reserving the labels *first* and *second* kinds to Dirichlet and Neumann conditions, respectively.
9. These equations are easily generalized to the "two-sided" 3-d case when the dividing surface is a plane described by two coordinates, y and z, say, that is orthogonal to the x-axis into the bulk phases. Eliminating the bulk fluxes with Fick's first law, we have:

$$\frac{\partial C^{(1)}}{\partial t} = D^{(1)}\left(\frac{\partial^2 C^{(1)}}{\partial x^2} + \frac{\partial^2 C^{(1)}}{\partial y^2} + \frac{\partial^2 C^{(1)}}{\partial z^2}\right) \quad \text{in bulk phase (1) } x < 0, \tag{1}$$

$$\frac{\partial C^{(2)}}{\partial t} = D^{(2)}\left(\frac{\partial^2 C^{(2)}}{\partial x^2} + \frac{\partial^2 C^{(2)}}{\partial y^2} + \frac{\partial^2 C^{(2)}}{\partial z^2}\right) \quad \text{in bulk phase (2) } x > 0, \tag{2}$$

$$\frac{\partial C^{(s)}}{\partial t} = D^{(s)}\left(\frac{\partial^2 C^{(s)}}{\partial y^2} + \frac{\partial^2 C^{(s)}}{\partial z^2}\right) + D^{(2)}\left.\frac{\partial C^{(2)}}{\partial x}\right|_{x=0^+} - D^{(1)}\left.\frac{\partial C^{(1)}}{\partial x}\right|_{x=0^-} \tag{3}$$

at the dividing surface $x = 0$. These conservation laws must be supplemented by coupling relations of the form (6.39) that describe surface–volume interactions:

$$-D^{(1)}\left.\frac{\partial C^{(1)}}{\partial x}\right|_{x=0^-} = k^{(1)}C^{(1)}|_{x=0^-} - C^{(s)}/\tau^{(1)}, \tag{4}$$

$$+D^{(2)}\left.\frac{\partial C^{(2)}}{\partial x}\right|_{x=0^+} = k^{(2)}C^{(2)}|_{x=0^+} - C^{(s)}/\tau^{(2)}. \tag{5}$$

As mentioned in Sect. 6.3, the generalization to curved dividing surfaces is not trivial, especially if these move.

References

1. D. McLean, *Grain Boundaries in Metals* (Oxford University Press, London, 1957).
2. *Interfacial Segregation*, W. C. Johnson and J. M. Blakely, Eds. (Amer. Soc. for Metals, Ohio 1979).
3. J. Cabané and F. Cabané, "Equilibrium Segregation in Interfaces," in *Interface Segregation and Related Processes in Materials*, J. Novotny, Ed., pp. 1–160 (Trans. Tech. Publ., Brookfield, 1991).

4. H. S. Carslaw, *Introduction to the Theory of Fourier's Series and Integrals*, 3rd ed. Macmillan, New York, 1930. (Reprinted by Dover, New York).
5. E. Isaacson and H. B. Keller, *Analysis of Numerical Methods* (John Wiley, New York, 1966).
6. R. H. Eddy and R. Fritsch, "An Optimization Oddity," *College Mathematics J.* **25**, 227 (1994).
7. R. Paré, "A Visual Proof of Eddy and Fritsch's Minimal Area Property," *College Mathematics J.* **26**, 43 (1995).
8. J. W. Gibbs, "On the Equilibrium of Heterogeneous Substances," *Trans. Conn. Acad.* (1875–1878). [Reprinted in *Scientific Papers*, Vol. 1 (Dover, New York, 1961).]
9. J. D. van der Waals, "The Thermodynamic Theory of Capillarity under the Hypothesis of a Continuous Variation of Density," originally published in Dutch (1893); translated by J. S. Rowlinson in *J. Statist. Phys.* **20**, 197 (1979).
10. J. W. Cahn and J. E. Hilliard, "Free Energy of a Nonuniform System," Parts I–III, *J. Chem. Phys.* **28**, 258 (1958); *ibid.* **30**, 1121 (1959); *ibid.* **31**, 688 (1959).
11. L. Prandtl, "Über Flüssigkeitsbewegung bei sehr kleiner Reibung," Proceedings 3rd Intern. Math. Congr., Heidelberg (1904), pp. 484–491. [See Collected Works, Vol. 2, pp. 575–584.]
12. H. Schlichting, "Boundary-Layer Theory," 6th ed. (McGraw-Hill, New York, 1968).
13. M. Van Dyke, *Perturbation Methods in Fluid Mechanics*, 2nd annotated edition (The Parabolic Press, Stanford, CA, 1975).
14. H. J. Queisser, K. Hubner, and W. Shockley, "Diffusion along Small-Angle Grain Boundaries in Silicon," *Phys. Rev.* **123**, 1245 (1961).
15. C. Lea and M. P. Seah, "Kinetics of Surface Segregation," *Phil. Mag.* **A35**, 213 (1977).
16. W. R. Tyson, "Kinetics of Temper Embrittlement," *Acta Met.* **26**, 1471 (1978).
17. R. T. P. Whipple, "Concentration Contours in Grain Boundary Diffusion," *Phil. Mag.* **45**, 1225 (1954).
18. A. D. Brailsford, "Surface Segregation Kinetics in Binary Alloys," *Surf. Sci.* **94**, 387 (1980).
19. P. Benoist and G. Martin, "Atomic Model for Grain Boundary and Surface Diffusion," *Thin Solid Films* **25**, 181 (1975); *errata, ibid.* **27**, L8 (1975).
20. G. Rowlands and D. P. Woodruff, "The Kinetics of Surface and Grain Boundary Segregation in Binary and Ternary Systems," *Phil. Mag.* **A40**, 459 (1979).
21. R. Ghez, "Irreversible Thermodynamics of a Stationary Interface," *Surf. Sci.* **20**, 326 (1970).

22. D. A. Edwards, H. Brenner, and D. T. Wasan, *Interface Transport Processes and Rheology*, (Butterworth-Heinemann, Boston, 1991).
23. S. R. de Groot and P. Mazur, *Non-Equilibrium Thermodynamics* (North-Holland, Amsterdam, 1962). [Reprinted by Dover, New York.]
24. H. S. Carslaw and J. C. Jaeger, *Conduction of Heat in Solids*, 2nd ed. (Oxford University Press, London, 1959).
25. W. H. Reinmuth, "Diffusion to a Plane with Langmuirian Adsorption," *J. Phys. Chem.* **65**, 473 (1961).
26. R. B. Bird, W. E. Stewart, and E. N. Lightfoot, *Transport Phenomena* (John Wiley, New York, 1960).
27. G. H. Gilmer, R. Ghez, and N. Cabrera, "An Analysis of Combined Surface and Volume Diffusion Processes in Crystal Growth," Parts I and II, *J. Cryst. Growth* **8**, 79 (1971); *ibid.* **21**, 93 (1974).

7

A User's Guide to the Laplace Transform

This chapter provides a compact, self-contained introduction to the Laplace transform, a powerful technique that can be applied with ease to a wide variety of time-dependent problems. Rather than strive for the greatest generality, this introduction addresses only those simplest properties that are most useful for the solution of diffusion problems. Only elementary methods of calculus are used, and integration in the complex plane is avoided altogether. It follows that questions regarding analyticity, the inversion integral, and asymptotic behavior cannot be completely answered, although some such properties will be demonstrated.[N1]

7.1. Motivation, Definition, and Some Examples

Rational operations, supplemented by raising to powers, do not exhaust all possible operations on functions. The basic operations of differentiation and integration are cornerstones of analysis. They are sometimes called transcendental operations because they are defined through limiting processes. It would be convenient to translate these operators into a form that resembles multiplication and division. Then, for example, repeated differentiation might correspond to repeated multiplication in the new language. The Laplace transform,[1–3] one of many useful *integral transforms*,[2,4] has precisely this property. Its systematic use, in the context of the diffusion equation, allows us to *reduce by one* the number of independent variables and thus to reduce a given problem to a far simpler one. We begin with the definition that follows.

DEFINITION: *Let $f(t)$ be a function (real or complex) of the real variable t. Its Laplace transform, denoted $\mathscr{L}\{f(t)\}$ or $\bar{f}(p)$, is defined by the integral*

$$\mathscr{L}\{f(t)\} = \bar{f}(p) = \int_0^\infty e^{-pt} f(t)\,dt. \tag{7.1}$$

The integral (7.1) clearly transforms functions of t (time, say) into new functions of a new variable p, the Laplace variable. That is to say, for each value of p for which the above integral converges, we get the corresponding value of the transformed function.

For example, the Laplace transform of an arbitrary exponential $f(t) = e^{\alpha t}$ is

$$\mathscr{L}\{e^{\alpha t}\} = \int_0^\infty e^{-(p-\alpha)t}\,dt = \frac{1}{p-\alpha}, \tag{7.2}$$

and the exponential function has been transformed into the new function $\bar{f}(p) = (p - \alpha)^{-1}$. This example, elementary though it may seem, can teach us quite a lot about the Laplace transform. First, we note that the transform of a transcendental function of t (the exponential) is a rational function (thus much simpler) in the Laplace variable p. Second, we carelessly performed the integration (7.2) as if this were always possible. But that integral will not converge if $p \leqslant \alpha$ (for real p and α), and the Laplace transform then cannot exist. In other words, the definition (7.1) requires a condition on the values of p (real or complex) that ensures convergence of the integral. For this first example, the condition on the real parts is $\Re(p) > \Re(\alpha)$. Third, we have just learned that the image function $\bar{f}(p)$ is generally a complex function of the complex variable p and that the defining integral (7.1) converges in some right-hand plane of the complex p-plane. Fourth, the original function $f(t)$ can also be complex. For example, taking $\alpha = i\omega$ in the exponential and separating the real and imaginary parts, we get

$$\int_0^\infty e^{-(p-i\omega)t}\,dt = \int_0^\infty e^{-pt}(\cos\omega t + i\sin\omega t)\,dt$$

$$= \frac{1}{p - i\omega} = \frac{p}{p^2+\omega^2} + i\frac{\omega}{p^2+\omega^2}, \tag{7.3}$$

from which we obtain the transforms of the trigonometric functions, which are defined for $\Re(p) > 0$. These transforms are listed in Table 7.1. Another special case of the exponential emerges when $\alpha = 0$. The function $f(t)$ is a constant everywhere equal to unity, and, from Eq. (7.2), we have $\mathscr{L}\{1\} = 1/p$.

Table 7.1. Short Table of Laplace Transforms[a]

Original function $f(t)$	Image function $\bar{f}(p)$	Conditions
1. $\cos\omega t$	$p/(p^2+\omega^2)$	
2. $\sin\omega t$	$\omega/(p^2+\omega^2)$	
3. $e^{\alpha t}$	$1/(p-\alpha)$	$\Re(p) > \Re(\alpha)$
4. 1	$1/p$	
5. $\delta(t)$	1	
6. t^n	$n!/p^{n+1}$	$n \geqslant 0$
7. $e^{\alpha t}t^\nu$	$\Gamma(\nu+1)/(p-\alpha)^{\nu+1}$	$\Re(p) > \Re(\alpha)$, $\nu > -1$
8. $\dfrac{a}{2\sqrt{\pi}\,t^{3/2}}e^{-a^2/4t}$	$e^{-a\sqrt{p}}$	$a > 0$
9. $\dfrac{1}{\sqrt{\pi t}}e^{-a^2/4t}$	$p^{-1/2}e^{-a\sqrt{p}}$	$a \geqslant 0$
10. $\operatorname{erfc}(a/2\sqrt{t})$	$p^{-1}e^{-a\sqrt{p}}$	$a \geqslant 0$
11. $(2\sqrt{t})^n\,\mathrm{i}^n\operatorname{erfc}(a/2\sqrt{t})$	$p^{-(1+n/2)}e^{-a\sqrt{p}}$	$a \geqslant 0$, $n \geqslant -1$
12. $e^{ha+h^2t}\operatorname{erfc}\left(\dfrac{a}{2\sqrt{t}}+h\sqrt{t}\right)$	$\dfrac{e^{-a\sqrt{p}}}{\sqrt{p}(\sqrt{p}+h)}$	$a \geqslant 0$, any h

[a] A very complete compilation is given in Ref. 15. In addition, almost every book on Laplace transforms contains an adequate table, e.g., Refs. 1–3. It is understood that all the original functions vanish identically for negative t, in the sense of Eq. (7.5). The third column indicates the existence conditions of transforms. Unless otherwise stated, $\Re(p) > 0$ is the condition for transforms that are rational in p. Only the negative real axis is excluded for transforms in $\sqrt{p}$: the phase of p can be anything at all in $(-\pi, \pi)$.

Table 7.1 is an illustration of the *mapping* provided by Eq. (7.1). To each function $f(t)$ for which that integral makes sense (i.e., converges) there corresponds a single image $\bar{f}(p)$. But is that correspondence one-to-one? Evidently not, because values of $f(t)$ for *negative* t's are irrelevant to the definition (7.1). In other words, two functions that differ only on the negative t-axis have the same image. For example, our exponential can be anything at all to the left of the origin. Because the Laplace transform was designed to deal with processes that develop from some initial condition, one simply restricts (or defines) the class of original functions $f(t)$ to those that are identically zero for $t < 0$.[N2]

Heaviside's function

$$H(t) = \begin{cases} 0 & \text{if } t < 0 \\ 1 & \text{if } t > 0 \end{cases} \tag{7.4a}$$

whose Laplace transform

$$\mathscr{L}\{H(t)\} = 1/p \tag{7.4b}$$

is the same as the unit function's, evidently falls into that class. It can be used to "truncate" any function that does not vanish for negative t's. Thus,

$$\bar{f}(p) = \mathscr{L}\{f(t)H(t)\} \tag{7.5}$$

defines this one-to-one correspondence, regardless of f's behavior to the left of the origin.[N3] Figure 7.1 shows the mapping $\mathscr{L}$ prescribed by Eq. (7.1), as well as its inverse $\mathscr{L}^{-1}$, whenever these can be reckoned.

7.2. Elementary Properties and Further Examples

The Laplace transform has many interesting and useful properties that we will exploit. Among these are linearity, change of scale, translations, the effect of differentiation and integration, convolution, and asymptotic behavior. These properties serve to derive new transform pairs from known ones. We shall survey only the most important ones for our purpose.

A. Linearity

This property is so evident from the definition (7.1) that we even failed to point it out in the derivation of Eqs. (7.3). By linearity of the

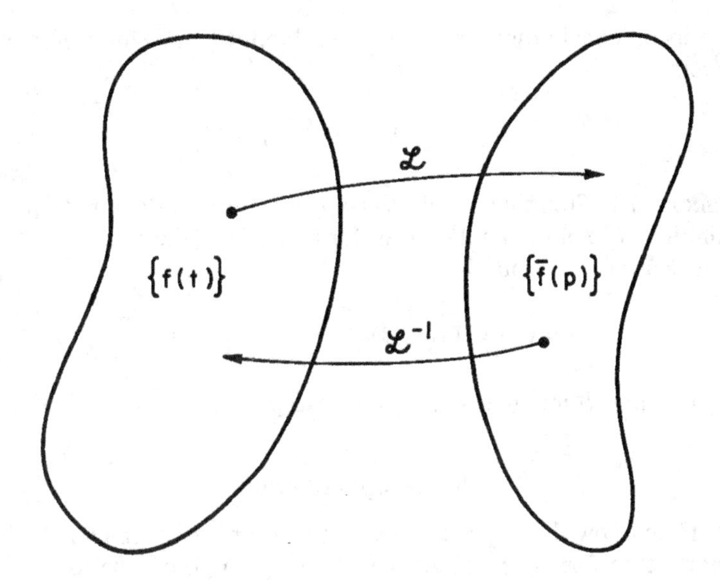

Fig. 7.1. Schematic representation of the Laplace transform and its inverse.

integral operator we have

$$\mathscr{L}\{\alpha f(t) + \beta g(t)\} = \alpha\mathscr{L}\{f(t)\} + \beta\mathscr{L}\{g(t)\} \tag{7.6}$$

for any scalars[†] α and β and any functions $f(t)$ and $g(t)$ for which the indicated operations make sense. The inverse operator $\mathscr{L}^{-1}$ is also linear. For example, expanding our exponential $e^{\alpha t}$ in a Taylor series around the origin, integrating term by term (without concern for rigor), and expanding the right-hand side of Eq. (7.2) in a geometric series, we get

$$\begin{aligned}\mathscr{L}\{e^{\alpha t}\} &= \int_0^\infty dt\, e^{-pt} \sum_{n=0}^{\infty} \frac{(\alpha t)^n}{n!} = \sum_{n=0}^{\infty} \frac{\alpha^n}{n!} \mathscr{L}\{t^n\} \\ &= \frac{1}{p-\alpha} = p^{-1} \sum_{n=0}^{\infty} (\alpha/p)^n.\end{aligned}$$

[†]In these sections, real variables or parameters (most often positive) are denoted by Latin letters, reserving Greek letters for complex quantities. The only notable exception is the Laplace variable p which can be complex.

Identifying equal powers of α, we then obtain the transform of integer powers of t:

$$\mathscr{L}\{t^n\} = n!/p^{n+1}, \qquad n \geqslant 0. \tag{7.7a}$$

EXERCISE 7.1. *Transform of arbitrary powers:* Show, directly from the definition (7.1) of the Laplace transform and from the definition (4.22) of the gamma function, that

$$\mathscr{L}\{t^\nu\} = \Gamma(\nu + 1)/p^{\nu+1}, \qquad \nu \geqslant -1. \tag{7.7b}$$

In particular, compute explicitly the cases $\nu = \pm\frac{1}{2}$. □

B. Change of Scale

If we know the Laplace transform of a function $f(t)$, can we then infer the transform of $f(at)$? As the following simple calculation shows, the answer is affirmative if a is positive. By definition

$$\mathscr{L}\{f(at)\} = \int_0^\infty e^{-pt} f(at)\, dt = a^{-1} \int_0^\infty e^{-(p/a)s} f(s)\, ds = \frac{1}{a}\, \bar{f}(p/a), \tag{7.8}$$

where the second equality results from the substitution $s = at$. The restriction on a follows because the dummy variables t and s must both run over the same half-axis $[0, \infty]$.

The next properties are equally easy to verify. Thus, they are noted without proof and left as Exercises 7.2 and 7.3 below.

C. Translations

The observant reader will have noticed that the transform of the exponential $e^{\alpha t}$ is obtained from that of Heaviside's function by a mere translation $p \to p - \alpha$ of the Laplace variable. This feature is quite general: If $\bar{f}(p)$ is the transform of $f(t)$, then

$$\mathscr{L}\{e^{\alpha t} f(t)\} = f(p - \alpha). \tag{7.9}$$

In other words, multiplication by exponentials in the original space of functions corresponds to translations of the independent variable p in

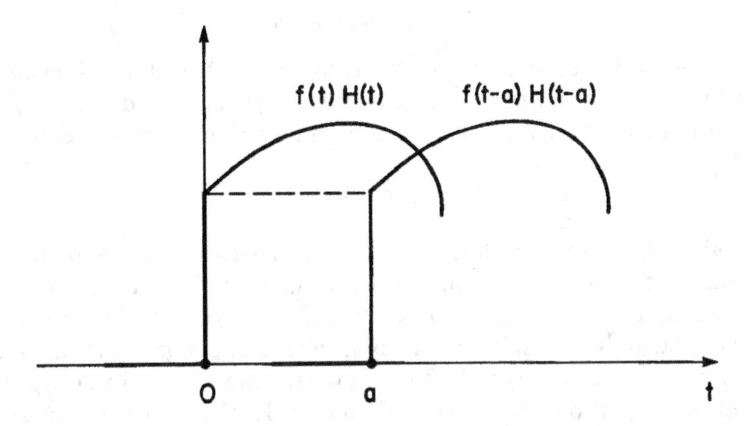

Fig. 7.2. An arbitrary function, and the same function with retarded argument.

image space. For example, from Eq. (7.7a) we immediately infer that $\mathscr{L}\{e^{\alpha t}t^n\} = n!/(p-\alpha)^{n+1}$.

There exists a second "shifting" theorem: Translations of the original variable t induce multiplication by exponentials in image space. Let $g(t) = f(t-a)$ be the "retarded" function $f(t)$, shifted a units to the right, with the understanding that $g \equiv 0$ for $t < a$. This situation is illustrated in Fig. 7.2. We then have

$$\mathscr{L}\{f(t-a)H(t-a)\} = e^{-ap}\bar{f}(p), \qquad a \geqslant 0. \tag{7.10}$$

For example, the Laplace transform of $H(t-a)$, Heaviside's function with a unit step at $t = a$, is simply $p^{-1}e^{-ap}$. Equations (7.9) and (7.10) illustrate a general feature of Laplace transforms, namely, that theorems often occur in "dual" pairs.

EXERCISE 7.2. *Simple proofs*: Prove the Properties (7.9)–(7.10). What can you say about these statements if α or a is negative? Can α be complex, and if so, what is the condition for convergence? [Hint: Assume that f is bounded.] □

We now come to the most useful feature of the Laplace transform: its effect on differentiation.

D. Differentiation

The basic question is: Can we compute the transform of df/dt if we know the transform of $f(t)$? Under fairly general conditions, if $\dot{f}$ is substituted in the definition (7.1), then integration by parts yields

$$\mathscr{L}\{\dot{f}(t)\} = p\mathscr{L}\{f(t)\} - f(0^+). \tag{7.11}$$

In other words, the transform of a function's derivative corresponds to a multiplication by p in image space, with the initial value thrown in as an added bonus. Note that the *value* of $f(t)$ at the origin is irrelevant; only its limit to the right enters into this equation. For example, if we take $f(t)$ to be Heaviside's function (7.4a), its derivative vanishes for all positive t. The left-hand side of Eq. (7.11) then also vanishes, and, in view of the result (7.4b), the right-hand side also vanishes identically because $H(0^+) = 1$. This example is telling because the value $H(0)$ is undefined.

The differentiation property (7.11) is essential for the application of the Laplace transform to the diffusion equation (1.14). If, now, we identify the variable t with time, and if we recognize that the definition (7.1) does not affect the space variables, $\mathscr{L}\{C_x\} = \partial(\mathscr{L}\{C\})/\partial x$—tersely, the $\mathscr{L}$-operator and x-derivatives commute—then the transformed concentration $\bar{C}(x, p) = \mathscr{L}\{C(x, t)\}$ obeys the *ordinary* differential equation

$$D\frac{d^2\bar{C}}{dx^2} - p\bar{C} = -C(x, 0^+), \tag{7.12}$$

which is surely much more easily solved than the original partial differential equation (1.14). These considerations apply regardless of the number of space dimensions and the types of coordinates used to describe them. In sum, the Laplace transform reduces by one the number of independent variables in any linear differential expression, and the Laplace variable p is then just a parameter. The main task, however, is the computation of the inverse transform that takes the solution $\bar{C}(x, p)$ of Eq. (7.12) into the space of original functions $C(x, t)$.

EXERCISE 7.3. *More simple proofs:* Supply the proof of the property (7.11), and think of an example where it might not apply. [Hint: What happens if f has jump discontinuities on the half axis $t > 0$?] □

The basic operation (7.11) can be applied to higher derivatives. For example, repeated application of that rule to the function $g = \dot{f}$ gives

$$\begin{aligned}\mathscr{L}\{\ddot{f}(t)\} &= \mathscr{L}\{\dot{g}\} = p\mathscr{L}\{\dot{f}(t)\} - \dot{f}(0^+)\\ &= p^2\mathscr{L}\{f(t)\} - pf(0^+) - \dot{f}(0^+).\end{aligned} \tag{7.13}$$

For example, we derive De Moivre's equation $e^{it} = \cos t + i \sin t$, which was surreptitiously used in the calculations leading to Eq. (7.3). This will also illustrate how we can easily solve any linear differential equation with constant coefficients. We consider the function $f(t) = e^{it}$. We verify that it satisfies the differential equation $\ddot{f} + f = 0$ and the initial conditions $f(0) = 1$, $\dot{f}(0) = i$. Applying Eq. (7.13) and the linearity property (7.6) to the differential equation, we immediately get an *algebraic* expression for the transform $\bar{f}(p) = (p + i)/(p^2 + 1)$. Hence, reading the first two entries of Table 7.1 "backward," we get the desired equation.

E. Integration

The operation inverse to differentiation is integration

$$g(t) = \int_0^t f(s)\, ds, \tag{7.14}$$

and one safely predicts that if differentiations on the original space of functions correspond essentially to multiplications by p on the image space, then integrations, such as Eq. (7.14), will correspond to divisions by p of their transforms. This result is an immediate consequence of the differentiation property (7.11) because $\dot{g} = f$ and $g(0^+) = 0$. Hence,

$$\mathscr{L}\{\dot{g}(t)\} = p\mathscr{L}\{g(t)\} - g(0^+) = p\mathscr{L}\{g(t)\},$$

and the desired result follows:

$$\mathscr{L}\left\{\int_0^t f(s)\, ds\right\} = \frac{\bar{f}(p)}{p}. \tag{7.15}$$

This equation is often used in reverse: We can evaluate the inverse transform of an expression such as $\bar{f}/p$ if we know the corresponding pair $f(t)$ and $\bar{f}(p)$. For example, the equation (7.7a) for powers of t follows by repeated integration of $H(t)$.

There are also properties dual to (7.11) and (7.15), namely, multiplication and division by t on the space of original functions correspond to differentiation and integration with respect to p on the space of image functions.[1-3] We shall not dwell on these properties, because they are rarely used in the context of linear diffusion.

Before continuing to examine transforms that are of interest for diffusion, we pause briefly to consider the *delta function* $\delta(t)$ and its transform. There are many ways to interpret this function, mainly because it is not a function at all in the usual sense, but rather a construct. We illustrate this with our exponential. Consider the family of functions

$$g_a(t) = ae^{-at}H(t), \qquad t > 0,\ a > 0, \tag{7.16}$$

that depends continuously on the real positive parameter a. As a function of t, each member of this family has an upper bound, a, that increases, and a "time constant," a^{-1}, that decreases with a. Nonetheless, the integral over all time of this family is a constant. Indeed, we easily verify that

$$\lim_{a\to\infty} g_a(t) = \begin{cases} 0 & \text{if } t > 0 \\ \infty & \text{if } t \to 0^+ \end{cases} \tag{7.17a}$$

and

$$\int_0^\infty g_a(t)\,dt = 1, \qquad \text{for all } a > 0. \tag{7.17b}$$

In other words, this family exhibits a singularity at the origin $t = 0$ if a is large, and that singularity is integrable. This should sound familiar from our work with the gaussian (4.3), which, we recall, is singular in the space variable x, time t playing the role of a parameter. Now the essential property of the delta function is its ability to select a particular functional value from a definite integral. This property is indeed shared by the family (7.16), as integrating by parts over an arbitrary function f,

$$\int_0^\infty g_a(t)f(t)\,dt = f(0^+) + \int_0^\infty e^{-at}\dot{f}(t)\,dt,$$

and taking the limit of large a, we get

$$\lim_{a \to \infty} \int_0^\infty g_a(t) f(t)\, dt = f(0^+). \tag{7.17c}$$

Consequently, there is some justification in setting

$$\lim_{a \to \infty} ae^{-at}H(t) = \delta(t). \tag{7.18a}$$

Further, since $\mathscr{L}\{g_a(t)\} = a/(p + a)$ follows from Eqs. (7.2) and (7.6), we then have

$$\mathscr{L}\{\delta(t)\} = \lim_{a \to \infty} \frac{a}{p + a} = 1. \tag{7.18b}$$

The transform of the delta function is thus *independent* of p. None of our other examples has this feature, and none ever will, because, again, the delta function is defined through a limiting process on a family of functions. The limit does not exist in the usual sense of a continuous function, as should be clear from Eq. (7.17a), but its effect on an integral is nonetheless well defined.[N4,N5]

EXERCISE 7.4. *Selection property of delta functions:* Show with any suitable family of functions that

$$\int_0^\infty f(t)\delta(t - a)\, dt = f(a) \tag{7.17d}$$

for any integrable function f and for any positive number a. □

EXERCISE 7.5. *More delta functions:* Consider the family of functions $g_\varepsilon(t) = 1/\varepsilon$ for $0 \leqslant t < \varepsilon$, which vanish elsewhere. Show that they have all the properties (7.17) in the limit $\varepsilon \to 0$. In what sense have we defined a derivative of $H(t)$? Can you generalize to other similar families that vanish outside an interval? □

7.3. Laplace Transforms in $\sqrt{p}$

We now have the basic tools to evaluate a large number of inverse transforms that may arise in practice. For example, if $\bar{f}(p)$ is a rational function of p, then a partial-fraction expansion, together with the translation property (7.9) and Eqs. (7.7), allow its inversion. Likewise, arbitrary inverse powers of p, perhaps multiplied by exponentials in p, can be inverted using Exercise 7.1 and the translation property (7.10).

Of major concern to us, however, are solutions of the transformed diffusion equation (7.12). Its fundamental solutions are of the form $\exp[\pm x\sqrt{(p/D)}]$, which are not yet in our lexicon of available transforms. We begin by stating and proving the lemma that follows.

LEMMA: *If we know the Laplace transform $\bar{f}(p)$ of the function $f(t)$, then the inverse transform of $\bar{f}(\sqrt{p})$ (i.e., the* same *function of the square-root argument) is given by the integral*

$$\mathscr{L}^{-1}\{\bar{f}(\sqrt{p})\} = \frac{1}{2\sqrt{\pi}\,t^{3/2}} \int_0^\infty s e^{-s^2/4t} f(s)\, ds. \tag{7.19}$$

We verify this lemma by first applying the $\mathscr{L}$-operator to both sides of Eq. (7.19). Then we interchange the s- and t-integrations, and perform the t-integration through the change of variables $t \to \xi = s/2\sqrt{t}$. Thus

$$\begin{aligned}
\mathscr{L}\left\{\frac{1}{2\sqrt{\pi}\,t^{3/2}} \int_0^\infty s e^{-s^2/4t} f(s)\, ds\right\} &= \int_0^\infty ds \frac{sf(s)}{2\sqrt{\pi}} \int_0^\infty \frac{dt}{t^{3/2}} e^{-pt} e^{-s^2/4t} \\
&= \int_0^\infty ds \frac{2f(s)}{\sqrt{\pi}} \int_0^\infty d\xi\, e^{-ps^2/4\xi^2} e^{-\xi^2} \\
&= \int_0^\infty e^{-s\sqrt{p}} f(s)\, ds,
\end{aligned}$$

where the integral over ξ is exactly the third result of Exercise 4.5. We now apply this lemma to a number of transforms that we have already encountered:

a. From the result (7.18b) and the second translation property (7.10), we know that $\mathscr{L}\{\delta(t-a)\} = e^{-ap}$. It follows immediately from the lemma (7.19), applied to the function $f(t) = \delta(t - a)$, and the selection property (7.17c) that

$$\mathscr{L}^{-1}\{e^{-a\sqrt{p}}\} = \frac{a}{2\sqrt{\pi}\,t^{3/2}} e^{-a^2/4t}. \tag{7.20}$$

Likewise, from Eq. (7.4b) for Heaviside's function, we have $\mathscr{L}\{H(t-a)\} = p^{-1}e^{-ap}$. Application of the lemma, once again, gives

$$\begin{aligned}\mathscr{L}^{-1}\{p^{-1/2}e^{-a\sqrt{p}}\} &= \frac{1}{2\sqrt{\pi}\,t^{3/2}} \int_a^\infty s e^{-s^2/4t}\,ds \\ &= \frac{1}{\sqrt{\pi t}} e^{-a^2/4t},\end{aligned} \tag{7.21a}$$

because we recognize that the integrand is

$$s e^{-s^2/4t} = -2t \frac{d}{ds} e^{-s^2/4t}. \tag{7.21b}$$

b. It should be noted that Eq. (7.21a) also results from Eq. (7.20) on integrating both sides of the latter with respect to the parameter a over the range $[a, \infty]$. This integration evidently commutes with the $\mathscr{L}$-operator. We thus have a general procedure for the computation of $\mathscr{L}^{-1}\{p^{-n/2}e^{-a\sqrt{p}}\}$. For example, from Eq. (7.21a) we have

$$\mathscr{L}^{-1}\{p^{-1}e^{-a\sqrt{p}}\} = \frac{1}{\sqrt{\pi t}} \int_a^\infty e^{-b^2/4t}\,db = \operatorname{erfc}(a/2\sqrt{t}), \tag{7.22}$$

where the last step results from a simple change of variables and the definition (4.10) of the complementary error function. We should now sense that these inverse transforms are related to the class of iterated error functions, which is the object of the next exercise.

EXERCISE 7.6. *Transforms of the iterated error functions:* By induction, show that

$$\mathscr{L}\{(2\sqrt{t})^n \mathrm{i}^n \operatorname{erfc}(a/2\sqrt{t})\} = p^{-(1+n/2)} e^{-a\sqrt{p}}, \tag{7.23}$$

for all integers $n \geqslant -1$. □

We note that Eqs. (7.21) and (7.22) are special cases of this result (7.23). Therefore, we know how to invert $e^{-a\sqrt{p}}$ divided by arbitrary *integer powers* of $\sqrt{p}$.[N6] There are cases, however, where one must learn to invert this exponential divided by *polynomials* in $\sqrt{p}$. The next example is of this type, and it requires a little more work.

c. We know from Eqs. (7.2), and (7.4) and (7.5) that

$$\mathscr{L}^{-1}\left\{\frac{1}{p(p+h)}\right\} = \mathscr{L}^{-1}\left\{\frac{1}{h}\left[\frac{1}{p} - \frac{1}{p+h}\right]\right\}$$

$$= \frac{1}{h}(1 - e^{-ht})H(t), \tag{7.24}$$

where h is any constant. From the second translation property (7.10), it follows that

$$\mathscr{L}^{-1}\left\{\frac{e^{-ap}}{p(p+h)}\right\} = \frac{1}{h}[1 - e^{-h(t-a)}]H(t-a),$$

for any positive constant a. Applying the lemma to this transform pair, we have

$$\mathscr{L}^{-1}\left\{\frac{e^{-a\sqrt{p}}}{\sqrt{p}(\sqrt{p}+h)}\right\} = \frac{1}{2h\sqrt{\pi}\,t^{3/2}}\int_a^\infty se^{-s^2/4t}[1 - e^{-h(s-a)}]\,ds,$$

and, with Eq. (7.21b), an integration by parts yields

$$\mathscr{L}^{-1}\left\{\frac{e^{-a\sqrt{p}}}{\sqrt{p}(\sqrt{p}+h)}\right\} = \frac{e^{ha}}{\sqrt{\pi t}}\int_a^\infty e^{-s^2/4t}e^{-hs}\,ds.$$

We then note that the exponents on the right-hand side can be written as a square. A first change of variables $\xi = s/2\sqrt{t}$ makes this quite obvious, and a second change $u = \xi + h\sqrt{t}$ brings the integral into the form (4.10) for the complementary error function. We get finally

$$\mathscr{L}^{-1}\left\{\frac{e^{-a\sqrt{p}}}{\sqrt{p}(\sqrt{p}+h)}\right\} = e^{ha+h^2t}\operatorname{erfc}\left(\frac{a}{2\sqrt{t}} + h\sqrt{t}\right), \tag{7.25}$$

where the parameter a must be nonnegative, but h can be anything, real or complex. This last result is very useful because many other transform pairs follow, either on setting the parameters a and h to special values or on differentiating or integrating with respect to these parameters.

EXERCISE 7.7. *Two useful transforms:* Find the inverse transforms of

$$\frac{e^{-a\sqrt{p}}}{\sqrt{p}+h} \quad \text{and} \quad \frac{e^{-a\sqrt{p}}}{p(\sqrt{p}+h)}$$

by differentiation and integration of Eq. (7.25) with respect to a. □

A whole cottage industry has grown out of the need to invert Laplace transforms. References 5–9 are representative.

7.4. A Sample Calculation

We apply the Laplace transformation to a simple problem that was solved earlier by direct methods: the impurity diffusion problem (4.8) in a half-space under constant surface conditions. This example illustrates how the systematic use of the Laplace transform reduces the search for solutions to mere algebraic juggling, rather than to detailed knowledge of the behavior of special functions. For convenience, we state the problem once more: to solve the diffusion equation

$$C_t = DC_{xx}, \qquad x > 0, \qquad t > 0, \tag{7.26a}$$

under the initial condition

$$C(x, 0^+) = C_\infty, \qquad x > 0, \tag{7.26b}$$

and the boundary conditions

$$C(0, t) = C_0, \qquad t > 0 \tag{7.26c}$$

$$C(\infty, t) = C_\infty, \qquad t > 0 \tag{7.26d}$$

where C_0 (called C_e in Chap. 4) and C_∞ are constants. Transforming the diffusion equation and using the initial condition (7.26b), we get the ordinary differential equation (7.12) with constant coefficients

$$D\bar{C}'' - p\bar{C} = -C_\infty, \tag{7.27}$$

and, because the boundary conditions are constant in time, they transform into

$$\bar{C}(0, p) = C_0/p, \tag{7.28a}$$

$$\bar{C}(\infty, p) = C_\infty/p, \tag{7.28b}$$

by virtue of Eq. (7.4b). The general solution of Eq. (7.27) is

$$\bar{C}(x, p) = p^{-1}C_\infty + Ae^{-x\sqrt{p/D}} + Be^{x\sqrt{p/D}}, \tag{7.29a}$$

where A and B are arbitrary constants with respect to x. They are, however, functions of the Laplace variable p. Indeed, the second transformed boundary condition (7.28b) shows that solutions must be bounded at large distances from the surface. This implies that $B = 0$ because, in general, $\Re(\sqrt{p}) > 0$ is required for transforms to exist. On the other hand, condition (7.28a) allows us to find A, and we obtain

$$\bar{C}(x, p) = p^{-1}C_\infty + p^{-1}(C_0 - C_\infty)e^{-x\sqrt{p/D}}. \tag{7.29b}$$

Hence, from transforms (7.4b) and (7.22), or reading the appropriate entries (4 and 10) in Table 7.1 "backward," we get the desired solution

$$C(x, t) = C_\infty + (C_0 - C_\infty)\operatorname{erfc}\left(x/2\sqrt{Dt}\right). \tag{7.30}$$

It is identical to Eq. (4.8b) if we note the relation (4.10a) between erf and erfc. Note also that the initial and boundary conditions need not be constants. This is illustrated in the next exercise.

Exercise 7.8. *Variable initial condition:* Assume that the initial condition of the above problem is an arbitrary function of x, i.e., Eq. (7.26b) is replaced by

$$C(x, 0^+) = \varphi(x), \qquad \text{with} \quad \lim_{x\to\infty} \varphi(x) = C_\infty.$$

Assume, also, that the boundary conditions (7.26c) and (7.26d) remain unaltered. Find the general solution in terms of a Green's function. [Hint: Solve the differential equation by the method of "variation of constants."] Compare your result with Eq. (4.5). □

The above procedure is quite general and can be applied to a wide variety of problems. To summarize: we apply the Laplace transform to any linear diffusion equation in 1-d and obtain an ordinary linear differential equation; the initial condition is, felicitously, "built in." We solve this equation with the *transformed* boundary conditions. Finally, using a table of transforms and, perhaps, some of the properties in Sect. 7.2, we find the solution in the space of original functions. But what should we do if the table does not contain the desired entries? We return to the general question of inversion in the last section of this chapter.

7.5. The Convolution Theorem

This property is next in importance to the Laplace transform's ability to translate differentiation and integration into rational operations in image space. To motivate convolutions, consider the differential equation of a damped linear oscillator. Its schedule $x(t)$ satisfies

$$m\ddot{x} + \mu\dot{x} + kx = f(t), \tag{7.31a}$$

under the initial conditions

$$x(0^+) = x_0 \qquad \text{and} \qquad \dot{x}(0^+) = v_0. \tag{7.31b}$$

Here, m, μ, and k are the particle's mass and its friction and restoring constants, respectively, and $f(t)$ is an arbitrary impressed force. The solution of this linear second-order differential equation (7.31a) is well known; we recall its structure: It consists of the sum of the general solution $x_h(t)$ of the *homogeneous* equation (i.e., when $f \equiv 0$) and *any particular* solution $x_p(t)$ of the full inhomogeneous equation. The general solution here depends on two arbitrary, multiplicative constants. But these solutions, while interesting, do not describe physical situations, unless the initial conditions (7.31b) can be met. These conditions can be distributed quite arbitrarily over x_h and x_p. We shall now see that it is best to attribute them to the solutions of the homogeneous equation. Indeed, if we take the transform of Eq. (7.31a) and use the differentiation properties (7.11) and (7.13), then we get

$$\bar{x}(p) = \frac{mx_0p + mv_0 + \mu x_0}{mp^2 + \mu p + k} + \frac{\bar{f}(p)}{mp^2 + \mu p + k}. \tag{7.32}$$

It is now evident that the first term of Eq. (7.32), i.e., the transform $\bar{x}_h(p)$, has been entirely specified by the initial conditions and that the second term $\bar{x}_p(p)$ stems from the *homogeneous* conditions

$$x(0^+) = \dot{x}(0^+) = 0. \tag{7.33a}$$

In addition, we recognize the characteristic polynomial of the differential equation (7.31a) in both denominators of Eq. (7.32). Now, $x_h(p)$ is a ratio of two polynomials in p, and it thus has a partial-fraction expansion. Consequently, its inverse $x_h(t)$ is available from our lexicon (see Exercise 7.9 below). The particular solution is another matter, because we might not know how to invert $\bar{f}(p)$, for arbitrary f's, divided by a second-degree polynomial. Let us assume for a moment that $\bar{f} = 1$. In this case the inversion of the second term would be similar to that of the first term. But its numerator corresponds precisely to the fifth entry $f(t) = \delta(t)$ in Table 7.1. As in the development of Eqs. (1.29) and (4.5), we call *this* particular solution the Green's function $G(t)$. Because any impressed force f can be decomposed according to

$$f(t) = \int_0^t f(s)\delta(t - s)\,ds, \tag{7.33b}$$

and because the differential equation (7.31a) is linear, it follows that the particular solution x_p, satisfying the homogeneous initial conditions (7.33a), is given by the integral

$$x_p(t) = \int_0^t f(s)G(t - s)\,ds. \tag{7.33c}$$

EXERCISE 7.9. *Explicit solution of Eqs. (7.31):* Cast the problem in dimensionless form by choosing appropriate time and distance scales. Find x_h and G explicitly. □

Results such as Eq. (7.33c) are familiar to us. They indicate that the effect of an arbitrary forcing term is the superposition, with retarded arguments, of elemental effects due to instantaneous point sources. This situation is so common that the right-hand side of Eq. (7.33c) is graced with a name: the Laplace *convolution* of the functions f and G, and it is denoted $f \star G$. More generally, one defines the convolution of two functions, f and g, as the integral

$$f \star g = \int_0^t f(s)g(t - s)\, ds. \tag{7.34}$$

It represents a "product" operation on the space of integrable functions, as shown in the next exercise.

EXERCISE 7.10. *Algebraic properties of convolutions:* Show that convolutions are commutative, associative, and distributive with respect to the addition of functions. (Such a set of operations goes a long way toward defining what is called an *algebra*.) □

Now what does all this have to do with Laplace transforms? The problem Eq. (7.31) could have been solved by direct methods. Nevertheless, the development that led from Eq. (7.32) to (7.33c) shows that there must be a connection between convolutions in the original space of functions and products in image space. Indeed, we have the so-called *convolution theorem*

$$\mathscr{L}^{-1}\{\bar{f}(p)\bar{g}(p)\} = f \star g, \tag{7.35}$$

which we now verify. We take the $\mathscr{L}$-operation of both sides and evaluate the left-hand side:

$$\bar{f}(p)\bar{g}(p) = \int_{\mathscr{D}_{uv}} e^{-p[u+v]} f(u)g(v)\, du\, dv = \int_{\mathscr{D}_{ts}} e^{-pt} f(s)g(t - s)\, ds\, dt$$
$$= \int_0^\infty dt\, e^{-pt} \int_0^t ds\, f(s)g(t - s) = \mathscr{L}\{f \star g\}.$$

The first integral is nothing but the product of the two Laplace transforms, written as a double integral over the dummy variables (u, v). The domain $\mathscr{D}_{uv}$ of these variables extends over the first

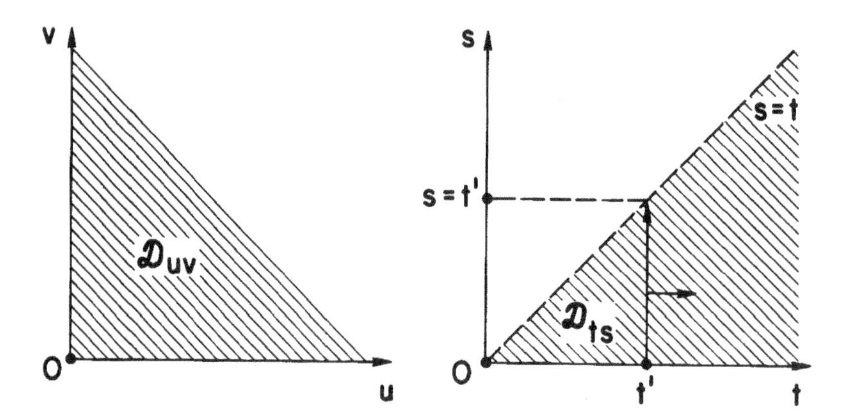

Fig. 7.3. Corresponding domains under the coordinate transformation required for the proof of the convolution theorem.

quadrant, as shown in Fig. 7.3. The change of variables $t = u + v$ and $s = u$ transforms this domain into the octant $\mathscr{D}_{ts}$. It is easily verified that the jacobian of the transformation is unity, whence the second integral. Finally, this octant is swept out by first integrating over s (from $s = 0$ to $s = t$), and then varying t over its whole range to yield the last integral.

This theorem has an immediate application to our sample problem of the last section. If, for example, the first boundary condition (7.26c) is an *arbitrary* function of time $C_0(t)$, then Eq. (7.29a) still holds, but the "constant" A will now be expressed in terms of the transform $\bar{C}_0(p)$. Thus, Eq. (7.29b) now reads

$$\bar{C}(x, p) = p^{-1}C_\infty + [\bar{C}_0(p) - C_\infty/p]e^{-x\sqrt{p/D}},$$

and, using the result (7.20), the convolution theorem (7.35), and the definition (4.10a), we get the desired solution

$$C(x, t) = C_\infty \operatorname{erf}(x/2\sqrt{Dt}) + \int_0^t C_0(s)\, \frac{xe^{-x^2/4D(t-s)}}{2\sqrt{\pi D}\,(t - s)^{3/2}}\, ds. \tag{7.36}$$

It should be noted that nowhere was it necessary to actually compute the transform $\bar{C}_0$ of the arbitrarily given concentration at the boundary $x = 0$.

EXERCISE 7.11. *To verify the boundary condition:* Can you merely put $x=0$ into Eq. (7.36) to verify the boundary condition $\lim_{x\to 0^+} C(x, t) = C_0(t)$? [Hint: Find a good change of variables for the integral.] □

The convolution theorem has many other applications. For example, it can be used to solve certain *integral equations*, as will be illustrated in the next chapter. Here, we will only mention its application to the evaluation of some definite integrals. The first of these defines the so-called *beta function*

$$B(\mu, \nu) = \int_0^1 (1 - s)^{\mu - 1} s^{\nu - 1}\, ds, \tag{7.37a}$$

where μ and ν are any real exponents greater than zero. This integral would be a convolution of the form (7.34) if the independent variable t were set to unity. The trick, therefore, is to embed it in the larger class of integrals of the form

$$f(t) = \int_0^t (t - s)^{\mu - 1} s^{\nu - 1}\, ds, \tag{7.37b}$$

whose Laplace transform is

$$\bar{f}(p) = \mathscr{L}\{t^{\mu - 1}\}\mathscr{L}\{t^{\nu - 1}\} = \frac{\Gamma(\mu)\Gamma(\nu)}{p^{\mu + \nu}}$$

according to the convolution theorem (7.35) and the seventh entry of Table 7.1. The inverse transform follows from this same entry, and we get

$$f(t) = \frac{\Gamma(\mu)\Gamma(\nu)}{\Gamma(\mu + \nu)} t^{\mu + \nu - 1}. \tag{7.37c}$$

Hence, setting $t = 1$, we have evaluated the beta function

$$B(\mu, \nu) = \Gamma(\mu)\Gamma(\nu)/\Gamma(\mu + \nu) \tag{7.37d}$$

entirely in terms of the gamma function.

EXERCISE 7.12. *The area of a circle:* Use the beta function to calculate the area of a circle. [Hint: Write the area as the integral $dx\,dy$ over the unit circle in the first quadrant. Then change variables to $\xi = x^2$ and $\eta = y^2$.] Can you generalize for the volume of a d-dimensional sphere? □

Another useful integral is that obtained by repeated "nested" integration (n-fold) of a given function

$$g_n(t) = \int_0^t dt_n \int_0^{t_n} dt_{n-1} \cdots \int_0^{t_2} dt_1 f(t_1), \tag{7.38a}$$

where $t > t_n > t_{n-1} > \cdots > t_1 > 0$. We wish to express it as a single integral. The previous discussion on transforms of integrals provides a clue because Eq. (7.14) is of this form. Further, it can be written symbolically $g_1 = f \star 1$, and its Laplace transform (7.15) is indeed given by the convolution theorem (7.35). Now, the multiple integral (7.38a) can be written as a recursion

$$g_n(t) = \int_0^t g_{n-1}(t_n)\, dt_n = g_{n-1} \star 1, \tag{7.38b}$$

and its transform is

$$\bar{g}_n(p) = \frac{1}{p}\bar{g}_{n-1}(p) = \frac{1}{p^2}\bar{g}_{n-2}(p) = \cdots = \frac{1}{p^n}\bar{f}(p), \tag{7.38c}$$

by repeated application of the integration property (7.15), and because $g_0 = f$. Inverting this last with the convolution theorem and the sixth entry of Table 7.1, we get the desired result

$$g_n(t) = \int_0^t \frac{(t-s)^{n-1}}{(n-1)!} f(s)\, ds. \tag{7.38d}$$

We see once again that the Laplace transform translates transcendental operations (7.38a,b) into simple divisions (7.38c) on the space of image functions.

7.6. A Few Words on Asymptotics

It may happen that we are only interested in the behavior of a diffusive system, either for short times or in the limit of long times. On the other hand, many problems are intractable analytically except within these limits. Consequently, we may ask if there are general methods to extract these asymptotic limits from a given problem's Laplace transform. The answer is affirmative, and the vast subject of asymptotics is related to the analytic properties of the Laplace integral (7.1) in the complex p-plane. For example, we have seen that this integral converges only in certain regions of that plane. In addition, the problem (7.31) of rectilinear motion shows that the complete solution (see Exercise 7.9) depends crucially on the partial-fraction expansion of the inverse of a second degree polynomial in p, i.e., on the *poles* of that function. Therefore, a full discussion of asymptotic behavior requires a good dose of complex variable theory. Here, in keeping with our elementary discussion of diffusion, we content ourselves with a few remarks.

The first thing we ought to note from the definition (7.1) of a Laplace transform is that $\bar{f}(p)$ vanishes when p is large: The original function $f(t)$ gets damped by the exponential for large positive values of p.† Indeed, all the tabulated transforms in this chapter vanish in this limit, except, of course, the transform of the delta function. But does that large-p limit correspond to anything of interest in the space of original functions? The sixth entry of Table 7.1 gives a clue: Convergence to zero of t^n corresponds to rapid decay for large p of its transform $n!/p^{-(n+1)}$. Therefore, we anticipate that the behavior of $f(t)$ on the t-axis should be, in some sense, "reciprocal" to the behavior of $\bar{f}(p)$ in the p-plane. This statement can be substantiated if the original function has a convergent Taylor expansion around $t = 0^+$. Then, as in the calculations leading to Eq. (7.7a), we get

$$\bar{f}(p) = \int_0^\infty dt\, e^{-pt} \sum_{n=0}^{\infty} \frac{t^n}{n!} f^{(n)}(0^+) = \sum_{n=0}^{\infty} \frac{f^{(n)}(0^+)}{n!} \mathscr{L}\{t^n\}$$

$$= \sum_{n=0}^{\infty} \frac{f^{(n)}(0^+)}{p^{n+1}}, \tag{7.39}$$

†More generally, legitimate Laplace transforms must vanish at infinity along any ray in the right-hand complex p-plane.

with the usual notation $f^{(n)}$ for the nth derivative of $f(t)$. In other words, the way in which $f(t)$ approaches its initial value is related to the way in which its transform $\bar{f}(p)$ behaves around the point at infinity. In particular, extracting the zeroth term from the sum, multiplying throughout by p, and taking the limit of large p, we get the so-called *initial-value* theorem

$$\lim_{p\to\infty} p\bar{f}(p) = \lim_{t\to 0^+} f(t). \tag{7.40}$$

EXERCISE 7.13. *Initial- and final-value theorems:* Use the differentiation property (7.11) to prove the initial-value theorem (7.40) directly. Likewise, prove the *final-value theorem*

$$\lim_{p\to 0} p\bar{f}(p) = \lim_{t\to\infty} f(t). \quad \square \tag{7.41}$$

These results call for several remarks. In the first place, although the theorems are correct, we must remember that the above manipulations do not constitute rigorous proofs, since we have freely interchanged limits. Next, the initial- and final-value theorems (7.40) and (7.41) must be understood as follows: *If* the limits $f(0^+)$ or $f(\infty)$ of the original function are otherwise known to exist, *then* we can compute their values from limits of Laplace transforms. To proceed in the opposite direction can sometimes yield absurd results. For example, applying the final-value theorem (7.41) to the first entry of Table 7.1 leads to a perfectly reasonable left-hand limit, $\lim_{p\to 0}(p^2/(p^2+\omega^2)=0$, but the right-hand side, $\lim_{t\to\infty}\cos\omega t$, does not exist. Last, the expansion (7.39) can be understood as asymptotic rather than convergent. Indeed, integrating the definition (7.1) by parts, we get

$$\bar{f}(p) = \int_0^\infty e^{-pt} f(t)\,dt = -\frac{1}{p} e^{-pt} f(t)\big|_{t=0}^{\infty} + \frac{1}{p}\int_0^\infty e^{-pt}\dot{f}(t)\,dt.$$

This process can be repeated to yield a series of the form (7.39), except that now we have not assumed Taylor convergence of f.

The previous considerations were all questions of the following type: Given the behavior of the original function around some value of t, can we infer the behavior of the image function around some "reciprocal" value of p? These considerations can be elaborated. For example, one can show (see Ref. 10, pp. 180–183) that the seventh entry of Table 7.1 holds asymptotically, namely, if

$$f(t) \sim e^{\alpha t} t^{\nu} \qquad \text{for } t \to 0^+ \text{ or } t \to \infty, \tag{7.42a}$$

then

$$\bar{f}(p) \sim \Gamma(\nu + 1)/(p - \alpha)^{\nu+1} \qquad \text{for } p \to \infty \text{ or } p \to \alpha^+. \tag{7.42b}$$

In addition, the power-series expansion (7.39) admits an important generalization known as Watson's lemma (see Ref. 11, p. 34): If the original function has an *asymptotic* expansion in arbitrary powers t^{ν} ($\nu > -1$), then its image function results by termwise transformation of these powers. Even this lemma can be generalized to asymptotic expansions in terms of "gauge functions" other than powers.[12]

The converse question, however, is more interesting: Given the image function, or at least, its asymptotic behavior, can we recover the original function of time? That question is related to the problem of inversion of Laplace transforms,[1–3] and, as was mentioned earlier, it cannot be addressed properly in the real domain. Here, the following remarks will suffice. First, if the transform $\bar{f}(p)$ admits a convergent series expansion in inverse powers (not necessarily integer) of p, then the original function can be recovered by formal termwise inversion (see Ref. 13, Chap. 8, especially p. 174, and Ref. 14, Theorem 2). Therefore, by construction, large-p expansions lead to small-t expressions of the original function. This procedure is sometimes applicable to convergent expansions in powers of functions other than $p^{-\nu}$ (see Ref. 3, pp. 198–200). Second, the theorem (7.42) can often be used in reverse, namely, (a) follows from (b), when α is a singularity of the transform (see Ref. 3, pp. 234–238). The precise conditions under which this is valid cannot be stated without a thorough examination of the properties of the "inversion integral." The interested reader will find more details in Ref. 13 (Chap. 7, especially pp. 150–154) and Ref. 14 (Theorem 3). Nevertheless, we will use or verify the previous statements in the examples to follow in the next chapter.

Notes

1. The Laplace transform was invented by Euler to solve differential equations. Laplace developed this technique further to solve difference and differential equations that occur in probability theory. Heaviside is the unsung hero of integral transforms (as well as of many other things), for he advocated their use in electromagnetic theory. See the remarkable article, "Oliver Heaviside" by P. J. Nahin, in *Scientific American* **262**, p. 122 (June, 1990).
2. There are other more subtle restrictions on the class of original functions that provide a one-to-one mapping. Two functions that differ by an amount whose integral (7.1) vanishes must be considered equal. Nevertheless, for functions f that are continuous on the positive real axis, except for countably many jump discontinuities, their Laplace transforms determine the f's uniquely, except for the values at the jumps.
3. The multiplicative function $H(t)$ in Eq. (7.5) is most often implied. This generally does not cause difficulties, except when one wishes to find transforms of functions with retarded arguments, in accordance with Eq. (7.10). For instance, see the calculation that follows Eq. (7.24).
4. The existence of many distinct families of functions of a real parameter a, like $g_a(t)$, each of which has a constant transform in the limit when that parameter a increases without bound (see Exercise 7.5), drives another nail in the coffin of the appellation delta "function." Strictly speaking, with the selection property (7.17d) taken as an equivalence relation, $\delta(t)$ is the "equivalence class" of such limits.
5. Those who believe that $\delta(t)$ holds no further mystery should consult the two short articles: "Zeroing in on the Delta Function" by J. R. Hundhausen, in *College Mathematics J.* **29**, 27 (1998), and "Dirac Deltas and Discontinuous Functions" by D. J. Griffiths and S. Walborn, in *Amer. J. Phys.* **67**, 446 (1999).
6. Inspection of Eq. (7.20) shows that we could extend equation (7.23) to integers $n < -1$ if we knew how to define the corresponding iterated error functions for such values of n. This is achieved by repeated differentiation of Eq. (7.21a) with respect to a. We obtain the class of Hermite polynomials $H_{-n-1}(a/2\sqrt{t})$ multiplied by gaussians, of which Eq. (7.20) is a member.

References

1. M. R. Spiegel, *Theory and Problems of Laplace Transforms* (Schaum Publishing Co., New York, 1965).
2. R. V. Churchill, *Operational Mathematics*, 3rd ed. (McGraw-Hill, New York, 1972).

3. G. Doetsch, *Introduction to the Theory and Application of the Laplace Transform* (Springer-Verlag, Berlin, 1974).
4. C. J. Tranter, *Integral Transforms in Mathematical Physics*, 3rd ed. (Methuen & Co. , Ltd, London, 1966).
5. L. A. Pipes, "The Summation of Fourier Series by Operational Methods," *J. Appl. Phys.* **21**, 298 (1950).
6. R. B. Hetnarski, "An Algorithm for Generating some Inverse Laplace Transforms of Exponential Type," *Z. angew. Math. Phys.* (ZAMP) **26**, 249 (1975).
7. F. T. Lindstrom and F. Oberhettinger, "A Note on a Laplace Transform Pair Associated with Mass Transport in Porous Media and Heat Transport Problems," *SIAM J. Appl. Math.* **29**, 288 (1975).
8. K. S. Crump, "Numerical Inversion of Laplace Transforms Using a Fourier Series Approximation," *J. Assoc. Comput. Mach.* **23**, 89 (1976).
9. P. Puri and P. K. Kythe, "Some Inverse Laplace Transforms of Exponential Type," *Z. angew. Math. Phys.* (ZAMP) **39**, 150 (1988); *errata, ibid.* **39**, 954 (1988).
10. D. V. Widder, *The Laplace Transform* (Princeton University Press, Princeton, NJ, 1941).
11. A. Erdélyi, *Asymptotic Expansions* (reprinted by Dover, New York, 1956).
12. A. Erdélyi, "General Asymptotic Expansions of Laplace Integrals," *Arch. Rational Mech. Anal.* **7**, 1 (1961).
13. G. Doetsch, *Handbuch der Laplace-Transformation*, Band II (Birkhäuser-Verlag, Basel, 1972).
14. R. A. Handelsman and J. S. Lew, "Asymptotic Expansion of Laplace Convolutions for Large Argument and Tail Densities for certain Sums of Random Variables," *SIAM J. Math. Anal.* **5**, 425 (1974).
15. A. Erdélyi, W. Magnus, F. Oberhettinger, and F. Tricomi, *Tables of Integral Transforms*, Vol. 1 (McGraw-Hill, New York, 1954).

8

Further Time-Dependent Examples

The constructive (or synthetic) methods of Chaps. 4–6 often require considerable ingenuity. Solving a given diffusion problem depends on our ability to match certain known elementary solutions of the diffusion equation to the initial and boundary conditions. It thus requires detailed knowledge of the behavior of these solutions. In contrast, this last chapter describes deductive (or analytic) methods, based mainly on the Laplace transform, that can be applied with ease to a wide variety of problems.†

8.1. Laser Processing

Sources of heat, such as lasers and electron beams, are commonly used to induce either structural modifications[1,2] or phase changes in engineering[3,4] and biological[5,6] materials. The many investigations of "laser annealing," especially for ion-implanted materials, have been excellently reviewed in Refs. 7 and 8. In this chapter, we use a simplified form of this process to investigate means of computing interactions between external sources of heat and a solid material whose properties we wish to modify.

We suppose that we have a laser beam impinging on such a material. We assume, in addition, that the solid is semi-infinite and that it is bounded by a planar surface. We then let $F_0(t)$ be the laser flux

†See also Appendix E for an introduction to Fourier methods.

incident on that surface†; its time-dependence is, for the moment, arbitrary. Not all the impinging power is effective; if R is the reflection coefficient of the material, then a fraction RF_0 is reflected by the solid. The remainder, $(1 - R)F_0$, transits into the solid, where it dissipates. The interaction of matter and light is a very complex process; it can be described phenomenologically, however, by introducing the notion of *skin depth* λ. According to this simple picture, the change in dissipated power P (per unit volume) is proportional to the power itself, the coefficient of proportionality being precisely $1/\lambda$. Thus, $dP/dx = -\lambda^{-1}P$ leads to

$$P(x) = P_0 e^{-x/\lambda}, \tag{8.1}$$

where P_0 is the power dissipated at the solid's surface $x = 0$. In the absence of diffusion processes, the total power dissipated in the solid is the integral over all x of expression (8.1). This must balance exactly the net power input $(1 - R)F_0$, and it follows that $P_0 = (1 - R)F_0/\lambda$.

It is easy to write the appropriate equation for temperature changes because we recognize that Eq. (8.1) is a source term (in the sense of Sect. 1.8) of the diffusion equation that follows from Eqs. (1.37). Thus, for the case of constant coefficients, we must solve

$$\rho c \frac{\partial T}{\partial t} = K \frac{\partial^2 T}{\partial x^2} + (1 - R)F_0(t)\lambda^{-1} e^{-x/\lambda}. \tag{8.2}$$

We suppose that the solid is initially at a constant temperature T_∞, which is somehow maintained far from the surface. Furthermore, we assume that the surface temperature is low enough so that radiation into the surrounding medium is negligible compared to the laser flux.‡ Therefore, we have the initial and boundary conditions

$$T(x, 0) = T_\infty, \tag{8.3a}$$

$$T(\infty, t) = T_\infty, \tag{8.3b}$$

$$\partial T/\partial x|_{x=0} = 0. \tag{8.3c}$$

†In this context, the word "flux" means the same thing as "areal power density"; both are measured in units of power per unit area. Here, we also assume that the beam has an infinite lateral extent.

‡According to the Stefan–Boltzmann σT^4 radiation law, where the constant σ has the value 5.64×10^{-12} W/cm^{2}K^4, an upper bound to the emitted flux from a silicon surface at its melting point, 1683 K, is only 45 W/cm^2. This should be compared to the power input, 0.1 J/cm^2 delivered in a nanosecond, of a typical pulsed laser.

The attentive reader may have noted the similarity between this problem and the heating problem Eq. (4.21). Here, the laser energy gives rise to a *distributed source* (8.1) that penetrates a distance λ into the solid, and the surface flux (8.3c) vanishes. On the other hand, energy input was simulated earlier through a *surface source* (4.21a), expressed as a boundary condition.† Are these distinct representations of the same physical process? One would hope not, and a clue emerges if we recognize that the power dissipation (8.1) is proportional to $\lambda^{-1}\exp(-x/\lambda)$. This, according to Eqs. (7.16)–(7.18), is nothing but $\delta(x)$ in the limit of small λ's, i.e., a unit surface source.

EXERCISE 8.1. *Heat balance and boundary conditions:* Compute an irradiated solid's change in total internal energy, and balance it against the power input from external sources. Use the heat equation (8.2), but retain the form (8.1) for the source. Discuss the boundary conditions when the skin depth is finite and when it vanishes. □

This problem is easily solved by a Laplace transformation. First, it is best to recast it in a dimensionless form that eliminates many of the physical constants. The skin depth λ evidently provides a length scale, and, consequently, its square divided by the thermal diffusivity $\kappa = K/\rho c$ is a unit of time. Further, if T_m denotes some characteristic temperature (such as the melting point) at which some desired change occurs, then, by inspection, we see that $(T_m - T_\infty)K/\lambda$ has units of flux. Introducing the new independent variables

$$\xi = x/\lambda \tag{8.4a}$$

and

$$\tau = \kappa t/\lambda^2, \tag{8.4b}$$

and the dimensionless temperature and flux

$$u(\xi, \tau) = \frac{T(x, t) - T_\infty}{T_m - T_\infty} \tag{8.5a}$$

and

$$\varphi(\tau) = \frac{\lambda(1 - R)}{K(T_m - T_\infty)} F_0(t), \tag{8.5b}$$

†Our former input flux F is exactly the unreflected portion $(1 - R)\,F_0$ considered here.

we transform Eq. (8.2) into the convenient dimensionless form

$$\frac{\partial u}{\partial \tau} = \frac{\partial^2 u}{\partial \xi^2} + \varphi e^{-\xi}. \tag{8.6a}$$

This we must solve under the conditions (8.3)

$$u(\xi, 0) = 0, \tag{8.6b}$$

$$u(\infty, \tau) = 0, \tag{8.6c}$$

$$u_\xi(0, \tau) = 0, \tag{8.6d}$$

expressed now in the new variables.

Taking the Laplace transform of Eq. (8.6a), defining $\bar{u}(\xi, p) = \mathscr{L}\{u(\xi, \tau)\}$ and $\bar{\varphi}(p) = \mathscr{L}\{\varphi(\tau)\}$ according to the definition (7.1), and using the linearity and differentiation properties (7.6) and (7.11), we get the ordinary linear differential equation

$$p\bar{u} = \frac{d^2 \bar{u}}{d\xi^2} + \bar{\varphi} e^{-\xi} \tag{8.6e}$$

because the initial condition (8.6b) vanishes. Its solution, under the boundary conditions (8.6c) and (8.6d) is easily seen to be

$$\bar{u}(\xi, p) = \frac{\bar{\varphi}(p)}{p - 1}\left[e^{-\xi} - \frac{e^{-q\xi}}{q}\right], \tag{8.7}$$

where $q = \sqrt{p}$, which we must invert to recover the temperature response of the irradiated material.

A formal inversion is not hard to obtain through the convolution theorem (7.35). In fact, each term of the image temperature (8.7) is available in Table 7.1 (entries 3 and 12) because

$$\frac{1}{p-1} = \frac{1}{q^2 - 1} = \frac{1}{2}\left(\frac{1}{q-1} - \frac{1}{q+1}\right).$$

If, however, we are only interested in the surface temperature T_0 (or its dimensionless embodiment u_0) rather than in the whole temperature distribution, then it suffices to evaluate (8.7) for $x = 0$. We then have

$$\bar{u}_0(p) \equiv \bar{u}(0, p) = \bar{\varphi}(p) \frac{1}{q(q+1)}, \tag{8.8}$$

which is easily inverted to give

$$u_0(\tau) = \int_0^\tau \varphi(\sigma) e^{(\tau-\sigma)} \operatorname{erfc}\sqrt{\tau - \sigma}\, d\sigma, \tag{8.9a}$$

or, reverting to the dimensional variables (8.4),

$$T_0(t) = T_\infty + \frac{1-R}{\rho c \lambda} \int_0^t F_0(s) e^{\kappa(t-s)/\lambda^2} \operatorname{erfc} \frac{\sqrt{\kappa(t-s)}}{\lambda}\, ds. \tag{8.9b}$$

There are two points to bear in mind regarding this last result. First, we evaluated the surface temperature *before* the Laplace inversion; second, it is valid for *arbitrary* laser pulses. Indeed, the coefficient of $\bar{\varphi}$ in the transformed solution (8.7) is the problem's (transformed) Green's function $\bar{G}(\xi, p)$, and we can write formally $u(\xi, \tau) = \varphi(\tau) \star G(\xi, \tau)$. This Green's function, in fact, contains *all* the information on the temperature's spatial distribution. [The reader should note the similarities and differences between this result and the problem leading to the integral (4.5).] We end this section with general remarks about short- and long-time behavior, and we defer to the next section some detailed calculations for specific laser pulse shapes.

The behavior for short times is easy to obtain from the Laplace transform (8.8) if we remember the discussion of Sect. 7.7. For $p \to \infty$ we get $\bar{u}_0 \sim \bar{\varphi}/p$. Consequently, with the integration property (7.15), we have

$$u_0 \approx \int_0^\tau \varphi(\tau)\, d\tau, \tag{8.10a}$$

or in dimensional form

$$T_0(t) \approx T_\infty + \frac{1-R}{\lambda \rho c} \int_0^t F_0(s)\, ds. \tag{8.10b}$$

In other words, the material system responds initially by *integrating* the laser pulse. Physically, in the beginning, essentially all the laser's energy goes into heating a surface layer of depth λ, while the rest of the sample appears adiabatically insulated from this surface layer. Indeed, neither thermal conductivity nor diffusivity appears in Eq. (8.10b). This result can also be verified directly from Eqs. (8.9) because $e^z \operatorname{erfc}\sqrt{z} \sim 1 - 2(z/\pi)^{1/2}$ by virtue of the expansion property (d) of Sect. (4.1) and the definition (4.10b). Consequently, "short" times mean that $\tau^{1/2}$ must be much less than unity, i.e., the diffusion length must be smaller than the skin depth. Taking $\kappa \simeq 0.2\,\text{cm}^2/\text{s}$ (for Si at 800°C), and $\lambda \simeq 1\,\mu\text{m}$ (for an argon-ion laser), we find from Eq. (8.4b) that τ must be much less than 50 ns for us to observe initial transient behavior.

On the other hand, we now assume that the laser had been switched on for only a duration t_1 (the details of the pulse's shape are irrelevant). We ask what is the surface temperature long after the pulse has been switched off. By the mean-value theorem, there exists a time $t^* \in [0, t_1]$ (and corresponding scaled times τ) such that equation (8.9a) can be written

$$u_0(\tau) = \varphi(\tau^*) \int_0^{\tau_1} e^{(\tau-\sigma)} \operatorname{erfc}\sqrt{\tau - \sigma}\, d\sigma$$
$$= \varphi(\tau^*) \int_{\tau-\tau_1}^{\tau} e^{\sigma} \operatorname{erfc}\sqrt{\sigma}\, d\sigma, \qquad (8.11)$$

where the second equality follows from a simple change of variables. Since we are interested in the behavior for $t \gg t_1$, it suffices to examine the asymptotic behavior of the integrand, which, by virtue of Eqs. (4.11), is asymptotically equal to $(\pi\sigma)^{-1/2}$. Thus we have

$$u_0(\tau) \sim \varphi(\tau^*) \int_{\tau-\tau_1}^{\tau} \frac{d\sigma}{\sqrt{\pi\sigma}} = \frac{2\varphi(\tau^*)}{\sqrt{\pi}}\left(\sqrt{\tau} - \sqrt{\tau - \tau_1}\right) \sim \varphi(\tau^*)\tau_1/\sqrt{\pi\tau}, \qquad (8.12a)$$

where, applying the binomial theorem to $\sqrt{(\tau - \tau_1)}$, the second equality holds when $\tau \gg \tau_1$. We also get its dimensional form

$$T_0(t) \sim T_\infty + \frac{(1-R)F_0(t^*)t_1}{K}\left(\frac{\kappa}{\pi t}\right)^{1/2}. \qquad (8.12b)$$

Therefore, for long times, we find that the *average energy* $F_0(t^*)t_1$ contained in the laser pulse causes a temperature decay proportional to $t^{-1/2}$. Contrary to the short-time behavior (8.10b), this result is independent of skin depth, but it does depend on κ and K. Physically, this simply means that the laser's energy decays into the sample by diffusion and that all "memory" of this energy's provenance is lost after a sufficiently long time.

EXERCISE 8.2. *More asymptotics:* Show that the result (8.12b) can also be obtained directly from the Laplace transform (8.7). With the same parameters as in the case of short times, what does a "long" time mean? Carry out the expansions (8.10) and (8.12) to higher orders. Can you also discuss these asymptotic limits of the full temperature distribution $T(x, t)$? □

8.2. Laser Processing: Pulse Shapes

Equations (8.9), it is recalled, hold for any pulse shape. In particular, if we have a cw laser, then the input flux is simply proportional to the Heaviside step function (7.4a). We then have

$$F_0(t) = F_1 H(t), \tag{8.13a}$$

where F_1 is the constant flux delivered by the laser. Defining the corresponding dimensionless flux φ_1, we obtain the surface temperature rise

$$u_0(\tau) = \varphi_1 \int_0^\tau e^\sigma \operatorname{erfc}\sqrt{\sigma}\, d\sigma. \tag{8.13b}$$

But this integral can be evaluated explicitly by parts, because, using the differentiation formula (c) of Sect. (4.2), we have

$$\frac{d}{dz}\left(e^z \operatorname{erfc}\sqrt{z}\right) = e^z \operatorname{erfc}\sqrt{z} - \frac{1}{\sqrt{\pi z}}. \tag{8.13c}$$

Therefore,

$$u_0(\tau) = \varphi_1\left(e^\tau \operatorname{erfc}\sqrt{\tau} + 2\sqrt{\tau/\pi} - 1\right), \tag{8.14a}$$

or, reverting to dimensional variables,

$$T_0(t) = T_\infty + \frac{\lambda(1-R)F_1}{K}\left[e^{\kappa t/\lambda^2}\operatorname{erfc}\frac{\sqrt{\kappa t}}{\lambda} + \frac{2}{\lambda}\left(\frac{\kappa t}{\pi}\right)^{1/2} - 1\right]. \tag{8.14b}$$

We may now ask what relation exists between these equations and our earlier calculation (4.21e). The second remark that follows Eqs. (8.3) suggests that we should indeed recover the heating problem Eq. (4.21) in the approximation $\lambda \to 0$. This means that the scaled time τ [see Eq. (8.4b)] is almost always very large, and we can use the asymptotic expansion of the error function in Eqs. (8.14) to get

$$u_0(\tau) \sim \varphi_1\left(\frac{1}{\sqrt{\pi\tau}} + 2\sqrt{\tau/\pi} - 1\right), \tag{8.15a}$$

which, for $\tau > 100$, reduces to

$$u_0(\tau) \sim \varphi_1 2\sqrt{\tau/\pi}. \tag{8.15b}$$

This is precisely our earlier result (4.21e) in dimensionless form. On the other hand, when $\tau \ll 1$ (i.e., if either the processing time is short or the skin depth is very large), we have the initial behavior

$$u_0(\tau) \sim \varphi_1\tau. \tag{8.16}$$

This follows either from the general result (8.10) or from a short-time expansion of the error function in the particular case (8.14).

The surface temperature (8.14) is plotted in Fig. 8.1 as a function of scaled time, together with its limiting behavior (8.16) and (8.15b) for short and long times, respectively. We see that the surface temperature is a monotonically increasing function of time and that it scales linearly with the laser flux F_1. This figure can be read in yet another way. From the definition (8.5a) of the scaled temperature it follows that

$$u_0(\tau_m) = 1 \tag{8.17a}$$

defines that time τ_m when the surface temperature will have first reached the melting or transition point T_m. Inserting this condition

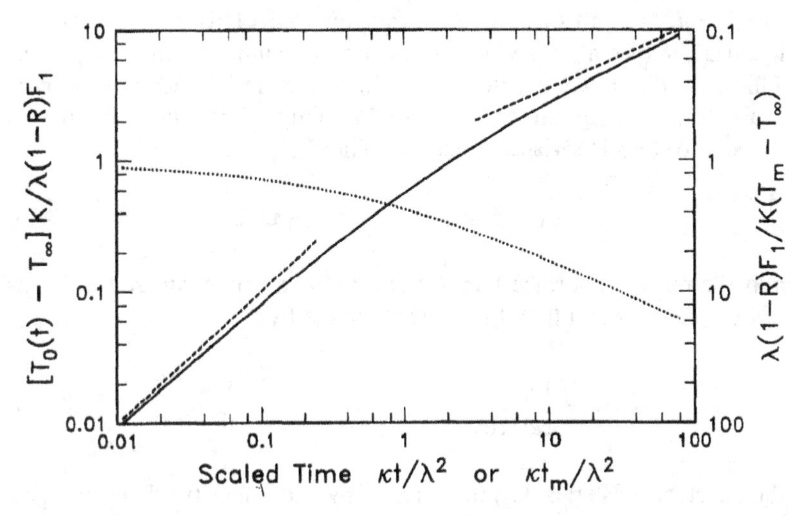

Fig. 8.1. Surface temperature (full line) as a function of time for the laser heating problem. The dashed lines correspond to the approximations (8.16) and (8.15b) for short and long times, respectively. The dotted line is the integrand of Eq. (8.13b). The use of the right-hand ordinate scale is explained in the text.

into Eq. (8.14a), we get

$$\varphi_1^{-1} = \left(e^{\tau_m} \operatorname{erfc}\sqrt{\tau_m} + 2\sqrt{\tau_m/\pi} - 1\right), \tag{8.17b}$$

and the log-log plot in Fig. 8.1 can be read "backward": A given flux F_1, if properly scaled (as shown on the right-hand ordinate), will produce a surface temperature T_m in a time t_m that can be read off directly from the abscissa.

EXERCISE 8.3. *Power required for melting:* Using the analytic expressions (8.15) and (8.16), estimate the time t_m as a function of laser power that is required to achieve a phase transformation. Carefully state the limits of validity of your estimates. □

Next in order of complexity, we can compute the effect of the rectangular laser pulse shown in Fig. 8.2. It is constant in the interval

$(0, t_1)$, and it vanishes outside this interval. This is evidently an idealization of real pulses, because rise rates cannot be infinitely large. This example, however, illustrates the principle behind any analytic evaluation of temperature response. We simply note that the pulse is the sum of two Heaviside functions. Thus,

$$\varphi(\tau) = \varphi_1[H(\tau) - H(\tau - \tau_1)], \tag{8.18}$$

and the required surface temperature (8.9) can be expressed in terms of our previous result (8.14). Indeed, define by

$$U(\tau) = \begin{cases} 0 & \text{if } \tau < 0 \\ e^{\tau} \operatorname{erfc}\sqrt{\tau} + 2\sqrt{\tau/\pi} - 1 & \text{if } \tau \geqslant 0 \end{cases} \tag{8.19a}$$

the effect of a Heaviside pulse. Then, by the linearity of the integral (8.9a), we must have the surface response

$$u_0(\tau) = \varphi_1[U(\tau) - U(\tau - \tau_1)]. \tag{8.19b}$$

This is the sum of two shifted functions (8.19a), each of which picks up at a break point of the laser's flux schedule (8.18).

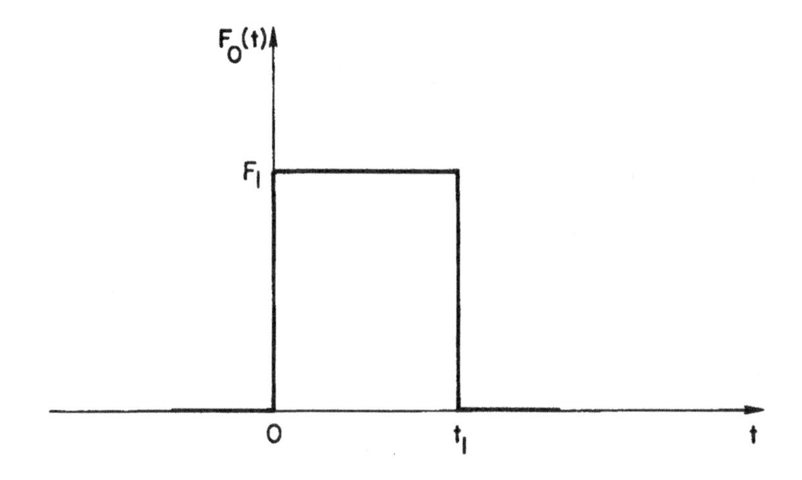

Fig. 8.2. Rectangular laser pulse of duration t_1 and of intensity F_1.

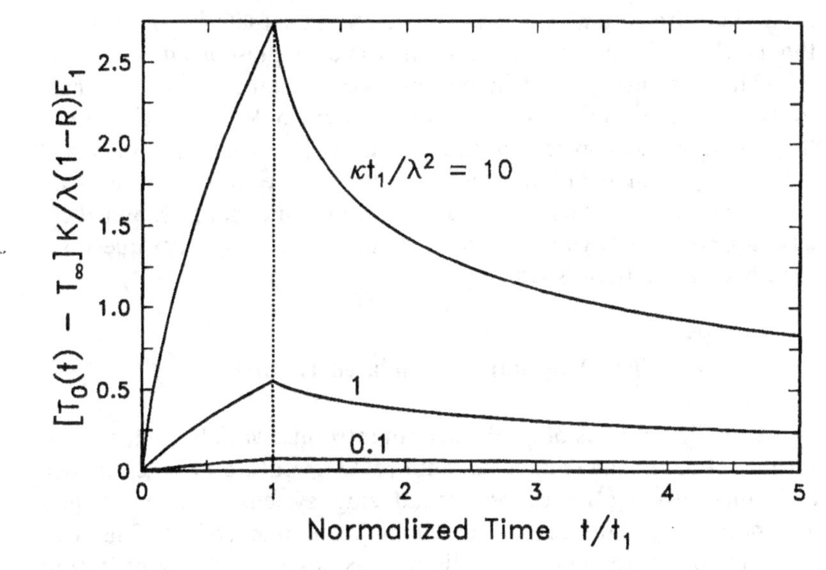

Fig. 8.3. Surface temperature response to a rectangular laser pulse. Note that this system, as all diffusion systems, has no inertia, i.e., the temperature drops in phase with the laser power.

EXERCISE 8.4. *Calculation using the Laplace transform:* Obtain this last result (8.19) from the general Laplace transform (8.8). □

Figure 8.3 shows the surface temperature as a function of time, scaled to the rectangular pulse length t_1. We see that the temperature rises as long as the laser is on, but that it drops immediately after the power drops to zero. Thus, the maximum temperature $T_{\max}$ occurs exactly at the pulse's end, $t = t_1$. This can be easily proved analytically, because the derivative of the U-function (8.19a) is $e^{\tau} \operatorname{erfc}\sqrt{\tau}$ [see Eq. (8.13b)], which never vanishes.† Consequently, as in the previous case (8.17), $u_0 \geqslant 1$ implies the condition

$$\varphi_1(e^{\tau_1} \operatorname{erfc}\sqrt{\tau_1} + 2\sqrt{\tau_1/\pi} - 1) \geqslant 1 \tag{8.20}$$

†One shows that there is no maximum in the *open* intervals $(0, t_1)$ and (t_1, ∞). Therefore, $T_{\max}$ occurs exactly at $t = t_1$. For more smoothly varying laser schedules (even triangular ones), one shows that $T_{\max}$ always occurs *after* the maximum of the laser pulse.

for a phase transition to occur. This relation characterizes a region in the (t_1, F_1)-plane between the laser flux and its pulse length.

There are many other interesting questions about heat and mass transport induced by laser irradiation. For example, we have assumed that the flux was constant in the *lateral* dimensions, a generally bad assumption. Moreover, cw lasers (or other sources of heat) often scan the material. Consequently, we often have moving sources, and these have approximately gaussian radial shapes. These and other questions have been actively pursued.[1,7–10]

8.3. Thermally Stimulated Diffusion

Even linear diffusion problems can introduce variable diffusivities. In Sect. 4.5, we saw that programmed temperature changes induce additional driving forces on crystallizing systems. There we had assumed that the diffusivity was temperature-independent. This may be a good approximation in liquids as long as the temperature excursions are moderate. In solids, however, diffusivities are often activated for reasons explained in Sect. 3.2. Consequently, they are strong functions of temperature and therefore of time. To motivate the following treatment of such cases, we turn to an old but popular method of surface analysis—*flash desorption*.

The flash desorption method consists, essentially, of placing a clean sample in an evacuated ampule, where it is thoroughly degassed. A known amount of foreign gas is then introduced into the chamber, and it is allowed to equilibrate with the sample. Thus, we know the initial adsorbed concentration $C_{s,e}$ of that foreign gas on our sample. If, now, the ampule is again evacuated, then that surface population is no longer in equilibrium with the gas phase, and desorption will occur. This can be monitored with pressure gauges or mass spectrometers, and the kinetics of desorption, i.e., essentially the decay of the adsorbed concentration $C_s(t)$, can be followed. This method yields insight into the forces that bind molecules to surfaces. Nevertheless, its analysis depends on the solution of models for the desorption process.[11,12] It is often suggested that a rate equation of the form

$$\dot{C}_s = -KC_s^{\nu}, \tag{8.21a}$$

with $C_s(0) = C_{s,e}$ holds. Here, K and ν are the rate constant and the order of the (desorption) reaction, respectively. This differential equation is separable

$$dC_s/C_s^\nu = -K\,dt \tag{8.21b}$$

as long as K does not depend on concentration. If K is also time-independent, then Eq. (8.21b) is easily solved:

$$Kt = \begin{cases} \ln(C_{s,e}/C_s) & \text{if } \nu = 1, \\ (\nu - 1)(C_s^{1-\nu} - C_{s,e}^{1-\nu}) & \text{if } \nu \neq 1. \end{cases} \tag{8.22}$$

The important point to note is that the right-hand side of this equation consists of pure functions of C_s, while the left-hand side depends exclusively on time. Consequently, if K is any function of temperature T, and if the temperature is programmed to change, e.g., linearly[†]

$$T(t) = T_0 + \alpha t, \tag{8.23}$$

then the kinetics (8.22) are still true by merely substituting

$$\int_0^t K[T(s)]\,ds \tag{8.24}$$

for the left-hand side Kt, as should be clear from Eq. (8.21b).

The analysis of these kinetics is not our concern here. It suffices to indicate that there are, again, two competing forces: (i) a temperature ramp, which tends to *increase* K and, with it, the rate of desorption; (ii) the natural (isothermal) tendency of $\dot{C}_s$ to *decrease* with time according to Eqs. (8.21). Therefore, the desorption rate exhibits peaks, whose analysis provides information on surface binding and structure. This experimental method is called *thermal stimulation*, and it has been applied to a wide variety of other physical systems, e.g., thermally stimulated electronic or ionic currents in dielectrics and semiconductors.[13]

Exactly the same situation holds for the diffusion equation (1.14) or (1.33) when the diffusivity depends on time through programmed

[†]Since positive temperature ramps are more common in thermally stimulated processes, here we take the ramp rate α to be positive for heating. The opposite sign was used in Eq. (4.35a) for crystallization. We note that the following considerations are valid for *any* given temperature schedule $T(t)$.

temperature changes (8.23). If λ is some constant distance scale, then it is easy to see that the change of variables†

$$\xi = x/\lambda \tag{8.25a}$$

and

$$\tau = \lambda^{-2} \int_0^t D[T(s)]\, ds \tag{8.25b}$$

transforms the diffusion equation (1.14) into $C_\tau = C_{\xi\xi}$, namely, an equation whose coefficients are *constant*. The same is evidently true in several space dimensions. Consequently, *any* solution obtained for constant D becomes formally valid by the mere substitution of ξ and τ for x and t. This method has been used to study impurity diffusion kinetics if the sample's surface does not accumulate these impurities.[14–16] The spectra exhibit peaks, just as in the case of flash desorption, because, on the one hand, D rises with temperature (thus causing larger fluxes) and, on the other hand, the amount of diffusant in any finite sample decreases with time, which tends to decrease the surface flux.

Now, in particular, Eq. (8.25b) is a relation between physical time t and "stretched" time τ, and it is this relationship that we must explore. For definiteness, let us suppose (see Chap. 3) that D has an Arrhenius dependence

$$D(T) = D_\infty e^{-E/kT} \tag{8.26}$$

on T, where E and k are, as usual, an activation energy and Boltzmann's constant, respectively, and where D_∞ is a constant "pre-exponential factor" that represents D at infinite temperature. Together with the ramp (8.23), its substitution into Eq. (8.25b) yields

$$\tau = (D_\infty/\lambda^2) \int_0^t e^{-E/k(T_0 + \alpha s)}\, ds. \tag{8.27a}$$

The exponent shows, as in statistical mechanics, that E/kT is a natural variable for this problem. Therefore, the change of variables

†It should be noted that $\partial/\partial t = (D/\lambda^2)\partial/\partial\tau$.

$s \to \beta = E/kT(s)$ transforms the integral into

$$\tau(t) = (ED_\infty/\alpha k\lambda^2) \int_{E/kT(t)}^{E/kT_0} e^{-\beta}\beta^{-2}\,d\beta. \tag{8.27b}$$

This integral, unfortunately, is not expressible in terms of elementary functions, but it is the difference between two integrals of the form

$$I(z) = \int_z^\infty e^{-\beta}\beta^{-2}\,d\beta, \tag{8.28a}$$

which is closely related to the *exponential integral.*[17] Thus, we have

$$\tau(t) = (ED_\infty/\alpha k\lambda^2)\{I[E/kT(t)] - I(E/kT_0)\}, \tag{8.29a}$$

and we are left wondering how to compute these integrals. In general, it is true, we must either compute the integrals numerically or use tables provided in Ref. 17. On the other hand, unless the activation energy E is very small, the argument E/kT (and *a fortiori* E/kT_0) is a very large number. For example, if the final temperature is 1000°C and if the activation energy is of the order of 1 eV, then that argument's value is approximately 10. Consequently, we are only interested in the asymptotic behavior of the integral (8.28a).

Exactly as with the error function in Eqs. (4.11) and Appendix D, repeated integration by parts produces the asymptotic expansion

$$I(z) \sim \frac{e^{-z}}{z^2}\left(1 - \frac{2}{z} + \frac{2 \times 3}{z^2} - \cdots\right), \tag{8.28b}$$

and when the argument is positive and large we retain only the first term. Therefore, the relation (8.29a) becomes asymptotically

$$\tau(t) \propto [E/kT(t)]^{-2}e^{-E/kT(t)} - (E/kT_0)^{-2}e^{-E/kT_0}. \tag{8.29b}$$

This function is plotted in Fig. 8.4 for various values of the initial energy ratio E/kT_0, together with an approximation, proved in the next exercise, that holds for moderate temperature excursions.

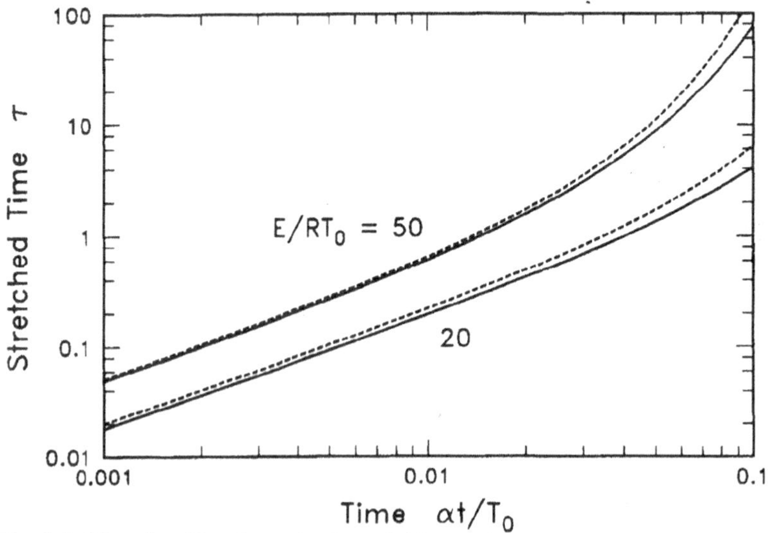

Fig. 8.4. The time-like (stretched) variable τ when the diffusivity depends explicitly on time.

EXERCISE 8.5. *Short-time expansion:* Expand the exponent in Eq. (8.27a) and find an approximation valid for short times. Invert for $t(\tau)$ in that approximation. What is an appropriate length scale λ? □

8.4. Green's Functions and the Use of Integral Equations

Green's functions are often cloaked in mystery, which is unfortunate because their motivation is so simple. Recalling Eqs. (1.29b), (4.5), (7.33c), (7.36), and (8.9), we remember that in all of these, solutions to given problems are expressed as convolution integrals over some *a priori arbitrary condition.* That condition might be an initial condition, as in Eqs. (1.29b) and (4.5), or an inhomogeneous forcing term, as in Eqs. (7.33c) and (8.9), or yet a Dirichlet condition, as in Eq. (7.36). The functions with which these conditions are convolved are precisely Green's functions. We have verified (often by their very construction) that each of these is a solution of the *same* underlying differential equation and that it satisfies some *homogeneous* auxiliary condition. For example, the function that multiplies the arbitrary surface concentration $C_0(t)$ in Eq. (7.36) is a solution of the diffusion equation, and it vanishes at both ends of the interval $[0, \infty]$. In general, each given problem, expressed through a differential equation *and* auxiliary conditions, has a Green's function G with these properties. A thorough investigation of Green's functions belongs in texts on differential equations. Here, we shall continue to construct them, almost automatically, through Laplace transforms. The following example illustrates the procedure.

Once again, we consider impurity diffusion in a half-space under the usual initial and far-field conditions

$$C(x, 0) = C_\infty \tag{8.30a}$$

and

$$C(\infty, t) = C_\infty. \tag{8.30b}$$

Rather than specify either concentration or flux at the surface $x = 0$, let us impose a radiation boundary condition

$$D\frac{\partial C}{\partial x} = k(C - C_e) \qquad \text{at } x = 0 \tag{8.31}$$

that was introduced in Sect. 6.6. Here k is some positive number, and C_e is any given positive function of time. In other words, we specify a linear combination of flux and field at the boundary. The radiation boundary condition reduces to a Dirichlet condition $C_0 = C_e$ when the parameter k is very large, and to a (homogeneous) Neumann condition $J_0 = 0$ when $k \to 0$. Physically, it merely states that the surface flux must accommodate any deviation from equilibrium at the surface, k being a "reaction constant." We have seen it in at least two different contexts: the boundary condition (2.9a) for oxidation kinetics, and the exchange-flux expression (2.28b) for the BCF theory. Its origin in elementary surface processes was discussed in Chap. 6.

The diffusion problem stated above is linear. It can be solved either directly (see Eqs. 6.47) or by straightforward application of the Laplace transform. But we would like this example to buy us a method of solution applicable to *nonlinear* boundary conditions as well. That application will be found in the next section. Here we solve our problem by reducing it to the solution of an integral equation. The procedure is very simple. Rather than solve the stated problem, let us suppose that the condition (8.31) is replaced by an arbitrary Neumann condition. In other words, we assume that the surface flux $-D\partial C/\partial x|_{x=0}$ is a given but yet unknown function $J_0(t)$ of time.

The Laplace transform of the diffusion equation, under the conditions (8.30), yields

$$\bar{C}(x, p) = p^{-1}C_\infty + Ae^{-x\sqrt{p/D}}, \tag{8.32a}$$

exactly as in Sect. 7.4. The "constant" A can be expressed in terms of this flux, because $\bar{J}_0 = -D\bar{C}'|_{x=0}$. Therefore, we have

$$\bar{C}(x, p) = \frac{C_\infty}{p} + \bar{J}_0 \frac{e^{-x\sqrt{p/D}}}{\sqrt{pD}}. \tag{8.32b}$$

The first term's inverse transform is just the constant C_∞ because of entry 4 of Table 7.1, and the second term is the product of two factors. The second factor's inverse transform appears in entry 9 of Table 7.1, and, using the convolution theorem (7.35), we get

$$C(x, t) = C_\infty + \int_0^t J_0(s) \frac{e^{-x^2/4D(t-s)}}{\sqrt{\pi D(t-s)}} ds. \tag{8.33}$$

This equation expresses the concentration distribution essentially as a convolution of an arbitrary flux and the gaussian (4.3).† In other words, should we ever find the flux J_0, then the full concentration distribution would be completely determined by the single quadrature (8.33). In particular, setting $x = 0$, we obtain the surface concentration

$$C(0, t) = C_\infty + \int_0^t J_0(s) \frac{ds}{\sqrt{\pi D(t-s)}}. \tag{8.34}$$

We note that, until now, the boundary condition (8.31) has not been used at all. If, however, we recognize that its left-hand side must be equal and opposite to J_0, then, inserting Eq. (8.34) into Eq. (8.31), we produce the equation

$$-J_0(t) = k\left[C_\infty - C_e + \int_0^t J_0(s) \frac{ds}{\sqrt{\pi D(t-s)}}\right]. \tag{8.35}$$

†By now, the reader surely knows that gaussians satisfy the diffusion equation and that their gradient vanishes at the position of their maximum i.e., they are authentic Green's functions for this Neumann problem.

This expression is called an *integral equation*, because the unknown function $J_0(t)$ appears both inside and outside an integral. In many ways, such expressions are the natural analog (for functions) of linear equations (for algebraic quantities). The method of solution that we shall explore in the next section will emphasize that analogy.

There are many types of integral equations. Their classification is beyond our scope, but we note that this one is linear in the unknown function and that the range of integration is limited to the interval $[0, t]$. Technically, it is called a "Volterra integral equation of the second kind, with a weakly singular, displacement kernel."[18] To elaborate somewhat, the function $(t - s)^{-1/2}$ that multiplies $J_0(s)$ is called the *kernel* of this integral equation. It is of *displacement* type because it depends exclusively on the retarded argument $t - s$. This kernel is also called *weakly singular* because its singularity is integrable. This last remark is of some importance for any numerical method of solution, as we shall soon see. But are numerical methods required for this simple problem? One would hope not, because, as mentioned earlier, linear diffusion problems under radiation boundary conditions can be solved directly.

Indeed, because Eq. (8.35) is linear, and because of its displacement kernel, we can apply the Laplace transform and its convolution theorem. If the equilibrium concentration C_e is constant, then

$$-\bar{J}_0 = k[p^{-1}(C_\infty - C_e) + \bar{J}_0(pD)^{-1/2}], \tag{8.36a}$$

because $\mathscr{L}\{(\pi t)^{-1/2}\} = p^{-1/2}$. This is merely a linear equation in the transform $\bar{J}_0$, which is trivially solved:

$$\bar{J}_0(p) = -kD\frac{C_\infty - C_e}{\sqrt{Dp}\left(k + \sqrt{Dp}\right)}. \tag{8.36b}$$

Its inversion is equally simple if we use entry 12 of Table 7.1 and the change of scale property (7.8), and we get the time dependence of the flux:

$$J_0(t) = -k(C_\infty - C_e)e^{k^2t/D}\,\mathrm{erfc}\left(k\sqrt{t/D}\right). \tag{8.37}$$

This procedure deserves a few remarks. First, it is clear that we have completely solved our diffusion problem, because, as mentioned earlier, insertion of the flux (8.37) into the Green's function representation (8.33) of the concentration yields a closed-form expression for $C(x, t)$.

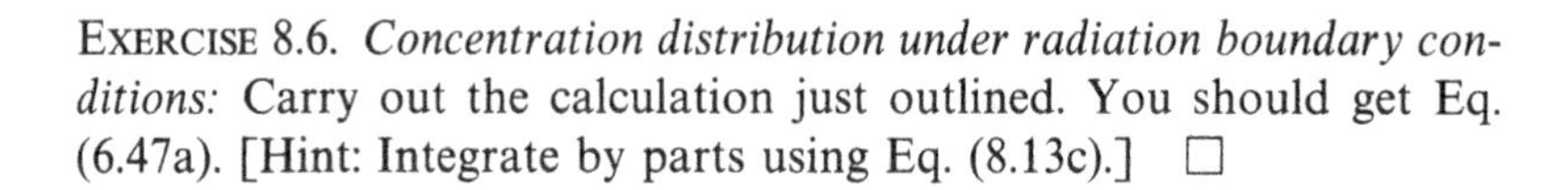

EXERCISE 8.6. *Concentration distribution under radiation boundary conditions:* Carry out the calculation just outlined. You should get Eq. (6.47a). [Hint: Integrate by parts using Eq. (8.13c).] □

Second, there are circumstances in which C_e is a given function of time; the effect of temperature ramps, discussed in Sect. 4.5, is a case in point. The previous problem can still be solved analytically, as is shown in the next exercise.

EXERCISE 8.7. *Exact solution with a variable forcing function:* Assume that the equilibrium concentration is a variable, but known, function of time. Solve the integral equation (8.35). □

Third and most important, we have, in effect, decomposed the given problem in the two independent variables, x and t, into two separate problems, each carrying over a *single* independent variable. Indeed, the integral equation (8.35) involves only the time t, and the quadrature (8.33) is again over the time-like variable s. Fourth, there exist circumstances in which one may be concerned only with the surface flux or the surface concentration, because, for example, the whole concentration distribution is unobservable. Integral-equation methods then provide a direct route to the quantities of interest. Fifth, in contrast to partial differential equations, integral equations are very easily solved numerically with great accuracy, as we will soon see. Finally, the reduction of a given problem to an integral equation often provides a means of studying the analytic properties and asymptotic behavior of solutions.[19–21]

Before going on to the final sections of this book, let us note that, from our work with lasers, we are already familiar with the particular function of time (8.37). Indeed, this very function, albeit with a different time-scaling, appears in Eqs. (8.9). It is even plotted in Fig. 8.1 as a dotted line, and it would represent the surface temperature response to a laser pulse in the form of a delta function of time.

8.5. Application to Nonlinear Surface Conditions

Are we to lose hope if faced with a problem where the boundary condition (8.31) is *nonlinear*? For example, chemical reactions are often *not* first order in the concentrations, and for thermal problems radiation involves the fourth power of the surface temperature. Indeed, the study of chemical boundary layers[22] requires similar analyses. Let us explore ways of solving the same diffusion problem as in the last section, but where the boundary condition (8.31) is replaced by

$$D\frac{\partial C}{\partial x} = K(C - C_e)^\nu \qquad \text{at } x = 0, \tag{8.38}$$

in which ν is an arbitrary power, and the reaction constant is now called K to distinguish it from the linear case. The method of integral equations quickly leads us to a formal solution because the development from Eqs. (8.32) to Eq. (8.34) is independent of this boundary condition. Consequently, it suffices merely to replace Eq. (8.35) by

$$-J_0(t) = K\left[C_\infty - C_e + \int_0^t J_0(s)\frac{ds}{\sqrt{\pi D(t-s)}}\right]^\nu. \tag{8.39}$$

This nonlinear integral equation must be solved numerically. Once this is accomplished, the complete concentration distribution again results from the quadrature (8.33), which now must be performed numerically.

Let us first reduce the integral equation to a standard dimensionless form. Taking a cue from the solution (8.37) of the linear radiation problem, we see that $\tau = k^2t/D$ and $J_0/k(C_\infty - C_e)$ are dimensionless because $[k] = \text{cm/s}$. By inspection, we see that $K(C_\infty - C_e)^{\nu-1}$ also has dimensions of velocity. Consequently, if we define a scaled time and flux

$$\tau = K^2(C_\infty - C_e)^{2(\nu-1)}t/D \tag{8.40a}$$

and

$$\varphi(\tau) = -\frac{J_0(t)}{K(C_\infty - C_e)^\nu}, \tag{8.40b}$$

and if we take the νth root of Eq. (8.39), then the integral equation takes the dimensionless form

$$[\varphi(\tau)]^{1/\nu} = 1 - \int_0^\tau \varphi(\sigma)\frac{d\sigma}{\sqrt{\pi(\tau-\sigma)}}. \tag{8.41}$$

It is just as easy to solve this integral equation when the constant, unity, on its right-hand side is replaced by an arbitrary but known forcing function f of time (see Exercise 8.7). Consequently, let us discuss the solutions, numerical and analytic, of the following standard integral equation:

$$[\varphi(t)]^{1/\nu} = f(t) - \int_0^t \varphi(s) \frac{ds}{\sqrt{\pi(t-s)}}, \tag{8.42}$$

where, for purely typographical reasons, we now call the dimensionless independent variables t and s.

The numerical solution of integral equations depends on methods for estimating the integral. In other words, we need a suitable quadrature formula. Let us seek a solution if $t \in [0, T]$, where T is any preassigned final value of t. If

$$t_0 = 0, t_1, \ldots, t_n, \ldots, t_N = T \tag{8.43a}$$

is any subdivision of this interval, then, just as in our discussion in Sect. 1.4, "seeking a solution" means that we wish to find the values of the unknown function

$$\varphi(t_n) \equiv \varphi_n \tag{8.43b}$$

at these nodes. At first glance, we might think that even the trapezoidal rule for computing integrals would be adequate. A moment's reflection, however, shows that such a procedure cannot succeed, because the kernel is singular whenever the integration variable s passes through any of the nodes (8.43a).

If, on the other hand, we are able to represent the unknown function $\varphi(t)$ as a linear combination of simple integrable functions—call them $S_n(t)$—and if these "basis functions" select the proper values φ_n at the nodes t_n, then our problem, in principle, will be solved. Therefore, let us write

$$\varphi(t) = \sum_{n=0}^{N} \varphi_n S_n(t), \tag{8.44}$$

where the S's are unit height "tents" (or "hats"), centered at each node. They are represented in Fig. 8.5, with the understanding that each $S_n(t)$ vanishes identically outside the interval $[t_{n-1}, t_{n+1}]$. Therefore, the

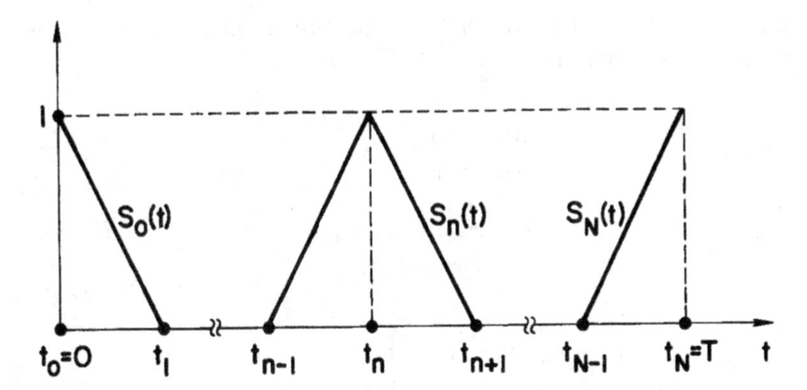

Fig. 8.5. Tent functions (finite elements) for the solution of integral equations.

sum (8.44) clearly satisfies the "collocation" condition (8.43b) at each of the nodes (8.43a). Indeed, that sum is simply a linear interpolation of adjacent functional values of φ.

Introducing Eq. (8.44) into the integral of Eq. (8.42) and requiring that the integral equation hold at any node t_n, we obtain

$$\varphi_n^{1/\nu} = f_n - \sum_{m=0}^{N} A_{nm}\varphi_m \qquad (n = 0, 1, \ldots, N), \tag{8.45a}$$

where, as in Eq. (8.43b), the f_n are the node values of $f(t)$, and the matrix elements are

$$A_{nm} = \int_0^{t_n} S_m(s) \frac{ds}{\sqrt{\pi(t_n - s)}}. \tag{8.45b}$$

This is an $(N + 1) \times (N + 1)$ system of *algebraic* equations in the unknown functional values φ_n. To solve this system, we need expressions for the matrix elements (8.45b), and to perform these integrals, we now need analytic expressions for the basis functions.

First, note that the $N + 1$ nodes (8.43a) are not necessarily equidistant. The following calculations, however, are easiest if we maintain equidistance, namely,

$$t_n = n\Delta t \qquad (n = 0, 1, \ldots, N), \tag{8.46}$$

where $\Delta t = T/N$. Since each basis function is piecewise linear, we easily get the equations

$$S_0(t) = \begin{cases} 1 - t/\Delta t & 0 \leqslant t \leqslant t_1, \\ 0 & \text{elsewhere}, \end{cases} \tag{8.47a}$$

$$S_n(t) = \begin{cases} 1 - |t - t_n|/\Delta t & t_{n-1} \leqslant t \leqslant t_{n+1}, \quad (n = 1, 2, \ldots, N-1), \\ 0 & \text{elsewhere}, \end{cases} \tag{8.47b}$$

$$S_N(t) = \begin{cases} 1 - (t - T)/\Delta t & t_{N-1} \leqslant t \leqslant T, \\ 0 & \text{elsewhere}. \end{cases} \tag{8.47c}$$

The integrals (8.45b) are now convergent. They can be performed explicitly, and we find

$$A_{00} = 0, \tag{8.48a}$$

$$A_{nm} = 0 \qquad \text{if } m > n, \tag{8.48b}$$

$$A_{nn} = \tfrac{4}{3}\sqrt{\Delta t/\pi} \qquad (n = 1, 2, \ldots, N), \tag{8.48c}$$

$$A_{n0}/A_{nn} = (n-1)^{3/2} - n^{1/2}(n - \tfrac{3}{2}) \qquad (n = 1, 2, \ldots, N), \tag{8.48d}$$

$$A_{nm}/A_{nn} = (n - m - 1)^{3/2} + (n - m + 1)^{3/2} - 2(n - m)^{3/2}$$
$$\text{if } 0 < m < n = 2, 3, \ldots, N. \tag{8.48e}$$

After this algebraic spell, let us attempt to interpret these results. First, Eq. (8.48a) is obvious because we are integrating over an interval of length zero. Second, Eq. (8.48b) also makes sense because we are integrating a null function. Therefore, the matrix A_{nm} is triangular with zero entries above its principal diagonal. This is a feature of all Volterra integral equations. Third, all elements (8.48c) of the principal diagonal are equal, except for the first, which is zero. Finally, all the remaining matrix elements depend exclusively on the difference $n - m$, because the kernel itself depends on the difference $t - s$. Thus, all elements of a given subdiagonal are equal, except for the first, A_{n0}, which is given by Eq. (8.48d). It follows that all the information about this matrix resides in its two first columns, A_{n0} and A_{n1}.

With this information, we can rewrite the system (8.45a) in the form

$$\varphi_0^{1/\nu} = f_0, \tag{8.49a}$$

$$\varphi_n^{1/\nu} + A_{nn}\varphi_n = f_n - A_{n0}\varphi_0 - \sum_{m=1}^{n-1} A_{n-m+1,1}\varphi_m \quad (n = 1, 2, \ldots, N), \tag{8.49b}$$

where the first and last terms have been extracted from the sum. This is a recursion whose first member, explicitly given by Eq. (8.49a), is simply $\varphi_0 = f_0^\nu$. The remaining expressions (8.49b) are algebraic equations in the current unknown φ_n in terms of a right-hand side that is a linear combination of all the previous φ's. Therefore, each step of this recursion involves, at most, the solution of a single algebraic equation, and this is very easily handled by standard numerical root-finding algorithms such as Newton's.

If the exponent ν is either 2 or 0.5, then the φ's are merely solutions of quadratic equations. Equations (8.49) are evidently linear when $\nu = 1$. The solutions of Eq. (8.41) when the forcing function f is unity are shown in Fig. 8.6 for the exponents $\nu = 0.5$, 1, and 2. The case $\nu = 1$, as we saw in the last section, has the analytic solution $e^t \operatorname{erfc}\sqrt{t}$. This provides a means to test the accuracy of this numerical method. If Δt is of the order 10^{-2}, then the numerical solution has at least three correct digits, except for the first two time steps.

The numerical method (8.49) rests crucially on the representation (8.44) of the unknown function in terms of very simple basis functions $S_n(t)$. As noted, this representation is, in fact, a linear interpolation between nodes. By choosing other basis functions, one can obtain higher-order schemes that are also more accurate.[23,24] Indeed, the tent functions (8.47) are continuous but not continuously differentiable, so that equation (8.44) would be inadequate if estimates of derivatives of $\varphi(t)$ were needed. The numerical scheme (8.49) is nonetheless very powerful because it holds, almost unchanged, for nonlinearities other than powers. In fact, if the left-hand side of the integral equation (8.42) is replaced by an *arbitrary* function F of φ, i.e., if we wish to solve

$$F[\varphi(t)] = f(t) - \int_0^t \varphi(s)\frac{ds}{\sqrt{\pi(t-s)}}, \tag{8.50}$$

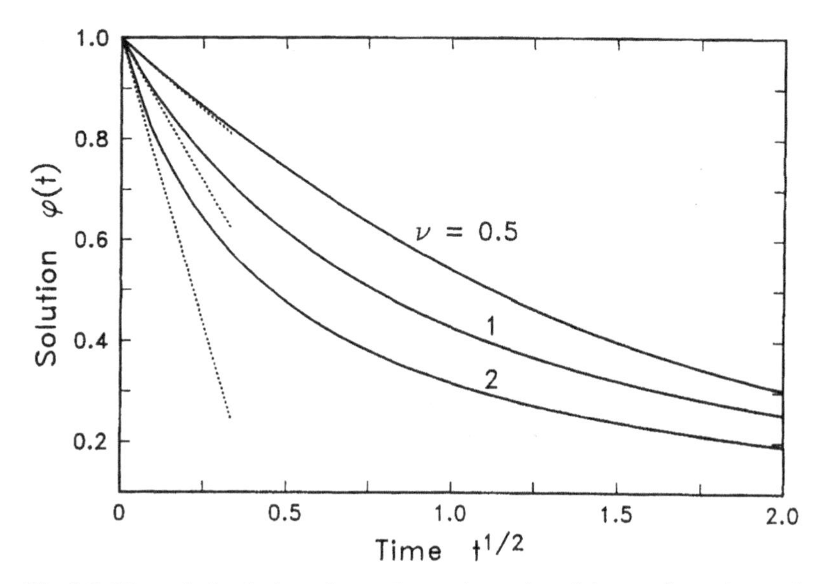

Fig. 8.6. Numerical solutions, for various values of ν, of the nonlinear integral equation (8.41). Dotted lines represent the starting solution (8.54).

then it suffices to replace $\varphi_n^{1/\nu}$ on the left-hand sides of Eqs. (8.49) by the functional values $F(\varphi_n)$ at the nodes (8.46).

8.6. Starting Solutions and an Epilogue

There is a last question of some interest in connection with the nonlinear equation (8.42). We might only be interested in the unknown function's behavior for short times. This would be the case when the time constant (see Eq. 8.40a) of a given physical system is very large. Alternatively, the forcing function $f(t)$ may have very rapid rates of change around $t = 0$, and any reasonable subdivision of time (8.46) would be inadequate. We will now show that it is very easy to find the analytic behavior around the origin.

We assume that the forcing function has a known expansion

$$f(t) = \sum_{n=0}^{\infty} f_n t^{n/2} \tag{8.51a}$$

in powers of $\sqrt{t}$.† The constant term f_0 must not vanish, as will soon be clear. We then seek a similar expression for $\varphi(t)$. If it exists, then its arbitrary power $\psi \equiv \varphi^{1/\nu}$ is similarly expressible. Thus, we write

$$\varphi(t) = \sum_{n=0}^{\infty} \varphi_n t^{n/2} \tag{8.51b}$$

and

$$\psi(t) \equiv [\varphi(t)]^{1/\nu} = \sum_{n=0}^{\infty} \psi_n t^{n/2}, \tag{8.51c}$$

and our problem reduces to the determination of the coefficients φ_n.

Inserting expansions (8.51) into the integral equation (8.42), performing the integrals using the convolution theorem (7.35) and the seventh entry of Table 7.1,[‡] and identifying equal powers of $\sqrt{t}$ produces the recursion

$$\psi_0 = f_0 = \varphi_0^{1/\nu}, \tag{8.52a}$$

$$\psi_n = f_n - c_{n-1}\varphi_{n-1} \qquad n \geqslant 1. \tag{8.52b}$$

Here, the second equality (8.52a) follows from the continuity of f, φ, and ψ around the origin, and the coefficients c_n are

$$c_n = \frac{\Gamma(1 + n/2)}{\Gamma(3/2 + n/2)}. \tag{8.52c}$$

EXERCISE 8.8. *Relations for the c's:* Verify that $c_0 = 2/\sqrt{\pi}$ and that the coefficients (8.52c) obey the recursions

$$c_n c_{n-1} = 2/(n + 1) \qquad \text{and} \qquad c_n = n c_{n-2}/(n + 1).$$

Hence, compute the first few c's. □

It remains to express the coefficients of ψ in terms of those of φ. If, for convenience, we momentarily set $\mu = 1/\nu$, application of the

[†] Here, the index n labels the various powers of $\sqrt{t}$. It must not be confused with that label for nodes in the last section. The odd coefficients f_{2n+1} would all vanish for functions analytic in t.

[‡] This is equivalent to the calculation (7.37) of the beta function.

binomial theorem yields

$$\begin{aligned}\psi(t) &= [\varphi_0 + \varphi_1\sqrt{t} + O(t)]^{\mu} \\ &= \varphi_0^{\mu}[1 + \mu(\varphi_1/\varphi_0)\sqrt{t}] + O(t)\end{aligned} \tag{8.53a}$$

to first order in $\sqrt{t}$. Consequently, we have ψ's first two coefficients

$$\psi_0 = \varphi_0^{\mu}, \qquad \psi_1/\psi_0 = \mu\varphi_1/\varphi_0, \tag{8.53b}$$

which, when inserted into the recursion (8.52), produce the solution

$$\varphi_0 = f_0^{\nu}, \qquad \varphi_1/\varphi_0 = \nu f_0^{-1}(f_1 - c_0\varphi_0). \tag{8.54a}$$

This first-order starting solution

$$\varphi(t) = \varphi_0 + \varphi_1\sqrt{t} \tag{8.54b}$$

is drawn as dotted lines in Fig. 8.7.

The computations illustrated in Eqs. (8.53) and (8.54) can evidently be carried out to any order, and, for example, one can show that each coefficient φ_n is an nth-degree polynomial in ν. The reader is encouraged to work out the second-order terms in order to appreciate the need for tools that are more efficient than binomial expansions. Fortunately, there are formal procedures for finding the coefficients of arbitrary powers of power series,[25,26] and these investigations perhaps deserve more attention from scientists curious about diffusion processes. Finally, we are not limited to powers, and the nonlinear generalization (8.50) also has starting solutions as shown in our final exercise.

EXERCISE 8.9. *Starting solutions for Eq. (8.50):* Find a procedure to calculate the coefficients ψ_n for general nonlinear functions $\psi = F[\varphi]$. Hence, find the second-order starting solution of the integral equation (8.50). [Hint: Calculate the successive derivatives of ψ.] □

The last few sections of this book are paradigmatic, for they show the wide variety of methods that can be brought to bear on particular diffusion problems. They also show that it is quite artificial to oppose numerical to analytical methods. Indeed, in this instance, analysis allows the reduction of a nonlinear problem to an integral equation. For this, a clear understanding of Laplace transforms and Green's functions is necessary. On the other hand, numerics are required to find solutions of this integral equation. Then, to come full circle, analysis permits the establishment of starting solutions that are often useful, if not vital. Consequently, analytic and numerical methods appear as complementary tools for the solution of diffusion problems. An introduction to, and the interplay of, these methods was my intention. I have also tried to indicate that the solution of a given diffusion problem, like any problem, requires a firm grasp of its physicochemical background. Models abound, but few survive, often because some basic principle, such as mass balance, is violated.

References

1. L. G. Pittaway, "The Temperature Distributions in Thin Foil and Semi-infinite Targets Bombarded by an Electron Beam," *Brit. J. Appl. Phys.* **15**, 967 (1964).
2. D. Maydan, "Micromachining and Image Recording on Thin Films by Laser Beams," *Bell Syst. Tech. J.* **50**, 1761 (1971).
3. R. J. von Gutfeld and P. Chaudhari, "Laser Writing and Erasing on Chalcogenide Films," *J. Appl. Phys.* **43**, 4688 (1972).
4. R. A. Ghez and R. A. Laff, "Laser Heating and Melting of Thin Films on Low-Conductivity Substrates," *J. Appl. Phys.* **46**, 2103 (1975).
5. R. Srinivasan, "Ablation of Polymers and Biological Tissue by Ultraviolet Lasers," *Science* **234**, 559 (1986).
6. J. Marshall, S. Trokel, S. Rothery, and R. P. Krueger, "Photoablative Reprofiling of the Cornea using an Eximer Laser," *Lasers in Ophthalmology* **1**, 21 (1986).
7. I. W. Boyd and J. I. B. Wilson, "Laser Processing of Silicon," *Nature* **303**, 481 (1983).
8. I. B. Khaibullin and L. S. Smirnov, "Pulsed Annealing of Semiconductors. Status Report and Unsolved Problems," *Fiz. Tekh. Poluprovodn.* **19**, 569 (1985). [Engl. transl. in *Sov. Phys. Semicond.* **19**, 353 (1985).]
9. D. Rosenthal, "The Theory of Moving Sources of Heat and its Application to Metal Treatments," *Trans. ASME* **68**, 849 (1947).
10. H. E. Cline and T. R. Anthony, "Heat Treating and Melting with a Scanning Laser or Electron Beam," *J. Appl. Phys.* **48**, 3895 (1977).
11. G. Ehrlich, "Kinetic and Experimental Basis of Flash Desorption," *J. Appl. Phys.* **32**, 4 (1961).

12. P. A. Redhead, "Thermal Desorption of Gases," *Vacuum* **12**, 203 (1962).
13. *Thermally Stimulated Relaxation in Solids*, P. Bräunlich, Ed. (Springer-Verlag, Berlin, 1979).
14. G. Carter, B. J. Evans, and G. Farrell, "Gas Evolution from a Solid following De-trapping and Diffusion during Tempering," *Vacuum* **25**, 197 (1975).
15. M. A. Frisch, "Studies on Non-equilibrium Reactions by Knudsen Effusion Mass Spectrometry," in *Adv. in Mass Spectrometry* **8A**, pp. 391–401, A. Quayle, Ed. (Heyden, London, 1980).
16. J. D. Fehribach, R. Ghez, and G. S. Oehrlein, "Asymptotic Estimates of Diffusion Times for Rapid Thermal Annealing," *Appl. Phys. Lett.* **46**, 433 (1985).
17. M. Abramowitz and I. A. Stegun, *Handbook of Mathematical Functions*, Chap. 3 (Reprinted by Dover, New York, 1965).
18. F. G. Tricomi, *Integral Equations* (Interscience, New York 1957). [Also reprinted by Dover, New York, 1987.]
19. W. R. Mann and F. Wolf, "Heat Transfer between Solids and Gasses under Nonlinear Boundary Conditions," *Quart. Appl. Math.* **9**, 163 (1951).
20. R. Ghez and J. S. Lew, "Interface Kinetics and Crystal Growth under Conditions of Constant Cooling Rate," *J. Cryst. Growth* **20**, 273 (1973).
21. W. E. Olmstead and R. A. Handelsman, "Diffusion in a Semi-infinite Region with Nonlinear Surface Dissipation," *SIAM Review* **18**, 275 (1976).
22. P. L. Chambré and A. Acrivos, "On Chemical Surface Reactions in Laminar Boundary Layers," *J. Appl. Phys.* **27**, 1322 (1956).
23. C. Wagner, "On the Numerical Solution of Volterra Integral Equations," *J. Math. & Phys.* **32**, 289 (1954).
24. P. Linz, "Numerical Methods for Volterra Integral Equations with Singular Kernels," *SIAM J. Numer. Anal.* **6**, 365 (1969).
25. J. C. Jaeger, "Conduction of Heat in a Solid with a Power Law of Heat Transfer at its Surface," *Proc. Cambr. Phil. Soc.* **46**, 634 (1950).
26. A. C. Norman, "Computing with Formal Power Series," *ACM Trans. Math. Software* **1**, 346 (1975).

Appendices

Appendix A

Random Walks in Higher Dimensions

The considerations of Sects. 1.1–1.2 are extended here to higher dimensions. Particular attention is given to the notion of flux. Although, for simplicity, we emphasize derivations in two space dimensions, these carry over to any number of dimensions. Some familiarity with tensor algebra[1–3] would be helpful.

We suppose that a list of sites requires *pairs* (i, j) of integers and that $N_{i,j}(t)$ represents the number of particles that reside at the subscripted site at any time t. We also assume that *two* distinct transition frequencies, Γ_1 and Γ_2, are required to describe jumps along the ith and jth directions, respectively. The two-dimensional lattice of integer points is represented in Fig. A.1, and, exactly as in Eqs. (1.1)–(1.3), we find that the particle flows in these directions are

$$
\begin{aligned}
J_{i+1/2,j} &= \tfrac{1}{2}\Gamma_1(N_{i,j} - N_{i+1,j}),\\
J_{i-1/2,j} &= \tfrac{1}{2}\Gamma_1(N_{i-1,j} - N_{i,j}),
\end{aligned}
\tag{A.1a}
$$

and

$$
\begin{aligned}
J_{i,j+1/2} &= \tfrac{1}{2}\Gamma_2(N_{i,j} - N_{i,j+1}),\\
J_{i,j-1/2} &= \tfrac{1}{2}\Gamma_2(N_{i,j-1} - N_{i,j}).
\end{aligned}
\tag{A.1b}
$$

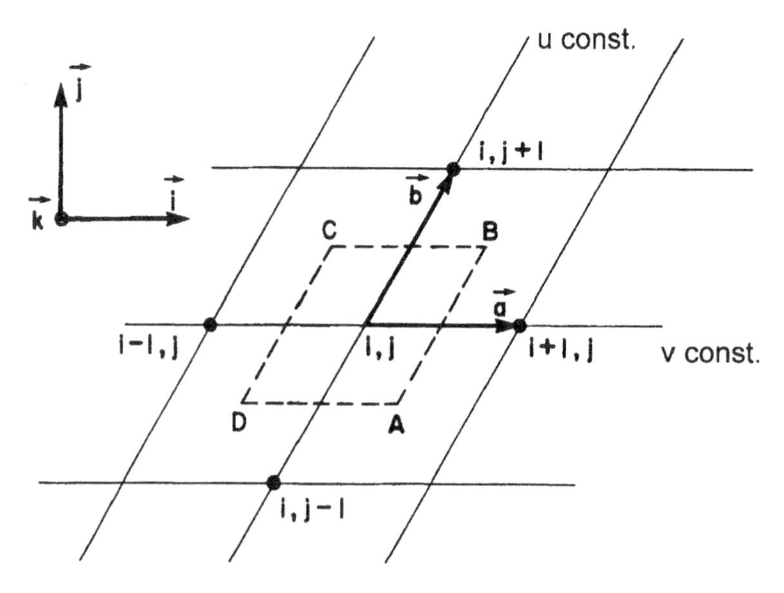

Fig. A.1. Oblique lattice for the 2-d random walk.

The time variation of the particle distribution is

$$\dot{N}_{i,j} = \tfrac{1}{2}\Gamma_1(N_{i+1,j} + N_{i-1,j} - 2N_{i,j}) + \tfrac{1}{2}\Gamma_2(N_{i,j+1} + N_{i,j-1} - 2N_{i,j}), \tag{A.2}$$

which can also be expressed in "divergence" form

$$\dot{N}_{i,j} = -(J_{i+1/2,j} - J_{i-1/2,j}) - (J_{i,j+1/2} - J_{i,j-1/2}). \tag{A.3}$$

To find the continuum form of these equations, we must take coordinate transformations somewhat seriously. The integer sites (i, j) can be viewed either as integer points of the real R^2-plane [namely, the set of real number pairs (u, v)] or as specified points (x_i, y_j) of our physical euclidean space E^2. The specification follows if we impose a one-to-one mapping,

$$x = x(u, v), \qquad y = y(u, v), \tag{A.4}$$

between these two spaces. Because E^2 is endowed with an orthonormal base $(\mathbf{i}, \mathbf{j})$, and thus with a metric, it follows that an arbitrary point has the vector representation $\mathbf{x} = x\mathbf{i} + y\mathbf{j}$, such that $x = \mathbf{x}\cdot\mathbf{i}$ and $y = \mathbf{x}\cdot\mathbf{j}$. In view of the mapping (A.4), its differential takes either one of the two forms

$$d\mathbf{x} = \mathbf{i}\,dx + \mathbf{j}\,dy = (x_u\mathbf{i} + y_u\mathbf{j})\,du + (x_v\mathbf{i} + y_v\mathbf{j})\,dv, \tag{A.5}$$

and the coefficients of du and dv define new basis vectors $\mathbf{a}$ and $\mathbf{b}$, tangent to the coordinate lines $v =$ const. and $u =$ const., respectively, which are generally neither orthogonal nor of unit length, nor yet of constant length as one travels along coordinate lines. If, however, the mapping (A.4) is linear (thus with constant coefficients), then $\mathbf{a}$ and $\mathbf{b}$ maintain constant length, and we have defined a (generally oblique) crystal lattice. These are the *primitive vectors* of crystallography. The correspondence between integer points of R^2 and lattice points of E^2 is complete if we take arbitrary integers for du and dv in the "differential map" (A.5) to get integral linear combinations of the primitive vectors $\mathbf{a}$ and $\mathbf{b}$.

Now, exactly as in Sect. 1.2, we introduce a sufficiently smooth function $\hat{N}(u, v, t)$ that interpolates the discrete functions $N_{i,j}(t)$ at integer sites $u = i$ and $v = j$. Taylor's expansion yields

$$N_{i\pm 1,j} = N_{i,j} \pm \left.\frac{\partial \hat{N}}{\partial u}\right|_{i,j} + \frac{1}{2}\left.\frac{\partial^2 \hat{N}}{\partial u^2}\right|_{i,j} + \cdots \tag{A.6}$$

and a similar expression in the v-direction. Insertion into Eq. (A.2) then produces the partial differential equation

$$\frac{\partial \hat{N}}{\partial t} = \tfrac{1}{2}\Gamma_1 \frac{\partial^2 \hat{N}}{\partial u^2} + \tfrac{1}{2}\Gamma_2 \frac{\partial^2 \hat{N}}{\partial v^2}, \tag{A.7}$$

which is "diagonal" in the oblique, lattice coordinates (u, v). But we are more interested in its form in physical space E^2. To find it efficiently requires a condensed notation.

Rename coordinates and basis vectors according to the scheme

$$x = x^1, \qquad y = x^2, \qquad u = u^1, \qquad v = u^2, \tag{A.8a}$$

and

$$\mathbf{i} = \mathbf{e}_1, \qquad \mathbf{j} = \mathbf{e}_2, \qquad \mathbf{a} = \mathbf{a}_1, \qquad \mathbf{b} = \mathbf{a}_2. \tag{A.8b}$$

Then, with Einstein's summation convention (over the range 1, 2) for diagonally placed Greek indices, Eqs. (A.5) read simply

$$d\mathbf{x} = \mathbf{e}_\alpha \, dx^\alpha = \mathbf{a}_\alpha \, du^\alpha, \tag{A.9a}$$

where the primitive vectors are

$$\mathbf{a}_\alpha = \frac{\partial x^\beta}{\partial u^\alpha} \mathbf{e}_\beta. \tag{A.9b}$$

The four coefficients $\partial x^\beta/\partial u^\alpha$ are the components of the primitive vectors $\mathbf{a}_\alpha$ in the euclidean basis $\mathbf{e}_\beta$, and they also form the jacobian of the mapping (A.4). Indeed, these components can be computed from the scalar products

$$\partial x^\beta/\partial u^\alpha = \mathbf{a}_\alpha \cdot \mathbf{e}_\beta \tag{A.10}$$

in the euclidean metric $\mathbf{e}_\alpha \cdot \mathbf{e}_\beta = \delta_{\alpha,\beta}$, and they are constants if the mapping (A.4) is linear, as will henceforth be assumed. Finally, we recall that an arbitrary vector field $\mathbf{V}$ of E^2 has the two decompositions†:

$$\mathbf{V} = V^\alpha(x) \, \mathbf{e}_\alpha = V^\alpha(u) \, \mathbf{a}_\alpha(u), \tag{A.11a}$$

which define its (contravariant) components in each basis. With the transformation law (A.9b) for basis vectors, we get the transformation law

$$V^\alpha(x) = \frac{\partial x^\alpha}{\partial u^\beta} V^\beta(u) \tag{A.11b}$$

for components. We shall not dwell further on notions of covariance, metric tensor, and Christoffel symbols, because they are not needed in the present context of rectilinear lattices. They are required, however, for a study of *strained* lattices.

†Note the further condensation in notation $(x) \equiv (x^1, x^2)$ and $(u) \equiv (u^1, u^2)$.

Returning to random walks, we can write Eq. (A.7) as

$$\frac{\partial \hat{N}}{\partial t} = D^{\alpha\beta}(u) \frac{\partial^2 \hat{N}}{\partial u^\alpha \partial u^\beta} \tag{A.12a}$$

in the condensed notation, where

$$D^{\alpha\beta}(u) = \frac{1}{2}\begin{pmatrix} \Gamma_1 & 0 \\ 0 & \Gamma_2 \end{pmatrix} \tag{A.12b}$$

is a diagonal matrix formed from the transition frequencies. We now show that it is also a second-order tensor, the *diffusivity tensor* $\mathbf{D}$, whose components in the lattice coordinates (u) are given precisely by Eq. (A.12b).

We first note that Eq. (A.12a) can also be written in divergence form

$$\frac{\partial \hat{N}}{\partial t} = -\frac{\partial}{\partial u^\alpha} \hat{J}^\alpha(u), \tag{A.13a}$$

if the flux components are

$$\hat{J}^\alpha(u) = -D^{\alpha\beta}(u) \frac{\partial \hat{N}}{\partial u^\beta}. \tag{A.13b}$$

Then, expressing the interpolating function $\hat{N}$ in terms of the euclidean coordinates (x), the mapping (A.4) transforms Eq. (A.12a) into

$$\frac{\partial \hat{N}}{\partial t} = \frac{\partial x^\lambda}{\partial u^\alpha} \frac{\partial x^\mu}{\partial u^\beta} D^{\alpha\beta}(u) \frac{\partial^2 \hat{N}}{\partial x^\lambda \partial x^\mu} \tag{A.14a}$$

if the transformation coefficients are constants. It follows that Eqs. (A.12a) and (A.14a) have the same form if the diffusivity components in the euclidean base are

$$D^{\lambda\mu}(x) = \frac{\partial x^\lambda}{\partial u^\alpha} \frac{\partial x^\mu}{\partial u^\beta} D^{\alpha\beta}(u), \tag{A.14b}$$

which is merely the law of transformation of a second-order contravariant tensor. It is easily seen that this tensor is symmetric.

Exercise A.1. *Connection with discrete fluxes (A.1):* Equations (A.13b) are a recipe for the computation of fluxes from the interpolating function $\hat{N}$. As in Exercise 1.2, show that this equation is optimal when derivatives are evaluated at midpoints of the integer lattice of R^2. □

We are now in a position to justify the diffusion equation in anisotropic media. To do so, however, requires inquiry into the physical dimensions of the quantities we have discussed. First, the numbers $N_{i,j}$ have no dimensions, and neither do the coordinates (u) of the R^2-plane. Indeed, the area of any region in this plane is merely a measure of the number of integer lattice points that it contains. Likewise, the interpolating function $\hat{N}(u, v, t)$ has no dimensions.† If, nonetheless, it is to represent its discrete counterpart, then not only must it "collocate" at integer points, but also its integral over a unit square $i - \frac{1}{2} \leqslant u \leqslant i + \frac{1}{2}$, $j - \frac{1}{2} \leqslant v \leqslant j + \frac{1}{2}$ in the R^2-plane (called "square" in the following integral) must match $N_{i,j}$. Thus,

$$\begin{aligned} N_{ij}(t) &= \int_{\text{square}} \hat{N}(u, v, t)\, du\, dv \\ &= \int_{ABCD} \hat{N}(x, y, t)\, |\partial(u, v)/\partial(x, y)|\, dx\, dy, \end{aligned} \tag{A.15a}$$

where $ABCD$ represents the interior of the contour thus named in Fig. A.1.‡ But, according to rules of calculus, the determinant of the jacobian that stands in the second integral is the reciprocal of that formed from the four coefficients in Eqs. (A.10), whose value is exactly the area formed by the primitive vectors. It follows that

$$\hat{N}(x, y, t)\, |\partial(u, v)/\partial(x, y)| = \hat{N}(x, y, t)/|\mathbf{a} \times \mathbf{b}| \tag{A.15b}$$

is precisely what we mean by the concentration distribution $C(x, y, t)$. Therefore, Fick's second law follows if we divide Eqs. (A.14a) and (A.14b) by the constant factor $1/|\mathbf{a} \times \mathbf{b}|$:

$$\frac{\partial C}{\partial t} = D^{\alpha\beta}(x)\, \frac{\partial^2 C}{\partial x^\alpha \partial x^\beta}. \tag{A.16}$$

†For clarity in the following calculations, we momentarily revert to the explicit notation of Eqs. (A.4)–(A.7).

‡This is, in fact, a body-centered unit cell.

On the other hand, analyzing fluxes is more delicate. Indeed, we now demonstrate that the quantities (A.13b) are *not* the components of a flux vector. For example, $\hat{J}^1(u, v, t) = -\frac{1}{2}\Gamma_1 \partial \hat{N}/\partial u$ is an optimal expression of $J_{i+1/2,j}$ if that partial derivative is evaluated at the midpoint of the segment $\overline{AB}$ in Fig. A.1. But the latter represents the *integrated* flux through that boundary. If **J** is the true flux vector, then

$$J_{i+1/2,j} = \mathbf{J} \cdot (\mathbf{b} \times \mathbf{k}), \tag{A.17}$$

where **k** is normal to the plane of Fig. A.1. Thus, **J** is dotted into the reciprocal vector to **b**, times the area $|\mathbf{a} \times \mathbf{b}|$. It follows that $\hat{J}^1/|\mathbf{a} \times \mathbf{b}|$ is the contravariant component J^1 (without a "hat") of **J**. The same reasoning holds for the other three discrete fluxes (A.1). We thus obtain the continuum analogs of Eqs. (A.1) and (A.3):

$$J^\alpha(x) = -D^{\alpha\beta}(x) \frac{\partial C}{\partial x^\beta}, \tag{A.18a}$$

$$\frac{\partial C}{\partial t} = -\frac{\partial}{\partial x^\alpha} J^\alpha(x). \tag{A.18b}$$

in physical space, and these expressions still hold in any number of dimensions by merely extending the range of the Greek superscripts†

EXERCISE A.2: *More dimensions:* Carefully discuss the dimensions of the diffusivity tensor and of the various flux components. [Hint: What are the dimensions of $\partial x^\alpha/\partial u^\beta$?] ☐

Appendix B

Space-Time Symmetries of the Diffusion Equation

It is likely that progress hinges on our notice of regularities or patterns in certain phenomena and geometric figures. The cyclical nature of days and nights, of seasons, and of other astronomical events

†The previous reasoning is nothing more than an elementary proof of the divergence theorem.

suggests basic periodicities, e.g., a certain phenomenon (such as the onset of heavy rains in the tropics) occurs at regular intervals of time. Likewise, certain geometric figures are "predictable" in the sense that they appear unchanged if one changes one's spatial point of view. For example, a perfectly smooth, unlabeled bottle of wine appears the same to all drinkers at a round table, and a single observer, walking around that bottle, would perceive no change in the bottle's appearance. A typical mathematician would say that seasons are invariant under translations of roughly 365 days in time and that our bottle is invariant under rotations of space in any plane perpendicular to its axis. Operations that cause no change in the appearance of a given object are called *symmetry operations*.

We might be interested in operations other than translations of time or rotations in space, and the objects we investigate may include things other than seasons or bottles. Here, we are particularly interested in the way certain functions and differential equations are modified—and perhaps even remain unchanged—under transformations of space and time. Thus, the description of symmetries in physical systems depends crucially on our understanding of transformations of space-time, and we will soon see that the study of symmetries is of more than mere esthetic appeal. In passing, let us mention that there are other transformations of interest in the physical sciences, e.g., permutations of particles and charge conjugation. In that connection, Chapter 52 in the first volume of Feynman's *Lectures on Physics*[4] is highly recommended, as is a wonderful little book by Weyl called simply *Symmetry*.[5]

B.1. Scalar Functions of One Variable and Transformations of the Real Line

Consider the simple function

$$y = x^2. \tag{B.1}$$

As is known, this is a mapping of the real number line R onto itself. Its graph, in the (x, y)-plane, is a certain curve, $\mathscr{C}$, called a parabola.

We wish to investigate how this function appears when the independent variable, x, is changed to a new variable $\bar{x}$. Let us examine three special cases:

$$\bar{x} = x + a, \qquad \text{where } a \text{ is a constant}, \tag{B.2a}$$

$$\bar{x} = 1/x, \tag{B.2b}$$

$$\bar{x} = -x. \tag{B.2c}$$

By substitution, we immediately obtain:

$$y = (\bar{x} - a)^2 = \bar{x}^2 - 2a\bar{x} + a^2, \tag{B.3a}$$

$$y = (1/\bar{x})^2 = 1/\bar{x}^2, \tag{B.3b}$$

$$y = (-\bar{x})^2 = \bar{x}^2. \tag{B.3c}$$

The first case, Eq. (B.2a), is called a *translation* for obvious reasons. Indeed, the parabola is transformed into a curve of identical shape even though its new analytic representation (B.3a) has changed. In the second case, the parabola has been transformed by the *reciprocal substitution* (B.2b) into an "inverse square law" (B.3b); the curve and its analytic representation have changed. The third case, Eq. (B.2c), is called a *reflection* in the y-axis, and Eq. (B.3c) shows that both the curve and its analytic representation are preserved. In high-school we call this phenomenon an "even" function. Here, we say that the parabola's equation (B.1) is *invariant* or *symmetric* under the reflection operation: the curve $\mathscr{C}$ has "bilateral symmetry."

These considerations can obviously be generalized. Let

$$y = u(x) \tag{B.4}$$

be a real function of the real variable x.† We also consider a point transformation of the independent variable

$$\bar{x} = f(x), \tag{B.5}$$

which we assume is invertible

$$x = \varphi(\bar{x}). \tag{B.6}$$

†Here, for once, it is essential to distinguish between the *mapping* u (i.e., the operations on the independent variable x) and its *values* y. Thus, we avoid the ambiguous way of writing $y = y(x)$.

Now, we merely substitute Eq. (B.6) into Eq. (B.4) to obtain

$$y = u(x) = u[\varphi(\bar{x})] \equiv \bar{u}(\bar{x}), \tag{B.7}$$

where the last equality *defines* the new function $\bar{u}$. But—and this is important—nothing is said about the functional form $\bar{u}$ as compared to u. *In general they will be quite different.*

For example, the transformations (B.2a) and (B.2b) change the analytic form of Eq. (B.1) into the distinct forms (B.3a) and (B.3b). Clearly, the new functions $\bar{u}$ are not u! On the other hand, with the transformation (B.2c) we note that $\bar{u}$ is the same function of $\bar{x}$ that u was of x, i.e., the equation (B.1) of the parabola $\mathscr{C}$ is symmetric under that transformation.

B.2. Scalar Functions of Two Variables and Transformations of the Real Plane

As we are mainly interested in solutions of the one-dimensional diffusion equation, let us recall some facts about *point transformations* of the (x, t)-plane, but, for the moment, we are not attaching any physical meaning to these variables. Incidentally, there are other interesting kinds of transformations of variables in mathematics, such as "Legendre transformations," which are widely used in thermodynamics and analytical mechanics.

Call the real number pair (x, t) a *point P*. Geometrically, we represent it in a plane whose coordinate axes we choose, for example, to be rectilinear and orthogonal. Units for the two axes are also specified. To avoid the clumsy expression, "the (x, t)-plane," let us call it simply R^2, namely, the two-dimensional, real number plane of these variables.

We now suppose that we are given two functions

$$\bar{x} = f(x, t), \qquad \bar{t} = g(x, t), \tag{B.8}$$

which we assume sufficiently smooth and also invertible in some domain of R^2:

$$x = \varphi(\bar{x}, \bar{t}), \qquad t = \psi(\bar{x}, \bar{t}). \tag{B.9}$$

Of course, this means nothing at all until we specify the *values* that these functions (B.8) can take. Are the "points" $(\bar{x}, \bar{t})$ to be matrices, colors, or elephants? Obviously here we will require that the "images" $\bar{x}$ and $\bar{t}$ be real numbers. In other words, the functions f and g map real number pairs, one-by-one, onto other real number pairs. For short, we say that Eqs. (B.8) and (B.9) define a *one-to-one point transformation*, and that the pairs (x, t) and $(\bar{x}, \bar{t})$ are the "old" and "new" variables, respectively. There are two analytically equivalent, yet physically distinct interpretations:

(*a*) A given point $P_0 = (x_0, t_0)$ of R^2 is carried by Eqs. (B.8) into a *new* point $\bar{P}_0 = (\bar{x}_0, \bar{t}_0)$, also in R^2, where

$$\bar{x}_0 = f(x_0, t_0), \qquad \bar{t}_0 = g(x_0, t_0). \tag{B.10}$$

Here we emphasize that all the numbers x_0, t_0, $\bar{x}_0$, $\bar{t}_0$, represent coordinates in the *same* space R^2; see Fig. B.1. This is sometimes called an "active" point of view because points get displaced, as in the "Lagrangian" description of elasticity or hydrodynamics.

(*b*) Equations (B.8) and (B.9) define a (one-to-one) mapping from R^2 onto *another* number plane $\bar{R}^2$, also referred conventionally to rectilinear and orthogonal axes; see Fig. B.2. Here a particular point P_0 suffers no displacement at all. It is merely represented by two

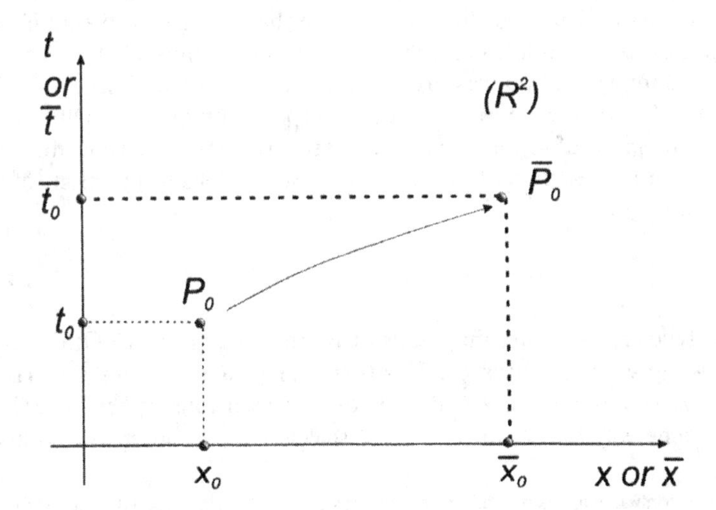

Fig. B.1. Schematic representation of the "active" interpretation of a point transformation (B.8) and (B.9).

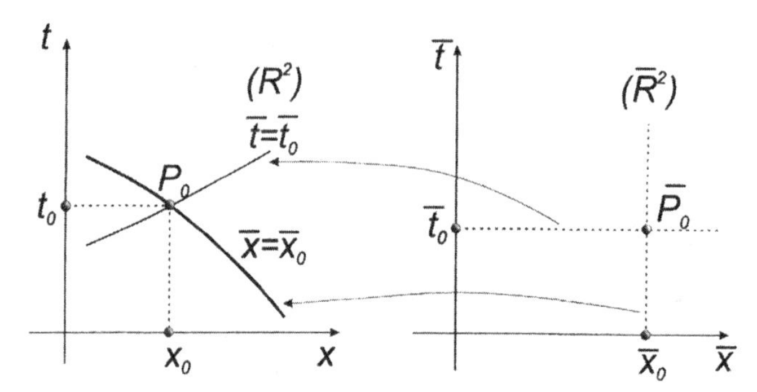

Fig. B.2. Schematic representation of the "passive" interpretation of a point transformation (B.8) and (B.9).

distinct sets of coordinates: those in R^2, itself, and those of its image $\bar{P}_0$ in $\bar{R}^2$. The coordinates of $\bar{P}_0$ are calculated according to the same equations (B.10) as for interpretation (a). This point of view is sometimes called "passive," or a "coordinate change," or a "change of reference frame." Indeed, the term "curvilinear coordinates" comes about because the straight lines $\bar{x} = \text{const.}$ and $\bar{t} = \text{const.}$ of $\bar{R}^2$ generally map onto curves of R^2 according to Eqs. (B.9).

To summarize briefly, point transformations of the plane R^2 can be understood in two different ways: either as displacements in that plane or as coordinate changes that cause no motion whatsoever.† The second interpretation was preferred in Appendix A and Sect. 1.6. Here and in Chap. 5, we adopt the first interpretation almost exclusively.

In spite of appearances, we are still interested in diffusion. Therefore, let us apply the previous considerations to functions of two variables

$$C = u(x, t). \tag{B.11}$$

Analytically, these are mappings of R^2 into the "C-axis." Graphically, they represent surfaces $\mathscr{S}$. In other words, a given function (B.11) selects a set of points, $\mathscr{S}$, from the three-dimensional space $\{x, t, C\}$, in the same way that a function (B.4) selects a set of points, $\mathscr{C}$ (a curve),

†Much confusion arises because, when the transformations (B.8) and (B.9) are linear, the second interpretation can also be viewed as an affine deformation of the coordinate axes of R^2.

from the two-dimensional space $\{x, y\}$. The result, in either case, is what we call a geometric *figure*.

What is the relation of functions to figures? We have all been so thoroughly brainwashed by analytic geometry that we often fail to distinguish between a geometric figure and its analytic representation through equations, finite or differential. For example, the equation $(x+3)^2 + (y-4)^2 - 25 = 0$ represents a circle of radius 5 that is centered at the point $(-3, 4)$ in the (x, y)-plane. But, this analytic representation contains a lot of useless information regarding our circle. If we are interested only in its geometric properties (in the sense of euclidean geometry), then the location of its center in space and, perhaps, also the magnitude of its radius are completely irrelevant. Also implicit in the analytic viewpoint is the disposition and orientation of the coordinate axes (with respect to the "fixed stars," maybe?), and that, too, is completely irrelevant to the definition and properties of circles.[†]

The analytic point of view is invaluable, however, because it can be extended to spaces of higher dimension where geometric visualization fails for most humans, and because the tools of calculus are then available. One might say that the study of point transformations and of symmetries is an attempt to regain the pure geometric (or "intrinsic") point of view.

The ordinary functions of one or several variables, to which we are accustomed, are called *scalar* in the sense that their *values* do not depend on the method (coordinates) for describing the arguments $x, \ldots, t$ in space-time. For example, the values of temperature in a given region of space must be independent of the spatial coordinate system that one chooses.[‡] Therefore, by the mere *substitution* of Eqs. (B.9) into Eq. (B.11), we get

$$C = u(x, t) = u[\varphi(\bar{x}, \bar{t}), \psi(\bar{x}, \bar{t})] \equiv \bar{u}(\bar{x}, \bar{t}), \qquad \text{(B.12)}$$

where the last equality *defines* the new function $\bar{u}$. Again, in general, $\bar{u} \neq u$.

[†]Some may remember that even parabolas [see Eq. (B.1)] have an intrinsic definition (unrelated to cartesian axes, etc.), namely, the locus of points in the plane that are at equal distance from a given point and a given line.

[‡]The values of T *do* depend on dimensions; after all, degrees Fahrenheit and Celsius are not quite the same. In this appendix we do not consider dimensional changes, although they are of the form (B.8) and (B.9).

Returning to the transformation equations (B.5) and (B.6) or (B.8) and (B.9), we note that the active interpretation (a) means that each curve or surface is generally displaced or deformed into a new curve $\bar{\mathscr{C}} \subset \{x, y\}$ or surface $\bar{\mathscr{S}} \subset \{x, t, C\}$, whereas the passive interpretation (b) does nothing at all to the given geometric figures; they are merely represented by generally different analytic forms in the new variables.

We are now in a position to define notions of symmetry that were introduced rather vaguely at the very beginning... remember seasons and bottles? Indeed, we had said, intuitively, that a "thing" is symmetric if specified changes of our point of view do not change its appearance. The kinds of things that we have in mind are geometric figures and their representation through equations. All the previous examples have illustrated that an object's symmetry (or lack thereof) involves two distinct notions. *First, we need a way to represent that object, most often by using coordinates. Second, we characterize an operation on that object (or a change of point of view) by introducing a specified point transformation of these coordinates. The object in question is said to be "symmetric" or "invariant" under that operation (which is then called a "symmetry operation") if its analytic representation remains formally unchanged.*

B.3. Symmetries of the Diffusion Equation

We finally come to the "meat" of this appendix. Among the objects whose symmetries one can discuss is the diffusion equation

$$\frac{\partial u}{\partial t} = D\frac{\partial^2 u}{\partial x^2}, \tag{B.13}$$

written here for constant diffusivity, D, and in one space dimension. But what does it mean to seek symmetries of a partial differential equation such as Eq. (B.13)? After all, it is hardly a bottle. To answer that question, we must remember what it means to "solve" that equation: Under given initial and boundary conditions,[†] find the function $C = u(x, t)$, defined on a certain domain of $R^2 = \{x, t\}$ of the independent variables and with values in $\{C\}$, that identically satisfies Eq. (B.13). Since Eq. (B.11) represents a surface $\mathscr{S} \subset \{x, t, C\}$, a solution is sometimes called a "solution (or integral) surface." Now for

[†] Recall that these are always *equations* that hold on particular lines of R^2.

an important remark: *Since this solution exists and is unique,*[†] *giving the diffusion equation, together with its initial and boundary conditions, is entirely equivalent to giving its solution surface* $\mathscr{S}$. We have thus "geometrized" our problem: To discuss symmetries of the diffusion equation (B.13) under point transformations (B.8) and (B.9) is equivalent to discussing the symmetries of its solution surfaces $\mathscr{S}$, described by Eq. (B.11).

Equation (B.12) describes the transformed shape of the solution surface, $\bar{\mathscr{S}}$, in the new variables $\bar{x}$ and $\bar{t}$. We now ask the question: What is the partial differential equation in these new variables (and the corresponding initial and boundary conditions) whose solution surface is precisely $\bar{\mathscr{S}}$? The answer is usually a sorry mess. We would have to calculate the partial derivatives of u, using Eqs. (B.12) and (B.8), according to the theorem on derivatives of composite functions (the "chain rule"), and insert these into Eq. (B.13). The initial and boundary conditions must be similarly transformed. For example,

$$\frac{\partial u}{\partial x} = \frac{\partial \bar{u}}{\partial \bar{x}}\frac{\partial f}{\partial x} + \frac{\partial \bar{u}}{\partial \bar{t}}\frac{\partial g}{\partial x}, \tag{B.14a}$$

$$\frac{\partial u}{\partial t} = \frac{\partial \bar{u}}{\partial \bar{x}}\frac{\partial f}{\partial t} + \frac{\partial \bar{u}}{\partial \bar{t}}\frac{\partial g}{\partial t}, \tag{B.14b}$$

where the "coefficients" $\partial f/\partial x$, etc., must be expressed in terms of the new variables according to Eqs. (B.9). In general, the resulting partial differential equation for $\bar{u}(\bar{x}, \bar{t})$ and its new initial and boundary conditions bear little resemblance to the original diffusion problem.

So far, so bad. But, what if we had been very, very clever in our choice of transforms, and the surface $\bar{\mathscr{S}}$ *coincides* overall with $\mathscr{S}$? Analytically, this means that the function $\bar{u}$ *equals* the function u. Then, from the string of equations (B.12), we have $C = u(x, t) \equiv \bar{u}(\bar{x}, \bar{t}) = u(\bar{x}, \bar{t})$, and thus, using Eqs. (B.8), the condition

$$C = u(x, t) = u[f(x, t), g(x, t)]. \tag{B.15}$$

In addition, because of the uniqueness theorem, the partial differential equation associated with $\bar{u}$ must be formally unchanged,

$$\frac{\partial \bar{u}}{\partial \bar{t}} = D\frac{\partial^2 \bar{u}}{\partial \bar{x}^2}, \tag{B.16}$$

[†] And these are *theorems*, not mere verbal statements!

as compared to Eq. (B.13) associated with u, and that is what we mean when we say that the diffusion equation is symmetric or invariant under a point transformation (B.8) and (B.9).

It should be appreciated that the previous operations are not mere substitutions of *names*. After all, by giving the old variables any new names, such as $u \to \heartsuit$, $x \to \flat$ $t \to \odot$, we could write the diffusion equation (B.13) as

$$\frac{\partial \heartsuit}{\partial \odot} = D \frac{\partial^2 \heartsuit}{\partial \flat^2}, \tag{B.17}$$

which is not quite the sequence of steps that led to Eq. (B.16). That, perhaps, is the difference between science and sloganeering.

We easily imagine that the problem of finding the most general transformation of coordinates (B.8) and (B.9) that leaves the diffusion equation and its initial and boundary conditions invariant is overly ambitious. Here we put aside the question of these conditions, and we ask instead the question: If u is solution of Eq. (B.13), then, given a coordinate transformation, is $\bar{u}$ [calculated according to Eq. (B.12)] a solution of Eq. (B.16)? We now examine several cases of physically significant transformations; in each case, we ask for the additional restrictions on Eqs. (B.8) and (B.9) (if any) that are required for invariance of the diffusion equation.

1. Translations in Space-Time. The transformation

$$\bar{x} = x + x_0 \qquad \text{and} \qquad \bar{t} = t + t_0, \tag{B.18}$$

where x_0 and t_0 are constants, displaces all points of R^2 by a constant "vector" (x_0, t_0) in the active interpretation.† We apply Eqs. (B.14), which, although trivial in this case, should nonetheless be done carefully. We get

$$\frac{\partial u}{\partial x} = \frac{\partial \bar{u}}{\partial \bar{x}}, \tag{B.19a}$$

†In the passive interpretation, the old origin (0, 0) of space-time coordinates has the new label (x_0, t_0).

hence a similar relation for second derivatives, and

$$\frac{\partial u}{\partial t} = \frac{\partial \bar{u}}{\partial \bar{t}}. \tag{B.19b}$$

Thus, Eq. (B.16) is satisfied: if $u(x, t)$ is a solution of the original diffusion equation (B.13), then so is $u(x + x_0, t + t_0)$, *regardless* of the values of x_0 and t_0. Although this result may appear trivial, it should be borne in mind that we have *proved* (rather than merely proclaimed) the independence of diffusion phenomena on the origin of space and time. Not only would the opposite be quite unfortunate (as, too, for any equation describing natural phenomena), but this result allows us to choose that origin of coordinates as conveniently as we please.

2. Reflections in Space. We have the transformation (B.2c) in mind. Equation (B.19b) holds since that transformation does not affect the time variable (formally, $\bar{t} = t$). The first spatial derivative changes sign because we get $\partial u/\partial x = -\partial \bar{u}/\partial \bar{x}$ instead of Eq. (B.19a). The second derivative sets things right, however, so that the diffusion equation is again invariant under reflections. Thus, if $u(x, t)$ is a solution, then $u(-x, t)$ is also a possible solution. In other words the orientation of the x-axis is irrelevant when setting up a diffusion problem. The extension to multidimensional problems is obvious.

3. Time Reversal. Here we encounter our first "failure." Indeed, the transformation

$$\bar{x} = x \qquad \text{and} \qquad \bar{t} = -t \tag{B.20}$$

now changes the sign of the time derivative, $\partial u/\partial t = -\partial \bar{u}/\partial \bar{t}$ [instead of Eq. (B.19b)], and we do *not* obtain the diffusion equation (B.16). In words: If $u(x, t)$ is a solution, then $u(x, -t)$ is *not* a solution! Here is the mathematical proof that diffusion is an irreversible process: Time travel in reverse will eventually get one into trouble (remember negative moduli in Table 1.1) or perhaps to the Big Bang.

4. Changes of Scale. Finally, we examine the way in which the diffusion equation behaves under changes of scale of both space and time. Setting

$$\bar{x} = \alpha x \qquad \text{and} \qquad \bar{t} = \gamma t, \tag{B.21a}$$

where α and γ are real constants, we calculate $\partial^2 u/\partial x^2 = \alpha^2 \partial^2 \bar{u}/\partial \bar{x}^2$ and $\partial u/\partial t = \gamma \partial \bar{u}/\partial \bar{t}$. Therefore, Eq. (B.16) is satisfied if and only if

$$\gamma = \alpha^2. \tag{B.21b}$$

We will use this important case to advantage in the next section. It should be noted that the coefficient γ can only be positive, in accord with our (negative) conclusion regarding time reversal.

B.4. A Basic Similarity Solution of the Diffusion Equation

Inserting the results (B.21) into Eq. (B.15), we have

$$C = u(x, t) = u(\alpha x, \alpha^2 t), \tag{B.22}$$

which must hold for all values of the parameter α. This type of relation for an unknown function u is called a "functional equation," and it can be solved analytically (meaning that one can find the form of u) by the so-called "method of characteristics." To not burden the reader with further mathematics, let us reason geometrically.

What are the features of the surface $\mathcal{S}$ that is described by Eqs. (B.22)? First, we note that Eqs. (B.21) imply $x/\sqrt{t} = \bar{x}/\sqrt{t} = \text{const.}$, regardless of the scale factor α. Second, according to Eq. (B.22), the corresponding points, (x, t) and $(\bar{x}, \bar{t})$ have the same "ordinate" C. Thus, in the active interpretation of the coordinate transformation (B.21), the level lines of $\mathcal{S}$ must all be parabolas $t \propto x^2$. Each parabola is invariant under that transformation. This suggests that we might be able to integrate the diffusion equation (B.13) by integrating "transversely" to these parabolic level lines.

This is precisely what Boltzmann did in 1894 when he introduced the notion of a *similarity variable*[†]:

$$\eta = \frac{x}{2\sqrt{Dt}}, \tag{B.23a}$$

which obviously has a constant value over each level line, together with his astute observation that the function u might be expressed only

[†]The factor $2\sqrt{D}$ is inserted to render η dimensionless. One finds other conventions in the literature.

in terms of that variable,

$$u = F(\eta), \tag{B.23b}$$

and not in terms of x and t separately. Equations (B.22) are then certainly satisfied.[†]

The rest is almost trivial. We calculate the derivatives of u by using Eqs. (B.23) and insert these into the diffusion equation (B.13). We then easily get

$$\frac{\partial u}{\partial t} = -\frac{\eta}{2t}\frac{dF}{d\eta} \quad \text{and} \quad \frac{\partial u}{\partial x} = \frac{1}{2\sqrt{Dt}}\frac{dF}{d\eta}. \tag{B.24}$$

and, hence, the *ordinary* differential equation

$$F'' + 2\eta F' = 0. \tag{B.25a}$$

Lo and behold! it is integrable, and we get

$$F(\eta) = A + B\int_0^{\eta} e^{-\xi^2}\,d\xi, \tag{B.25b}$$

which is a linear combination of a constant and the integral of a gaussian. For historical as well as practical reasons, one chooses the constants in this general solution so as to satisfy the boundary conditions: $F(0)=0$ and $F(\infty)=1$. It follows that $A=0$ and $B=2/\sqrt{\pi}$, and we recover the error function (4.7) that was obtained by entirely different means in Chap. 4. Then, with Eqs. (B.23), we get the important solution

$$C = u(x, t) = \mathrm{erf}\left(x/2\sqrt{Dt}\right) \tag{B.26}$$

of the diffusion equation (B.13). It is plotted in Fig. B.3, but most of its features are available from Fig. 4.3.

[†] Boltzmann's argument is certainly appealing. Just imagine that you have sliced up your surface $\mathscr{S}$, like a sausage, with the help of planes that are parallel to R^2. It is intuitively obvious that the whole surface can be reconstructed from a description of its level lines. After all, any scout can read a topographic map.

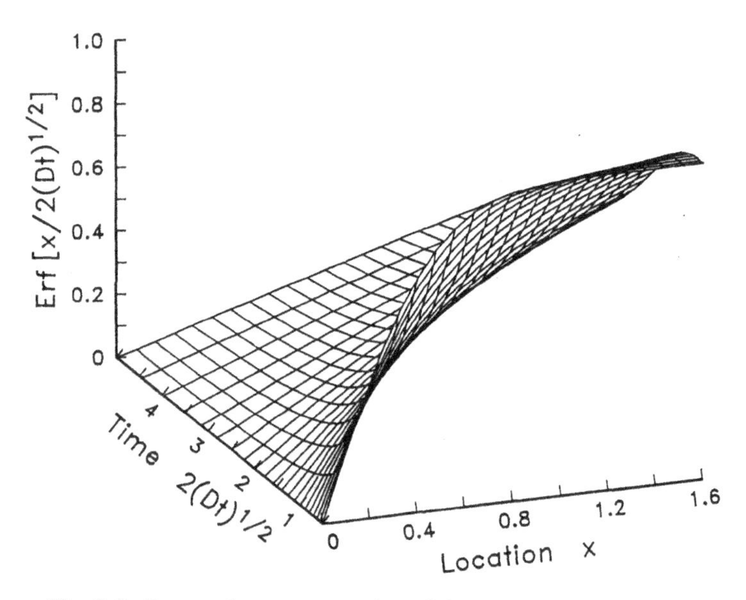

Fig. B.3. Perspective representation of the error function (B.26).

In sum, we used symmetry arguments (under scale changes) to find a similarity variable with which the diffusion equation (a partial differential equation) is reduced to an ordinary differential equation. Chapter 5 develops these questions further.

Appendix C

The Phase Rule and Some of Its Consequences

Some basic thermodynamic results are used to derive various forms of the phase rule[6,7] and their consequences for small systems.

Any fluidlike, 3-d phase is fully characterized by the $2 + c$ fields T, p, and μ_i $(i = 1, 2, \ldots, c)$, also called *thermodynamic intensive variables.* Their values are not all arbitrary, however, because the Gibbs–Duhem relation (2.14a) must hold at equilibrium. Therefore, only $1 + c$ of these values can be arbitrarily assigned. The generalization of this simple fact to a collection of phases is called the *phase rule*, and its derivation is easy.

Consider φ contiguous 3-d phases and assume, first, that they are all large in extent. This collection is characterized by the $\varphi(2 + c)$ fields

$$T^{(\alpha)},\ p^{(\alpha)},\ \mu_i(\alpha), \quad (i = 1, \ldots, c), (\alpha = 1, \ldots, \varphi), \tag{C.1}$$

which, at equilibrium, are constant over their respective domains $\mathscr{D}^{(\alpha)}$.† There are now φ Gibbs–Duhem relations, one for each phase, and equilibrium conditions at the $\varphi - 1$ pairs of phase boundaries

$$T^{(\alpha)} = T^{(\beta)}, \quad p^{(\alpha)} = p^{(\beta)}, \quad \mu_i^{(\alpha)} = \mu_i^{(\beta)}, \qquad (i = 1, \ldots, c), (\alpha \neq \beta), \tag{C.2}$$

for a total of $\varphi + (\varphi - 1)(2 + c)$ relations or "constraints." Therefore, the number of thermodynamic variables whose values can be arbitrarily prescribed is

$$f = c + 2 - \varphi. \tag{C.3}$$

By analogy with mechanical systems, this number f is called the thermodynamic system's number of *degrees of freedom*, and Eq. (C.3) is called the phase rule. This apparently trivial result will perhaps seem less so if we realize that, having experimentally chosen the values of f variables from the list (C.1), the remaining values, as well as the values of *all other* thermodynamic variables that depend on that list (such as the entropy densities and concentrations), are entirely determined when equilibrium is achieved.

This admittedly simple derivation of the phase rule glosses over several important issues that can cause confusion. First, it must be remembered that it holds for phases that are in global equilibrium. It *is* often used, however, in regions that are only in local equilibrium, as in the vicinity of phase boundaries. This is, in fact, an assumption about local kinetics. Next, the c components are assumed to be independent. If not (e.g., in the presence of chemical reactions), the number c in Eq. (C.3) must be reduced by the number of linearly independent constraints. Further, it is assumed that the list of variables (C.1) fully describes the phases' internal state, and this may not be true,

†Recall Note 1 in Chap. 2.

for instance for solids under anisotropic states of stress. Last, the phases are assumed to be large in extent, which means that capillarity effects are negligible. We now seek extensions of Eq. (C.3) when the phases are "small."

The general problem of surface effects is quite complex, because we must learn to characterize 2-d surface phases that constitute the boundaries of bulk 3-d phases. This forms the subject matter of surface thermodynamics, which we broach only through the notion of surface tension γ. In other words, save for γ, the phase boundaries are assumed to have no intrinsic physical properties such as adsorbed quantities, surface energies, and entropy. Therefore, almost all the examples in this book deal with physical situations in which the boundaries are not explicitly the seat of thermodynamic changes, the only notable exceptions being the considerations of Chap. 6 and the BCF theory of crystal growth treated in Sect. 2.6.[†] Enumeration of surface phases is another difficulty with the general theory, because they do not exist in splendid isolation from bulk phases. Since each of these (except for pathological examples like Möbius strips) must be in contact with at least two bulk phases and since the shape of all phases is incidental, it follows that the problem of enumeration is topological.

Because a system of *two* bulk phases in contact through their phase boundary is our main concern, we restrict our discussion to the case $\varphi = 2$ and to situations where γ entirely specifies the surface phase. Blithe application of the phase rule (C.3) would state simply that the number of degrees of freedom is equal to the number of independent components:

$$f = c. \tag{C.4}$$

But is this quite correct? Evidently not, because the second of the relations (C.2) must be replaced by Laplace's law (2.15b), and because we must enumerate all thermodynamic variables properly. (The next remarks are best read together with Sect. 2.4.) There are now $2(2 + c) + 1$ variables, because the mean curvature K must be appended to the list (C.1) to give the new list

$$T^{(\alpha)},\ p^{(\alpha)},\ \mu_i^{(\alpha)},\ K, \qquad (i = 1, \ldots, c),\ (\alpha = 1, 2). \tag{C.5}$$

[†]Indeed, surface properties are always implicit in kinetic processes that involve "surface rate limitations." Thus, oxidation and desorption phenomena fall into that class.

The equilibrium relations† then lead to the special phase rule

$$f = c + 1, \tag{C.6}$$

which, in fact, holds true even when other properties, in addition to the surface tension γ, hold at the boundary of a two-phase system.

A few words are in order regarding the role of the mean curvature as a thermodynamic variable. We first note that constant pressures $p^{(1)}$ and $p^{(2)}$ in the adjacent equilibrium bulk phases (over the domains $\mathscr{D}^{(1)}$ and $\mathscr{D}^{(2)}$, respectively; see Fig. 2.4), together with Laplace's law (2.15b), imply that K must be constant over the phase boundary Σ. Therefore, the *shape* of the surface is characterized by a constant mean curvature. The study of such shapes is known as the "problem of Plateau," and has been pursued intensively over the years.[8–10] Briefly, one shows that the sphere is the only connected and closed (therefore bounded) surface in 3-d space of constant and positive mean curvature. Open unbounded surfaces include the cylinder, with constant positive K, and the catenoid of revolution, with identically vanishing K. The general problem of open and bounded surfaces is both hard and interesting, as evidenced by the rich variety of soap bubbles.[8,11] From our point of view, namely kinetics, we must recognize that K is a field that varies over the phase boundary Σ if the adjacent phases are *not* in global equilibrium. In other words, the phase rules (C.4) and (C.6) apply to the small hatched domain in Fig. 2.4 if it is locally in equilibrium, no matter what kinetic processes occur in the core of the two adjacent bulk domains. The first, Eq. (C.4), holds locally when the surface element's curvature vanishes, while the second, Eq. (C.6), characterizes curved elements.

We now apply this apparatus to the estimation of a particular Gibbs–Thomson effect, namely, the influence of curvature on the melting or sublimation temperature of a pure solid. We now have $c = 1$ and, of course, two bulk phases (solid and liquid, or solid and vapor). Therefore, the phase rule (C.6) indicates that only two thermodynamic variables can be chosen at will from the seven-member list (C.5). As noted above, when the values of these two variables are chosen, all others on that list, as well as any others (such as the entropy densities $s^{(\alpha)}$) whose values depend explicitly on that list, are

†Two Gibbs–Duhem relations (2.14a) and the $2 + c$ relations (2.15), for a total of $4 + c$ relations.

fully determined. We choose a system such that the pressure $p^{(2)}$ of the second phase is maintained constant by contact with a "work reservoir" $p^{(r)}$. We then inquire into the functional dependence of the other variables—in particular, the melting or sublimation temperature—on the mean curvature K.

The calculation is formally simple, but it is fraught with logical shoals. Let us proceed with care as in Sect. 2.4. First, we have the two Gibbs–Duhem relations (2.14a), written here for pure systems

$$s^{(\alpha)}\, dT^{(\alpha)} - \Omega^{(\alpha)}\, dp^{(\alpha)} + d\mu^{(\alpha)} = 0, \qquad (\alpha = 1, 2), \tag{C.7}$$

where the atomic (or molar) volumes $\Omega^{(\alpha)}$ are the reciprocals of the total concentrations $C^{(\alpha)}$. Then, the equilibrium relation (2.15a) and the constancy of the T-fields demand equality of the temperatures in both phases: No matter what their functional dependence on other variables might be, their values must be equal at equilibrium. We call this common value T. By definition, it is the equilibrium temperature of the melting or sublimation process. The same remark holds, in view of Eq. (2.15c), for the chemical potentials. Laplace's law (2.15b), on the other hand, now reads simply $dp^{(1)} = d(\gamma K)$, because we have $dp^{(2)} = dp^{(r)} = 0$, regardless of the reasons for these variations. Subtracting the two relations (C.7) from one another, and taking the previous remarks into account, we get

$$(s^{(2)} - s^{(1)})\, dT + \Omega^{(1)}\, d(\gamma K) = 0. \tag{C.8a}$$

Now, the above difference in entropy densities is, by definition, the latent heat density L of the phase change divided by T.† The previous equation then becomes a differential expression that is trivially integrated from the state of zero curvature to one of finite K,

$$T(K) = T(0)\, e^{-\Omega^{(1)}\gamma K/L}, \tag{C.8b}$$

if both L and $\Omega^{(1)}$ are but weakly dependent on K. On the other hand, we note that γ can have *any* curvature dependence. In contrast to Eq. (2.17c) for curvature-induced shifts of the equilibrium concentration,

†For the latent heat to be positive, we must always choose phase 2 to be the "less condensed" phase, i.e., $s^{(2)} > s^{(1)}$.

the melting or sublimation point *decreases* over surface elements of positive mean curvature, i.e., over convex elements. The relation (C.8b), one of many regarding phase changes, is attributed to J. J. Thomson, who was not Lord Kelvin.†

EXERCISE C.1. *One of W. Thomson's results:* Consider a one-component, two-phase (liquid–gas or solid–gas) system at constant temperature. Discuss the phase rule (C.6), and compute the effect of curvature on the vapor pressure above a curved solid or liquid element. [Hint: Assume that the vapor phase is a perfect gas.] □

Appendix D

Series Can Be Useful!

We imagine being on some desert island (with nothing more than a solar-powered, four-function calculator and a bathing suit), and that there is an urgent need to calculate erf(1.5), for example.‡ Series expansions offer the easiest path, and they also offer convenient error bounds.

D.1. Expansion of erf(z) for "Small" z

Remembering the definition (4.7), we expand the exponential in a Taylor series around the origin and integrate term-by-term (ask yourself why is this allowed). We then have the power series expansion

$$\operatorname{erf}(z) = \frac{2}{\sqrt{\pi}} \int_0^z e^{-\xi^2}\, d\xi = \frac{2}{\sqrt{\pi}} \sum_{n=0}^{\infty} \frac{(-1)^n z^{2n+1}}{n!(2n+1)} \tag{D.1}$$

of the error function. All powers are odd, as they should be, since the error function itself is odd. In addition, we easily check that the power

†See Note 8 in Chap. 2.

‡This little scenario is designed to dramatize that the evaluation of any transcendental (i.e., nonalgebraic) function requires some analysis,[12] even for the smartest computer.

series (D.1) is everywhere convergent, which means that it can be used, in principle, to calculate this function for *all* values of z. The practical question, however, is: *how many terms of this series are required for a given level of accuracy*?

As usual, we write the series as a finite sum S_N (the so-called "partial sum") consisting of exactly N terms, plus its "remainder" R_N:

$$\operatorname{erf}(z) = \underbrace{\frac{2}{\sqrt{\pi}}\left[z - \frac{z^3}{1!3} + \frac{z^5}{2!5} - \cdots + \frac{(-1)^{N-1}z^{2N-1}}{(N-1)!(2N-1)}\right]}_{S_N(z)}$$

$$+ \underbrace{\frac{2}{\sqrt{\pi}}\sum_{n=N}^{\infty}\frac{(-1)^n z^{2n+1}}{n!(2n+1)}}_{R_N(z)}. \tag{D.2}$$

The finite sum is all that we *can* calculate by arithmetic means. The remainder contains an infinite number of terms, and must be estimated by other means. Now, the series (D.1) is "alternating," and calculus teaches that the remainder is then bounded by the first term *omitted* from S_N. Thus

$$|R_N(z)| \leqslant \frac{2}{\sqrt{\pi}}\frac{|z|^{2N+1}}{N!(2N+1)}. \tag{D.3}$$

The rest is easy. If we suppose that we tolerate an absolute error ε in the estimation of erf(z), it means that $|\operatorname{erf}(z) - S_N(z)| \leqslant \varepsilon$. But that difference is exactly the absolute value of the remainder, and Eq. (D.3) shows that the condition is surely satisfied if

$$\frac{2}{\sqrt{\pi}}\frac{|z|^{2N+1}}{N!(2N+1)} \leqslant \varepsilon. \tag{D.4}$$

The inequality (D.4) defines a region of the (N, z)-plane in which the error does not exceed ε. For example, three correct decimal places are assured if $\varepsilon = \frac{1}{2} \times 10^{-3}$.† Figure D.1 shows this region for this level

†A bit of thinking will convince you that, in general, p correct decimal places (after rounding) are assured if $\varepsilon = \frac{1}{2} \times 10^{-p}$. See Smith's entertaining article.[13]

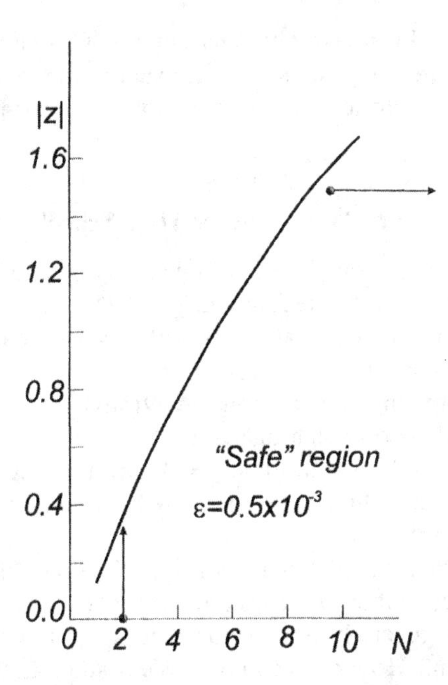

Fig. D.1. Region of the (N, z)-plane in which the inequality (D.4) is satisfied.

of accuracy. Thus, using $S_2(z)$ (two terms, only) is "safe" as long as $|z| < 0.338$. This figure [or Eq. (D.4)] can also be used to estimate the number N of terms that would be required for a given value of z. For example, at least nine terms, i.e., the sum $S_9(1.5) = 0.966$, are required to estimate erf(1.5) to three decimal places; our urgent need is now satisfied.

Convergence of series, though familiar, is a tricky concept. Although, for given z, the remainder (D.2) and (D.3) *always* tends to zero as N tends to infinity (factorials eventually overpower any power), the manner in which the individual terms of the series (D.1) eventually tend to zero is not so straightforward. For example, we calculate

$$\mathrm{erf}(2) = \frac{2}{\sqrt{\pi}}[2 - 2.667 + \mathbf{3.200} - 3.048 + 2.370 - 1.552 + \cdots]. \quad \text{(D.5)}$$

The terms first *increase* in absolute value, and they begin to decrease only after the third term (in bold characters). This effect is even more dramatic for larger arguments. It is exactly the opposite of what happens with asymptotic series, to which we now turn.

D.2. Asymptotics in High School

We all have distant memories of linear asymptotes in high school. To refresh these, consider the function $y = 1/(1 + x)$, which represents a hyperbola whose "horizontal" asymptote, as x increases by positive values, is the axis $y = 0$.[†]

Then, multiplying the previous function by x, we consider the function $y = x/(1 + x)$, which again represents a hyperbola, but now its asymptote is the horizontal line $y = 1$. Multiplying again by x, we have $y = x^2/(1 + x)$, which is no longer a hyperbola, but it does have the linear asymptote $y = x - 1$.

We're on a roll, so let's finally consider the function $y = x^3/(1 + x)$. For large positive x, it approaches the parabola $y = x^2 - x + 1$, i.e., a *curve*. (This was generally not in our high school curriculum, but it should have been.) How do we know? We assume that the difference between our function and the asymptote $y = ax^2 + bx + c$ is vanishingly small for large values of x. A simple calculation of limits yields the coefficients.

We now ask: *What's the use*? In each of the four preceding cases we have managed to replace the original function, $f(x)$, by a simpler one, $\tilde{f}(x)$, in such a way that the "distance" $|f(x) - \tilde{f}(x)|$ becomes vanishingly small when x is large.[‡] Consequently, we can use the simpler function instead of the original (perhaps very complicated) function to examine its behavior and for numerical evaluation. Formally, we have

$$\lim_{x \to \infty} |f(x) - \tilde{f}(x)| = 0, \tag{D.6}$$

and we then write $f \sim \tilde{f}$.

[†] It also has a "vertical" asymptote at $x = -1$ and a left-hand branch, which are not our concern here. The reader is advised to draw schematic graphs.

[‡] Remember that, even here, polynomials are algebraically simpler functions than rational functions.

D.3. Expansion of erfc(z) for "Large" z

The condition (D.6) can be approximated by what is called an *asymptotic expansion.* Here, we are not concerned with the general theory,[14,15] but we do wish to examine, heuristically, how erfc(z) approaches zero for large values of z.[†]

As we know, this function is defined by the integral

$$\operatorname{erfc}(z) = \frac{2}{\sqrt{\pi}} \int_z^\infty e^{-\xi^2}\, d\xi. \tag{D.7}$$

Here's a neat trick: multiply and divide the integrand by ξ. The integrand is now $-(1/2\xi)(e^{-\xi^2})'$, an ideal form for integration by parts:

$$\operatorname{erfc}(z) = \frac{e^{-z^2}}{\sqrt{\pi}\, z} - \frac{1}{\sqrt{\pi}} \int_z^\infty \frac{1}{\xi^2} e^{-\xi^2}\, d\xi. \tag{D.8}$$

Both terms on the right-hand side tend to zero for large positive z. We will soon show that the second term is always smaller than the first, and so we can use this first term as a useful, simple approximation of the complementary error function for large values of its argument. We write

$$\operatorname{erfc}(z) \sim \frac{e^{-z^2}}{\sqrt{\pi}\, z} \tag{D.9}$$

to symbolize this approximation, and we say that the complementary error function decays asymptotically as does a gaussian divided by the first power of the argument.

How "good" is this approximation? For example, tables give us $\operatorname{erfc}(2) = 0.00468$ to five decimal places (rounded). On the other hand, expression (D.9) gives us the approximation $\operatorname{erfc}(2) \sim e^{-4}/2\sqrt{\pi} = 0.00517$, also to five decimal places. The corresponding results when $z = 2.5$ are $\operatorname{erfc}(2.5) = 0.00041$ and $\operatorname{erfc}(2.5) \sim 0.00054$. We are definitely on the right track.[‡]

Can we do better? Yes, of course, because the residual integral in Eq. (D.8) can be handled by parts in exactly the same way. In fact,

[†] Indeed, in the complex domain, for $\Re(z) \gg 1$.

[‡] Is 2 a "large" number? Yes, in this case, because of gaussian decay. Recall the mathematicians' joke: 2 is much greater than 1 for large values of 1.

N-fold integration by parts yields

$$\operatorname{erfc}(z) = \frac{e^{-z^2}}{\sqrt{\pi}}\left[\frac{1}{z} - \frac{1}{2z^3} + \frac{1\cdot 3}{2^2 z^5} - \cdots + (-1)^{N-1}\frac{1\cdot 3\cdots(2N-3)}{2^{N-1}z^{2N-1}}\right]$$
$$+ (-1)^N \frac{1\cdot 3\cdots(2N-1)}{\sqrt{\pi}\,2^{N-1}}\int_z^\infty \frac{1}{\xi^{2N}}e^{-\xi^2}\,d\xi. \tag{D.10}$$

We have thus expressed the complementary error function as a finite sum of N terms, call it $S_N(z)$, which we can conveniently rewrite as

$$S_N(z) = \frac{e^{-z^2}}{\sqrt{\pi}\,z}\left[1 - \frac{1}{2z^2} + \frac{1\cdot 3}{(2z^2)^2} - \cdots + (-1)^{N-1}\frac{1\cdot 3\cdots(2N-3)}{(2z^2)^{N-1}}\right], \tag{D.11}$$

and a residual integral, call it $R_N(z)$. Therefore, by analogy with Eq. (D.2), we are tempted to use the apparatus of convergent power series. But expressions (D.10) and (D.11) are of quite a different character. First of all, S_N is a product of the "core" approximation (D.9) times a finite sum of *inverse* powers of the argument z. Therefore, just as the power series (D.1) converges in the neighborhood of the origin, we might expect the partial sums (D.11) to converge around the "point at infinity." But this is simply not true! Any of the standard convergence tests will show that the series generated by the partial sums S_N *does not converge*. Further, one can easily show [see Eq. (D.15), below] that the residual integral R_N also blows up as N becomes large. In other words, for any prescribed positive value of z, we do have a decomposition $\operatorname{erfc}(z) = S_N(z) + R_N(z)$ of the complementary error function, but each of these expressions increases without bound as $N \to \infty$. What *is* this strange beast?

The expression (D.10) is called an "asymptotic expansion," and its properties, as hinted above, are somewhat disconcerting for the beginner. Nonetheless, it is vitally important for the accurate evaluation of functions in regions where computer results would be worthless.† We ask in what sense can we use the finite expression (D.11) to obtain numerical results?

†Nonconvergent expressions were introduced by Euler; their systematic investigation stems from Poincaré's work on celestial motion.

Exactly as in the natural sciences, mathematics can (and should) be approached experimentally. For example, let us use eight terms of the finite sum (D.11) to estimate erfc(2), again to 5 decimal places:

$$S_8(2) = 0.00517 \times [1 - 0.125 + 0.04687 - 0.02930 + \mathbf{0.02563} - 0.02884 + 0.03965 - 0.06444]. \quad \text{(D.12)}$$

We notice a strange phenomenon: The first five terms decrease in absolute value, but then increase markedly.† The smallest term is in bold characters. Let's decide to cut off this sum at that point:

$$\begin{aligned} S_5(2) &= 0.00517 \times [1 - 0.125 + 0.04687 - 0.02930 + \mathbf{0.02563}] \\ &= 0.00517 \times 0.91820 = 0.00475. \end{aligned} \quad \text{(D.13)}$$

This is clearly a far better approximation of erfc(2) than what we had obtained previously [see the estimates that follow Eq. (D.9)]. However—and this may appear strange—*for this value of z we cannot do any better by including more terms of Eq. (D.12).*

To understand this bizarre phenomenon, we return to Eq. (D.10) to find a bound for the remainder R_N. This is quite easy because $e^{-\xi^2} \leqslant e^{-z^2}$ when $\xi \geqslant z$. Thus

$$|R_N(z)| \leqslant \frac{1 \cdot 3 \cdots (2N-1)}{\sqrt{\pi}\, 2^{N-1}} e^{-z^2} \left| \int_z^\infty \frac{d\xi}{\xi^{2N}} \right|, \quad \text{(D.14)}$$

and this elementary integral has the value $1/(2N-1)|z|^{2N-1}$. Collecting terms, we find that $|R_N|$ is bounded by the absolute value of the last term that we *retain* in the finite sum S_N:

$$|R_N(z)| \leqslant \frac{1 \cdot 3 \cdots (2N-3)}{\sqrt{\pi}\, 2^{N-1} |z|^{2N-1}} e^{-z^2}. \quad \text{(D.15)}$$

Now, if we remember that the remainder is a measure of the error between the given function (here the erfc) and its approximation (here S_N), then the best possible approximation (for a given value of z) is clearly obtained when we reach the smallest term of the finite sum.

†A little thought will convince you that the increase is at least exponential.

This term is a bound for the remainder. Analytically, for given z, we must find the value of N that makes the expression

$$\frac{1 \cdot 3 \cdots (2N-3)}{(2z^2)^{N-1}}$$

as small as possible. In other words, the ratio of this term to its predecessor and successor must be as close to unity as possible. A short calculation shows that N, the number of terms that must be retained for a given value of z, must lie within the "band"

$$z^2 + \tfrac{1}{2} \leqslant N \leqslant z^2 + \tfrac{3}{2}, \tag{D.16}$$

and the reader is encouraged to check that $N = 5$ [see Eq. (D.13)] indeed lies in that band for this example.

D.4. Application to Finite-Difference Calculations

Many diffusion problems, especially in supersaturated media, begin with an impulsive "shove." Initial gradients are then very large, and we ask how it might be possible to "smooth" these out. For definiteness, consider a typical problem (see Exercise 1.11) that has been nondimensionalized:

$$u_t = u_{xx}, \qquad \text{in } \{0 < x < 1; t > 0\}, \tag{D.17}$$

together with the initial and boundary conditions

$$u(x, 0) = 0, \qquad 0 < x < 1, \tag{D.18a}$$

$$u(0, t) = 1, \qquad u(1, t) = 0, \qquad t > 0, \tag{D.18b}$$

the origin being clearly a point of discontinuity.

Intuitively, the effect of this discontinuity at $x = 0$ will not be significant at the far boundary $x = 1$ for short enough times. The medium then appears to be semi-infinite, and we know that the corresponding analytic solution is

$$u_s(x, t) = \operatorname{erfc}(x/2\sqrt{t}). \tag{D.19}$$

It is called a "starting solution" for reasons that will now become clear, therefore the subscript 's' to distinguish it from the full solution of Eqs. (D.17) and (D.18).

Let us estimate how short is "short" by evaluating u_s at the far boundary, $u_s(1, t) = \operatorname{erfc}(1/2\sqrt{t})$, which we want negligibly small. Since finite-difference calculations are rarely performed with more than two decimal-place precision, let us impose the condition

$$\operatorname{erfc}(1/2\sqrt{t}) \leqslant \tfrac{1}{2} \times 10^{-2} = 5 \times 10^{-3}. \tag{D.20a}$$

But, from our estimate (D.13), we know that this condition is surely satisfied when the argument of the error function exceeds 2. Thus, we can use the starting solution (D.19) for times $t \leqslant t_0$ such that

$$1/2\sqrt{t_0} = 2 \Rightarrow t_0 = 1/16 = 0.0625. \tag{D.20b}$$

The starting solution (D.19), evaluated when $t = t_0$, can now serve as an initial condition for further finite-difference calculations, whence its name. The advantage of this method over the blind application of the initial condition (D.18a) is that the initial (infinite!) gradient has been smoothed out.

We still do have gradients when $t = t_0$, and these can be used to advantage in choosing a mesh size Δx intelligently. Indeed, the largest gradient occurs at the near boundary $x = 0$, and, using Eqs. (D.1) or (D.2), we have the Taylor expansion

$$\operatorname{erfc}(\Delta x/2\sqrt{t_0}) = 1 - \frac{2}{\sqrt{\pi}}\left[\frac{\Delta x}{2\sqrt{t_0}} - \frac{1}{3}\left(\frac{\Delta x}{2\sqrt{t_0}}\right)^3 + \cdots\right], \tag{D.21a}$$

evaluated at the first interior point $x = x_1 = \Delta x$. As shown in Fig. D.2, to get a reasonable evaluation of this largest gradient, we require that the distance $\overline{PQ}$ be negligible. But this distance is merely the remainder of the previous alternating series, itself bounded by the cubic term. Remembering estimate (D.20b) for this level of accuracy, we have

$$\overline{PQ} < \frac{2}{3\sqrt{\pi}}\left(\frac{\Delta x}{2\sqrt{t_0}}\right)^3 \leqslant \tfrac{1}{2} \times 10^{-2} \Rightarrow \Delta x < \tfrac{1}{2}[(3\sqrt{\pi}/4) \times 10^{-2}]^{1/3} = 0.128. \tag{D.21b}$$

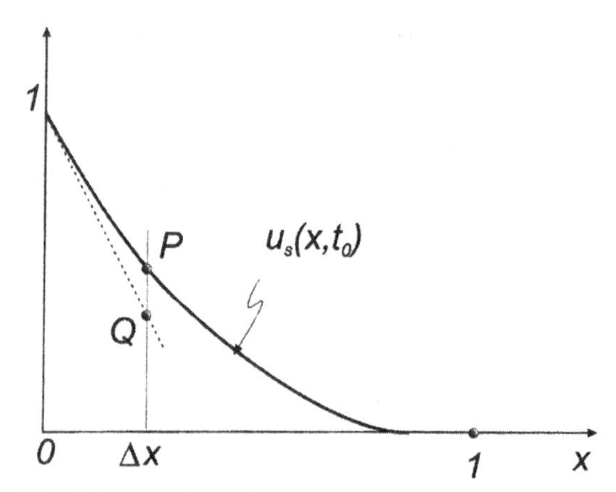

Fig. D.2. Profile of the starting solution that can be used to choose a spatial mesh size.

In other words, 10 equal space intervals are sufficient to obtain two decimal-place precision if one has had the good sense to use the starting solution (D.19) and (D.20). Other diffusion problems can be analyzed similarly. Note that the time step Δt is not yet determined; it depends on one's choice for the modulus λ.

Appendix E

Diffusion in Bounded Regions

Bodies that support diffusion are never infinite in extent, although they may so appear for short enough times. This means that we ought to know what conditions the unknown field satisfies on its free surfaces. Such problems are known as "boundary-value problems," and this appendix provides a systematic technique that is easy to apply when the boundary conditions are of Dirichlet-type, namely, when the concentration, temperature, etc, are *given* functions of time on the body's surfaces.

E.1. Finite Fourier Transforms

There are essentially two closely related techniques for the solution of linear boundary-value problems: the method of *separation of variables* and that of *integral transforms*. The first method is well known, and was illustrated in Sect. 5.3 (the "interior" problem for a sphere). It is admirably suited to handle arbitrarily complicated initial conditions, but its application to anything but the simplest boundary-value problems requires a fair amount of skill. On the other hand, the second class of methods transform a 1-d diffusion problem into a trivial ordinary differential equation. Chapter 7 is devoted to one such method, the Laplace transform, which is suited to independent variables that vary on the half-line $t > 0$ or $x > 0$. On the other hand, it is not directly applicable to bounded regions. Here, we restrict ourselves to one of the simplest of integral transforms that can handle such regions: the *finite Fourier sine transform*.[16–18]

This complicated name hides nothing more than a fancy way of using certain Fourier series. Consider a function $f(x)$ defined on an interval $[0, L]$. Under "sufficient conditions of regularity" (a standard phrase that avoids telling exactly *what* they are), this function can be expanded in a Fourier series

$$f(x) = \frac{2}{L} \sum_{n=1}^{\infty} \bar{f}_n \sin \frac{n\pi x}{L}. \tag{E.1}$$

The set of its coefficients $\{\bar{f}_n\}$, also called the "spectrum" or "transform" of f, can be calculated explicitly from the equation

$$\bar{f}_n = \int_0^L f(x) \sin \frac{n\pi x}{L} \, dx \qquad (n = 1, 2, \ldots). \tag{E.2}$$

Indeed, if we are not concerned with rigor, then multiplying Eq. (E.1) by $\sin(m\pi x/L)$, integrating over x from 0 to L, and using the "orthogonality property"

$$\int_0^L \sin \frac{n\pi x}{L} \sin \frac{m\pi x}{L} \, dx = \frac{L}{2} \delta_{n,m}, \tag{E.3}$$

we get precisely Eqs. (E.2).

Those who have never met a Fourier series might be too stunned to respond. Those who *have* might at least ask the question: Why only use sines? To answer that question, we observe, first, that Eqs. (E.1) and (E.2) apply to functions f that are defined on a finite interval $[0, L]$. The behavior of these functions elsewhere is completely irrelevant. If, however, we insist on extending the domain of definition of these functions, then, because the "basis functions" $\sin(n\pi x/L)$ are both $2L$-periodic and odd, these functions f must be extended in a way that is also $2L$-periodic and odd. For example, it is easy to see that the constant function $f = 1$ on $[0, L]$ has the Fourier sine expansion

$$1 = \frac{4}{\pi} \sum_{n=1}^{\infty} \frac{\sin[(2n-1)\pi x/L]}{2n-1}. \tag{E.4}$$

If, however, this series is evaluated outside of the fundamental interval $[0, L]$, then it converges to the "square-wave" represented on Fig. E.1. How this series converges, and how fast, is the stuff of lectures on Fourier series. Here, we simply note that the series (E.4) represents a *discontinuous* function, something that should surprise those who believe exclusively in Taylor series.

We return to our original question: why not use cosines or even exponentials? It all depends on the way we wish to extend the original

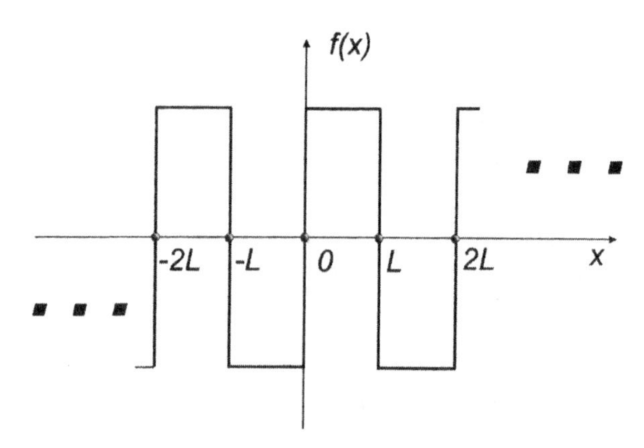

Fig. E.1. A square wave that is everywhere represented by the Fourier expansion (E.4).

function. If we extend it as an even function, then cosines are appropriate, and we can write expressions similar to Eqs. (E.1)–(E.3) but which involve the basis functions $\cos(n\pi x/L)$. Finally, if, for whatever reason, we need to extend f in a manner that is neither even nor odd, then we would need to use both sines and cosines to fully represent it.

But we haven't really answered the question: why sines? It turns out that many equations of mathematical physics are second- or fourth-order in spatial derivatives. The diffusion equation certainly is. We now investigate what happens to some such term. For simplicity, let us see how a second derivative f'' gets transformed. Inserting it into Eq. (E.2) and integrating by parts (twice), we get

$$\int_0^L f''(x)\sin\frac{n\pi x}{L}\,dx = f'(x)\sin\frac{n\pi x}{L}\Big|_0^L - \frac{n\pi}{L}\int_0^L f'(x)\cos\frac{n\pi x}{L}\,dx$$

$$= -\frac{n\pi}{L}f(x)\cos\frac{n\pi x}{L}\Big|_0^L - \left(\frac{n\pi}{L}\right)^2\int_0^L f(x)\sin\frac{n\pi x}{L}\,dx.$$

We observe that the second term on the right-hand side is proportional to $\bar{f}_n$, and the first term yields the boundary values of $f(x)$. Thus, we have the important formula

$$\int_0^L f''(x)\sin\frac{n\pi x}{L}\,dx = -\left(\frac{n\pi}{L}\right)^2\bar{f}_n + \frac{n\pi}{L}[f(0^+) - (-1)^n f(L^-)]. \quad \text{(E.5)}$$

What does this mean? By transforming a second derivative, we have managed to eliminate that derivative altogether, and Eq. (E.5) throws in the boundary values of the function as an extra bonus.†

EXERCISE E.1. *Summation of some Fourier series:* Finite Fourier sine expansions are easy to find when the original function $f(x)$ obeys a linear, second-order differential equation. Then, transforming that equation with Eq. (E.5), one gets the Fourier coefficients by purely algebraic means because the boundary values are known. Apply this apparatus to the functions $\alpha + \beta x + \gamma x^2$ and $\exp(\lambda x)$, and derive as many other results as possible by "playing" with the constants. □

†Just like the Laplace transform.

E.2. Application to Diffusion Problems

Consider one-dimensional diffusion problems for simplicity. We seek to express the concentration $C(x, t)$ as a Fourier series in the spatial variable x,

$$C(x, t) = \frac{2}{L} \sum_{n=1}^{\infty} \bar{C}_n(t) \sin \frac{n\pi x}{L}, \tag{E.6}$$

of the same form as Eq. (E.1). Its transform

$$\bar{C}_n(t) = \int_0^L C(x, t) \sin \frac{n\pi x}{L} \, dx \qquad (n = 1, 2, \ldots) \tag{E.7}$$

is no longer a set of constants but a denumberable set of functions of t. But how do we find these? By using the diffusion equation, of course.

Therefore, we multiply the diffusion equation (1.14) by $\sin(n\pi x/L)$, throughout, and integrate over x from 0 to L:

$$\int_0^L \frac{\partial C}{\partial t} \sin \frac{n\pi x}{L} \, dx = D \int_0^L \frac{\partial^2 C}{\partial x^2} \sin \frac{n\pi x}{L} \, dx.$$

On the left-hand side we note that t does not appear in the limits. The time-derivative thus commutes with the integration over space, and we recognize $d\bar{C}_n/dt$. For the right-hand side we use our result (E.5) because t is merely a parameter in those operations. Thus, we get the set of ordinary differential equations

$$\frac{d\bar{C}_n}{dt} + D\left(\frac{n\pi}{L}\right)^2 \bar{C}_n = D\frac{n\pi}{L}[C(0^+, t) - (-1)^n C(L^-, t)] \qquad (n = 1, 2, \ldots). \tag{E.8}$$

They are particularly simple because their coefficients are constants. General solution are available, and, in principle, we have completely solved our diffusion problem because the boundary values determine the inhomogeneous right-hand side.

Rather than write general solutions, let us solve—analytically—the problem that was proposed by finite differences in Exercise 1.12.[†] Here, to illustrate a common pitfall, we begin with a slightly more

[†] By linearity, one can subtract the constant C_L, which amounts to using that value as a baseline for concentrations.

complicated problem, namely,

$$C(0, t) = C_0 e^{-t/\tau} \qquad t > 0 \tag{E.9a}$$

and

$$C(L, t) = 0 \qquad t > 0, \tag{E.9b}$$

and let us include an arbitrary initial condition

$$C(x, 0) = \varphi(x), \qquad 0 < x < L. \tag{E.10}$$

In this case the differential equations (E.8) reduce to

$$\frac{d\bar{C}_n}{dt} + D\left(\frac{n\pi}{L}\right)^2 \bar{C}_n = DC_0 \frac{n\pi}{L} e^{-t/\tau} \qquad (n = 1, 2, \ldots), \tag{E.11}$$

and they do have explicit solutions

$$\bar{C}_n(t) = \alpha_n e^{-(n\pi/L)^2 Dt} + \beta_n e^{-t/\tau} \qquad (n = 1, 2, \ldots), \tag{E.12}$$

where the α's and β's are constants (with respect to time). The first term is the general solution of the homogeneous equation, and the second is any particular solution of the inhomogeneous equation. Inserting the latter into Eq. (E.11), we get

$$\beta_n = \frac{C_0(n\pi/L)}{(n\pi/L)^2 - (1/D\tau)}. \tag{E.13}$$

Now here is the pitfall. How do we determine the constants α_n? By using the initial condition, one would say. That is correct, but it is the *transformed* initial condition that we need because Eqs. (E.12) describe the transform of the concentration. Indeed, evaluating that equation at $t = 0$, we get

$$\bar{C}_n(0) = \bar{\varphi}_n = \alpha_n + \beta_n, \tag{E.14}$$

which is a set of equations for the α's. The problem is finished, in principle.

There is much more to solving problems "in principle." Indeed, we shall now see that there are serious *numerical* questions that require attention. After all, it is fine and good to write formal Fourier series; it is an altogether different matter to evaluate them accurately.

E.3. Analytical Solution of Exercise 1.12

We recall that the problem in Exercise 1.12 had demanded $\varphi(x) = 0$. In that case its transform is $\bar{\varphi}_n = 0$ for all n, Eqs. (E.14) imply that $\alpha_n = -\beta_n$, and, with Eqs. (E.6) and (E.13), we get the explicit solution

$$C(x, t)/C_0 = \frac{2}{\pi} \sum_{n=1}^{\infty} n \frac{e^{-t/\tau} - e^{-(n\pi/L)^2 Dt}}{n^2 - (L^2/\pi^2 D\tau)} \sin \frac{n\pi x}{L}. \tag{E.15}$$

We note the dimensional consistency of this expression. Indeed, all variables and parameters are expressed in the dimensionless combinations C/C_0, x/L, Dt/L^2, and $D\tau/L^2$.

Thus one might say, "all I have to do is program my trusty computer." Ah! that it were so simple. Should one do just that, one might need 10^4 terms of the series to get reasonable two-decimal precision—and that for only a single choice of your variables and parameters. Indeed, Eq. (E.15) is the sum of two series. The second generally offers no difficulties because the exponentials $\exp[-(n\pi/L)^2 Dt]$ tend to depress the successive terms quite fast. The first is problematic because a quick look shows that it converges as $n^{-1} \sin(n\pi x/L)$, and this is a very slowly converging series. We are fortunate, however, and in Exercise E.1, we showed that the first series can be summed:

$$\frac{2}{\pi} \sum_{n=1}^{\infty} \frac{n}{n^2 - (L^2/\pi^2 D\tau)} \sin \frac{n\pi x}{L} = \frac{\sin[(L - x)/\sqrt{D\tau}]}{\sin(L/\sqrt{D\tau})}. \tag{E.16}$$

Then, the expression

$$C(x, t)/C_0 = \frac{\sin[(L - x)/\sqrt{D\tau}]}{\sin(L/\sqrt{D\tau})} e^{-t/\tau} - \frac{2}{\pi} \sum_{n=1}^{\infty} \frac{n e^{-(n\pi/L)^2 Dt}}{n^2 - (L^2/\pi^2 D\tau)} \sin \frac{n\pi x}{L} \tag{E.17}$$

is usually quite well suited for computer calculations. We note that the first term of Eq. (E.17) satisfies both of the boundary conditions (E.9). The numerical results can be compared to the finite-difference solution that was proposed earlier.

There is one last case that does cause trouble, namely, that for "short" times. This generally means that $Dt/L^2 \ll 1$, and, numerically, the exponentials in the series (E.17) now do *not* damp significantly. There are several ways of addressing this problem, and perhaps the most compelling is through the use of Laplace transforms. Here we simply note that the short-time condition means that the concentration disturbance at $x = 0$ has not yet "propagated" to the other boundary $x = L$. In effect, the body *does* appear semi-infinite, and we must solve the diffusion problem under the sole boundary condition (E.9a). This we learned to do in Chap. 7 [see Eq. (7.36)]. However, if t is such that t/τ is also very small, then that boundary condition reduces approximately to $C(0, t) \approx C_0$, namely a constant. The corresponding solution (as we know all too well) is $C(x, t) \approx C_0 \operatorname{erfc}[x/2\sqrt{(Dt)}]$.

Appendix F

Moments of Distributions and Asymptotic Behavior

In several instances we have seen that mass balance leads to the diffusion equation at interior points of a mass-containing domain $\mathscr{D}$ and to boundary conditions on internal or external boundaries of $\mathscr{D}$. The total mass contained in $\mathscr{D}$,

$$M(t) = \int_{\mathscr{D}} C(\mathbf{x}, t)\, dV, \tag{F.1}$$

is a prototype of particular quantities called *moments*. Exactly as in probability theory, we shall see that the moments of the distribution $C(\mathbf{x}, t)$ characterize it entirely, including its asymptotic behavior.

Let us begin, however, by proving a general result concerning fluxes. We wish to know how $\mathbf{J}$ decays at large distances when $\mathscr{D}$ is unbounded. That question is easy to answer if we assume rotational symmetry far enough from the initial distribution. If R_0 is some large enough radius, then the change in mass exterior to R_0 can only be due to the flux J_0 on the surface $r = R_0$ that is received by the system.

Because one shows[19] (see also Exercise 7.12) that the area of a hypersphere in d dimensions is proportional to r^{d-1}, it follows that the volume element in Eq. (F.1) is

$$dV = \alpha r^{d-1}\, dr, \tag{F.2}$$

where α is a constant (that does depend on dimensionality d). Evaluating the time derivative of Eq. (F.1), both as the integrated flux $\alpha R_0^{d-1} J_0$ received by the system and by using the form (1.36) of the diffusion equation, we get the required asymptotic behavior

$$\lim_{r\to\infty} r^{d-1} J_r = 0. \tag{F.3}$$

For example, the gaussian (4.3) in 1-d,

$$G(x, t) = \frac{1}{2\sqrt{\pi D t}} e^{-x^2/4Dt}, \tag{F.4}$$

repeated here for convenience, evidently has this property. Indeed, not only do *all* its x-derivatives vanish for $x \to \pm\infty$, but they also vanish faster than any power of x. This is, in fact, a general property[20] of the diffusion equation's solutions: If the far field vanishes, i.e., if $\lim_{x\to\pm\infty} C(x, t) = 0$, then

$$\lim_{x\to\pm\infty} x^n \frac{\partial^m C}{\partial x^m} = 0, \qquad n, m \geqslant 0, \tag{F.5}$$

which includes the property (F.3) as a special case.

We recall how this gaussian serves to produce 1-d solutions of the diffusion equation for *arbitrary* initial conditions. In Chap. 4 we showed that the integral (4.5),

$$C(x, t) = \int_{-\infty}^{\infty} G(x - \xi, t)\varphi(\xi)\, d\xi, \tag{F.6}$$

also repeated for convenience, is precisely such a representation. Indeed, it satisfies the initial and boundary conditions

$$\lim_{t\to 0^+} C(x, t) = \varphi(x) \tag{F.7a}$$

and

$$\lim_{x \to \pm\infty} C(x, t) = 0. \tag{F.7b}$$

Moreover, by differentiation under the integral sign, we verify that the derivatives vanish at infinity, in accordance with Eq. (F.5). The integral representation (F.6) is sometimes called *Poisson's integral*, and, as with a similarly named integral in electrostatics and potential theory (e.g., see Ref. 21, pp. 343–346), it has an interpretation that we now explore.

For a given "observation point" x, the gaussian $G(x - \xi, t)$, as a function of ξ, peaks sharply at its maximum $\xi = x$. There, it samples the initial distribution φ, as would a filter of width $2\sqrt{(Dt)}$. Varying the "source point" ξ in Eq. (F.6) over the whole range $(-\infty, \infty)$, we get the additive effects of all points of φ, which is precisely the time-developed solution $C(x, t)$.

But what can we say about this solution if we are far enough (in space and time) from the initial distribution? The electrostatic analog gives a clue. If φ is a given charge distribution, then Poisson's integral (F.6) over an appropriate Green's function (the reciprocal distance $1/r$) yields the potential everywhere. Moreover, at large distances from φ, the potential appears *as if* caused by an equivalent point charge at the origin. "Equivalence" here means that the fictitious point charge and the real distributed charge density φ must have the same total charge, or "zeroth moment." Generally, one shows (Ref. 21, pp. 416–419) that the potential admits a "multipole expansion" in powers of $1/r$, whose coefficients, called moments, depend only on the given charge distribution. We now develop these considerations for diffusion problems.

We first define the moments of order n

$$M_n(t) = \int_{-\infty}^{\infty} x^n C(x, t)\, dx \qquad \text{for all } n \geqslant 0, \tag{F.8}$$

of the concentration distribution C[†]. Since C is some solution of the diffusion equation, we might anticipate that its moments also obey certain relations. This is indeed the case, and the proof is very simple:

[†]Readers familiar with probability theory will recognize that these moments are neither normalized nor centered.

We differentiate Eq. (F.8) with respect to time, and, as we have done many times, insert the diffusion equation (1.14) under the integral sign. We integrate twice by parts and use the property (F.5) to get the recursion

$$\dot{M}_n = n(n-1)DM_{n-2} \qquad \text{for } n \geqslant 2. \tag{F.9a}$$

Similar calculations for $n = 0$ and 1 yield simply

$$\dot{M}_0 = \dot{M}_1 = 0. \tag{F.9b}$$

Therefore, not only do we have differential equations for the moments, but these can also be integrated to give

$$M_0(t_2) = M_0(t_1) \equiv M_0, \tag{F.10a}$$

$$M_1(t_2) = M_1(t_1) \equiv M_1, \tag{F.10b}$$

$$M_2(t_2) - M_2(t_1) = 2M_0D(t_2 - t_1), \tag{F.10c}$$

where t_1 and t_2 are two arbitrary instants of time. In other words, the zeroth and first moments are constants "of the motion," and the second moment is linear in Dt. The recursion (F.9a) can evidently be carried forward, and it is clear that the time difference of moments, of order $2n$ and $2n + 1$, is an nth-degree polynomial in $D(t_2 - t_1)$. This property has been used to advantage for easy data-fitting of diffusion experiments.[22]

Exercise F.1. *Conditions for the moment equations:* Carefully carry out the derivation of Eqs. (F.9). What happens if C does not vanish at infinity? Interpret Eqs. (F.10) physically. □

Equations (F.10) are independent of the Poisson representation (F.6). In particular, however, they show that moments at time t are related to moments of the initial distribution φ. Consequently, we might expect that the full distribution $C(x, t)$ is expressible as a function of these initial moments. It is easy to find this expansion from Eqs. (F.4) and (F.6). For brevity, we put $z = (x - \xi)/2\sqrt{(Dt)}$, and expand the exponential $\exp(-z^2)$ in a Taylor series around the point $z_0 = x/2\sqrt{(Dt)}$. With the known property[23]

$$\frac{d^n}{dz^n} e^{-x^2} = (-1)^n H_n(z) e^{-z^2} \tag{F.11a}$$

of Hermite polynomials ($H_0 = 1$, $H_1 = 2z$, etc.), we have

$$e^{-z^2} = e^{-z_0^2} \sum_{n=0}^{\infty} (-1)^n \frac{(z - z_0)^n}{n!} H_n(z_0). \tag{F.11b}$$

Inserting this into Eqs. (F.4) and (F.6), and remembering the meaning of the abbreviations z and z_0, we easily get

$$C(x, t) = \frac{e^{-x^2/4Dt}}{2\sqrt{\pi Dt}} \sum_{n=0}^{\infty} \frac{H_n(x/2\sqrt{Dt})}{n!(4Dt)^{n/2}} M_n, \tag{F.12}$$

where $M_n \equiv M_n(0)$ are the moments (F.8) of the *initial* distribution φ. Thus, as promised, the initial moments entirely characterize the concentration distribution at time t. Indeed, for large t's and moderate x's, Eq. (F.12) reduces to its zeroth term, which is nothing but the gaussian (F.4) multiplied by the total (constant) mass M_0 in the system.

EXERCISE F.2. *Moments of gaussians:* The gaussian

$$(Q/\sigma\sqrt{\pi}) \exp[-(x - m)^2/\sigma^2]$$

depends on the three constants Q, m, and σ (see Exercise 4.2). Find its moments in terms of these constants and interpret. Is there a relation between moments? Why? □

The message of the expansion (F.12) should be clear: For long enough times in an infinite medium, any initial distribution will decay into a gaussian, proportional to Eq. (F.4), with corrections proportional to $t^{-n/2}$. Moreover, for the analogous electrostatic problem, it is recalled that an appropriate choice (at the center of charge) of the origin of coordinates cancels the "dipole" term. Likewise, here we may hope that there exists an origin of space and time, and a scale factor for C, such that some leading terms drop out of the moment expansion (F.12). This should be possible because our problem is translationally invariant in space and time, and it is also homogeneous (recall Properties C of Sect. 1.3). Physically, if we can find a leading gaussian that mimics the initial distribution (and the last exercise showed that there are precisely three parameters to choose from), then we expect, at least, "osculating" behavior of the series (F.12).

We assume, therefore, that C develops from some *fictitious* initial condition ψ that holds at some earlier time $t = -t_0$:

$$C(x, t) = \frac{1}{2\sqrt{\pi D(t + t_0)}} \int_{-\infty}^{\infty} e^{-(x-\xi)^2/4D(t+t_0)}\psi(\xi)\, d\xi. \qquad \text{(F.13)}$$

Consequently, the *real* initial distribution φ can also be expressed in terms of ψ by putting $t = 0$ in this Eq. (F.13):

$$\varphi(x) = \frac{1}{2\sqrt{\pi D t_0}} \int_{-\infty}^{\infty} e^{-(x-\xi)^2/4Dt_0}\psi(\xi)\, d\xi. \qquad \text{(F.14)}$$

Now, exactly as with Eqs. (F.11), we expand Eq. (F.13) around the as yet unspecified point $z_0 = (x - x_0)/[4D(t + t_0)]^{1/2}$ to get

$$C(x, t) = \frac{e^{-z_0^2}}{2\sqrt{\pi D(t + t_0)}} \sum_{n=0}^{\infty} \frac{H_n(z_0)}{n![4D(t + t_0)]^{n/2}} \int_{-\infty}^{\infty} (\xi - x_0)^n \psi(\xi)\, d\xi. \qquad \text{(F.15)}$$

Next, we define the moments

$$N_n = \int_{-\infty}^{\infty} x^n \psi(x)\, dx \qquad \text{(F.16)}$$

of the fictitious initial distribution. But we have no desire to characterize fictitious distributions. Our task, therefore, must be to express the integrals in the expansion (F.15) in terms of the moments M_n of the real initial distribution φ, and, along the way, to choose the undetermined constants x_0 and t_0 appropriately.

This is easily done because Eq. (F.14) connects φ to ψ, and their moments must also be related, in the spirit of Eqs. (F.10). Setting, $t_2 = 0$ and $t_1 = -t_0$, we have

$$M_0 = N_0, \qquad M_1 = N_1, \qquad M_2 = N_2 + 2M_0 D t_0, \ldots \qquad \text{(F.17)}$$

Let us calculate the leading integrals of the expansion (F.15). For $n = 0$ we have

$$\int_{-\infty}^{\infty} (\xi - x_0)\psi(\xi)\, d\xi = N_0, \tag{F.18a}$$

which, by the first equation (F.17), equals M_0. Therefore, the expansions (F.12) and (F.15) have the same total mass, a not altogether unexpected result. Next, for $n = 1$, we have

$$\int_{-\infty}^{\infty} (\xi - x_0)^1\psi(\xi)\, d\xi = N_1 - x_0 N_0 = M_1 - x_0 M_0 \tag{F.18b}$$

by the same arguments. This term vanishes if we choose $x_0 = M_1/M_0$ i.e., the origin must be at the "center of mass" of the real initial distribution, as expected from the electrostatic analog. Finally, for $n = 2$, we have

$$\int_{-\infty}^{\infty} (\xi - x_0)^2\psi(\xi)\, d\xi = N_2 - x_0^2 N_0 = M_2 - 2M_0 D t_0 - x_0^2 M_0, \tag{F.18c}$$

and this "moment of inertia" vanishes for a proper choice of t_0. In sum, in an infinite medium, the gaussian

$$\frac{Q}{2\sqrt{\pi D(t + t_0)}} \exp\left[-\frac{(x - x_0)^2}{4D(t + t_0)} \right] \tag{F.19a}$$

is an asymptotic representation of the solution C, up to terms of order $[4D(t + t_0)]^{-3/2}$, if

$$Q = M_0, \qquad x_0 = M_1/Q, \qquad t_0 = (M_2 - M_1^2/Q)/2DQ, \tag{F.19b}$$

and there are no other free parameters at our disposal. Indeed, when $t \to 0$, the choice (F.19b) forces the gaussian (F.19a) to match the real initial distribution φ through the equality of their first three moments. This gaussian, itself, is the result of a fictitious *point* source ψ, of strength Q, placed at the point $x = x_0$ when $t = -t_0$.

References

1. A. J. McConnell, *Applications of Tensor Analysis* (reprinted by Dover, New York, 1957).
2. A. Lichnerowicz, *Éléments de Calcul Tensoriel* 5e éd. (Armand Colin, Paris 1960). [Engl. transl. by J. W. Leech and D. J. Newman, (Methuen & Co., London, 1962.]
3. I. S. Sokolnikoff, *Tensor Analysis* 2nd ed. (John Wiley, New York, 1964).
4. R. P. Feynman, *Lectures on Physics* (Addison-Wesley, Massachusetts, 1963).
5. H. Weyl, *Symmetry* (Princeton University Press, Princeton, NJ, 1952).
6. H. B. Callen, *Thermodynamics and an Introduction to Thermostatistics* 2nd ed. (John Wiley, New York, 1985).
7. I. Prigogine and R. Defay *Thermodynamique Chimique* (Editions Desoer, Liège, 1950). [Engl. transl. by D. H. Everett, John Wiley, New York, 1954.]
8. R. Courant and H. Robbins, *What is Mathematics*? pp. 385–397 (Oxford University Press, London, 1941).
9. R. Osserman, *A Survey of Minimal Surfaces* (reprinted by Dover, New York, 1986).
10. W. H. Meeks III, "A Survey of the Geometric Results in the Classical Theory of Minimal Surfaces," *Bol. Soc. Bras. Mat.* **12**, 29 (1981).
11. E. V. Boys, *Soap Bubbles, Their Colours and the Forces Which Mold Them* (reprinted by Dover, New York, 1958).
12. R. P. Boas, "Partial Sums of Infinite Series, and How They Grow," *Amer. Math. Monthly* **84**, 237 (1977).
13. D. A. Smith, "What's Significant about a Digit," *College Mathematics J.* **20**, 136 (1989).
14. A. Erdélyi, *Asymptotic Expansions* (reprinted by Dover, New York, 1956).
15. N. G. de Bruijn, *Asymptotic Methods in Analysis* (reprinted by Dover, New York, 1981).
16. M. R. Spiegel, *Theory and Problems of Laplace Transforms*, Chaps. 6 and 8 (Schaum, New York, 1965).
17. C. J. Tranter, *Integral Transforms in Mathematical Physics*, 3rd ed., Chap. 6 (Methuen & Co., London, 1966).
18. R. V. Churchill, *Operational Mathematics*, 3rd ed., Chap. 11 (McGraw-Hill, New York, 1972).
19. C. Kittel, *Elementary Statistical Physics*, p. 37 (John Wiley, New York, 1958).
20. E. Goursat, *Cours d'Analyse Mathématique*, Tome III, pp. 287–322 (Gauthier-Villars, Paris, 1956).
21. A. N. Tikhonov and A. A. Samarskii, *Equations of Mathematical Physics* (Pergamon Press, Oxford, 1963). [Reprinted by Dover, New York.]

22. R. Ghez, J. D. Fehribach, and G. S. Oehrlein, "The Analysis of Diffusion Data by a Method of Moments," *J. Electrochem. Soc.* **132**, 2759 (1985).
23. N. N Lebedev, *Special Functions and their Applications*, R. A. Silverman, Ed., pp. 60–76 (reprinted by Dover, New York, 1972).

Index